THE HUMAN ORGANIZATION OF TIME

• Recently, there has been an explosion in research on time. This book provides a much needed summary of that work. *The Human Organization of Time* will prove a valuable resource to anyone interested in temporal research in organizations.

LESLIE PERLOW, *Harvard Business School.*

• Finally a masterful book about time. Bluedorn's work is comprehensive and cutting edge, laying out the interplay of time with fundamental aspects of organizations and individuals. It should be on every serious organizational scholar's bookshelf.

KATHLEEN EISENHARDT, *Department of Management Science and Engineering, Stanford University* Coauthor of *Competing on the Edge: Strategy as Structured Chaos*

• This is a wonderful and important book, full of fascinating information, insights, conjectures, and constructs. Bluedorn forges a compelling case for the importance of time, and of our roles as current stewards of the temporal commons. From the Big Bang to the Bolshevik revolution to the puzzles of Deep Time, from the social construction of zero to the theory of relativity, from the gates of Trenton State Prison to the gates of Dante's Inferno, *The Human Organization of Time* weaves a compelling fabric of temporal threads. Bluedorn has found power and poetry in time.

RAMON ALDAG, *Department of Management and Human Resources, University of Wisconsin*

• *The Human Organization of Time* is a broad look at how we truly think about time. It unifies the many human patterns of time-scale concepts and gives depth and perspective to a complex field. Thorough and insightful, it will become the standard work.

GREGORY BENFORD, *Department of Physics, University of California, Irvine* Author of *Deep Time*

• *The Human Organization of Time* stands to be a definitive source for those interested in temporality and time. Bluedorn's knowledge of diverse literatures and his attention both to historical perspectives as well as contemporary theorizing and research is noteworthy. Issues of time and temporality pervade the human experience; Bluedorn helps us to appreciate temporality as a social construction with very real consequences for organizations and their members.

JENNIFER M. GEORGE, *Jesse H. Jones Graduate School of Management, Rice University*

• A remarkable and original contribution to our understanding of the social construction of time and its effects on people and organizations. Playing off against a backdrop of work preoccupied with enduring and stable features of social life, Bluedorn underscores the importance of temporal features—pace, tempo, rhythm, entrainment, and historical turning points.

ALAN MEYER, *Lundquist College of Business, University of Oregon*

The

Human Organization

of Time

TEMPORAL REALITIES AND EXPERIENCE

Allen C. Bluedorn

STANFORD BUSINESS BOOKS
An Imprint of Stanford University Press

Stanford University Press
Stanford, California

© 2002 by the Board of Trustees of the
Leland Stanford Junior University

Printed in the United States of America
on acid-free, archival-quality paper

Library of Congress Cataloging-in-Publication Data

Bluedorn, Allen C.
    The human organization of time : temporal realities and experience /
Allen C. Bluedorn.
        p.   cm. — (Stanford business books)
    Includes bibliographical references and index.
    ISBN 0-8047-4107-7 (alk. paper)
    1. Time—Social aspects.   2. Time—Sociological aspects.   I. Title.
II. Series.
    HM656 .B58   2002
    304.2'3—dc21   2002001375

Original Printing 2002

Last figure below indicates year of this printing:
11   10   09   08   07   06   05   04   03   02

Designed by James P. Brommer
Typeset in 10.5/14.5 Caslon

To those who have brought such exquisite meaning to my times;
may their times be the best of times always:

To my wife, Betty;
To my sons, John and Nick;
To my brother, Ralph;
To my mother, Evelyn;
To my father, Rudolph, 1905–1988.

Contents

Preface

This book's road to publication was long and circuitous, and its history describes a course that only a chaos theorist could appreciate properly. That history will not be chronicled here; instead, I will acknowledge freely that like time itself, this book is truly a social construction, one whose existence is due to the thoughts and ideas, to the questions and research, and to the assistance and goodwill of many, many people. And as such, it is appropriate that I acknowledge as many of those people here as my memory permits.

I wrote this book in my office at home, which leads me to think of several home teams that have helped me with this project. And I will begin by acknowledging the team closest to home, my wife, Betty. Betty read *every* word in an earlier version of this manuscript and offered many cogent comments about both the ideas and the way they were presented. Betty is a registered nurse, a good one, and as one would expect from the ministrations of a good nurse, the "patient's" health definitely improved as a result of her efforts.

A second home team consists of several people at the University of Missouri-Columbia whose aid was invaluable. Two research collaborators, Steve Ferris and Gregg Martin, each gave me permission to conduct original analyses on data we had collected together and to publish the results in this book. I am grateful not only for their permission to do so but also for their enthusiastic interest in having me do so. Another member of this home team is Rhetta Standifer, who served as my research assistant during the last year leading up to the day that I put the final revised version of the manuscript into the mail to Stanford. Rhetta read the entire manuscript and helped me check the references, the accuracy of quotations, and so on. These jobs were extremely important, and Rhetta did them very well.

Jack and Cathleen Burns are colleagues here who helped in two very diverse ways. A critical series of conversations with Cathleen helped me clarify for myself that this was a project that I really should do, and Jack, Cathleen's astrophysicist husband, helped me avoid stepping into a theoretical black hole from which my credibility would never have reemerged. He was also kind enough to check the physics I present in Chapters 1 and 2 and to provide an introduction for me to his collaborator at the University of California, Irvine, Gregory Benford (results from this introduction appear in Chapters 5 and 7). In a similar vein, my friend Carol Ward, a physical anthropologist, was good enough to check an earlier version of Chapter 2 and provide some suggestions for upgrading the account of hominid evolution presented in it.

A special component of this home team is the staff of the University of Missouri-Columbia's library system. A large amount of library work takes place in any book project, but this is especially so when the project is cross-disciplinary and involves many historical materials. Although everyone I worked with at my campus' libraries was very helpful, I especially benefited from the assistance of librarians Gwen Gray, Michael Muchow, Nancy Turner Myers, Geoffrey Swindels, and John Wesselman. I should also thank the staffs at three other libraries: Cornell University, Rutgers University, and the University of Iowa. All three treated a visitor with exceptional courtesy and competence, and in the case of the University of Iowa, I discovered that I could go home again—having received all three of my degrees there—for late in the project I needed a particular edition of *Plutarch's Morals*, and they were kind enough to let me find it in their collection. Librarians are good people.

And so are graphic artists. The final versions of all of the figures in this book were prepared by Liz Priddy, a graphic artist in the Graphic Arts Services unit of the University of Missouri-Columbia's Academic Support Center. Not only did Liz prepare the final versions, but she also worked with me to help develop several of them conceptually. She even helped me with a decision about epigraphs!

Before moving beyond the boundaries of my university, I would like to acknowledge one more form of support it provided. I have conducted several original data analyses for this book and report the results in several chapters. All of the research projects from which data were used for these analyses were supported by one or more sources at either the University of Missouri System or the University of Missouri-Columbia, including a University of Missouri Research Board grant, a grant from the Research Council, an Alumni Association grant, Research Fellowships from the Center for the Study of Organizational Change, a grant from the Financial Research Institute, and several College of Business Summer Research Fellowships. This type of tangible support is greatly appreciated.

My extended home team begins with the people at my publisher, Stanford University Press. Let me begin with my editor, Bill Hicks. Bill has been an enthusiastic supporter of this project for many years, and his steadfast interest has spanned several salient events in both his career and mine. It is reasonable to say that without Bill this book would never have been written. For his enduring confidence and enthusiasm I shall be ever grateful. Another important member of this team is Kate

Wahl, my assistant editor. Not only was Kate tremendously helpful, but she would get things done and answer questions amazingly fast—a well-appreciated virtue in several situations. The devil is rumored to be in the details; if so, Kate kept the devil at bay. It was simply a delight to work with someone so competent, diligent, and enthusiastic. Such virtues also describe my production editor, Janna Palliser, and my copy editor, Mary Ray Worley.

Thousands of miles from Stanford are two colleagues who read the manuscript critically and helped me develop it. These colleagues are Ramon Aldag and Jean Bartunek. Not only did they give me good advice, but their overwhelming support for what I was doing was nothing less than inspirational. They have no idea how often I revisited their comments as a pick-me-up when the going was slow. Such support from such colleagues is truly manna from heaven.

And not only did Ray and Jean give me great advice and support, they also gave me permission to use some of the ideas in their comments and to cite them as the source of those ideas. So did several other colleagues and friends to whom I am very grateful and would like to say thank you to now for doing so: Robert Backoff, Gregory Benford, Betty Bluedorn, John Bluedorn, Nick Bluedorn, Jack Burns, Joseph Kasof, Carol Ward and Francis Yammarino. In a similar vein I would like to thank Cason Hall & Co. Publishers and Elliott Jaques for permission to use the quotation from *The Form of Time* as the epigraph for this book. It describes the point of the book perfectly (hence it is repeated in several of the chapters), and I am grateful they allowed me to use it. I would also like to thank Deloitte Consulting and Aileen Stacy for permission to quote their statement about multitasking.

Finally, there are the many people who sent me papers and articles, who answered my questions, and, often, who simply brought things to my attention because they knew I was interested in time and was writing a book about it. This is assistance it is impossible to buy, and it is amazing how many people helped me in this way. To the best that my memory and records allow me to acknowledge, I would like to thank all of the following people for helping me in this way: Yochanan Altman, Deborah Ancona, Neal Ashkanasy, Gregory Benford, Joel Bennett, Leonard Berry, Joseph Blackburn, Sally Blount, Betty Bluedorn, Evelyn Bluedorn, John Bluedorn, Nick Bluedorn, Cathleen Burns, Jack Burns, Todd Chiles, Yumiko Clevenger, Jeff Conte, T. K. Das, Paul Davis, Tom Dougherty, Randall Dunham, Ronald Ebert, Tracy Everbach, Michelle Fleig-Palmer, Fernando Galindo, Annabelle Gawer, Harrison Gough, Barbara Gutek, Mark Hayward, Stephen Horner, John Howe, Mary Hurt, Quy Nguyen Huy, Gregory Janicik, Joe Kasof, Carol Kaufman-Scarborough, Stephen Kern, Mark Kriger, Paul Lane, Herbert Lefcourt, Jay Lindquist, James March, Jim Mattingly, Marci McPhee, Kathy Neighbors,

Loren Nikolai, Ellen O'Connor, Greg Oldham, Wanda Orlikowski, David Palmer, J. P. Palmer, Leslie Perlow, Gil Porter, Lyman Porter, Michelle Reznicek, Bruce Rollier, Rakesh Sambharrya, David Schencker, Tom Slocombe, Rhetta Standifer, Scott Standifer, Kristine Stilwell, John Stowe, Paul Swamidass, Dan Turban, Jon Turner, Michael Tushman, Jean-Claude Usunier, Pierre Valette-Florence, Wende-lien van Eerde, Carol Ward, Francis Yammarino, and JoAnne Yates.

To all whom I have thanked here publicly, let me say thank you again. You have helped produce much of what is good in the pages that follow. To anyone I have omitted, I can offer only my apologies and the wish that my memory had been better. For any shortcomings or failings that occur in the following discourse, the responsibility is, of course, my own.

But allow me to comment on one failing I have tried to avoid. Over thirty years ago, E. P. Thompson commented about the writing on the sociology of in-dustrialization, describing it as "like a landscape which has been blasted by ten years of moral drought: one must travel through many tens of thousands of words of parched a-historical abstraction between each oasis of human actuality" (1967, p. 94). Whatever failings this volume may have, the kind of writing Thompson criticized should not be one of them. Instead, you will find real people in the pages that follow, occasionally speaking in the form of statistical aggregates, but much more frequently telling their stories from the pages of history; from newspapers, magazines, books, and journals; and from my own story as well. For what follows is the story of a human construction, time, and how it shapes the set of experi-ences we call life. And to tell this story requires human voices. They begin speak-ing on page 1.

In the form of time is to be found the form of living.

—Elliott Jaques, *The Form of Time*

All Times Are Not the Same

Que no son todos los tiempos unos.
(For all times are not the same.)
—Miguel de Cervantes Saavedra,
    Don Quixote de la Mancha

It was a foreboding hour early in the twentieth century. The date was the first of August in 1914, and because time was about to join the Allies' cause, the world would be changed forever. Anxiety about the German attacks scheduled for that evening led William II, emperor of Germany (kaiser), to propose a change of plans to his chief of staff, General Helmuth von Moltke. The kaiser proposed that Germany's plan of war be changed to sequential attacks: first an attack against Russia, followed—presuming a Russian defeat—by an attack against France, this rather than the anxiety-producing simultaneous two-front war specified in the original plan (Tuchman 1962, pp. 15, 93–104).

But the Kaiser failed to convince his subordinate. Von Moltke declined the change on the grounds that "once settled, it [the plan] cannot be altered" (Tuchman 1962, p. 100). And because the cold war took its origins from World War II, which took its origins from the outcome of the First World War, whose outcome was intimately linked to the Kaiser's decision that fateful August day, in a very real sense the entire direction of twentieth-century history turned on that strategic decision. Indeed, one could argue that the twentieth century truly began that day.

Could a skeptic dismiss this interpretation as just an exercise in hyperbolic history? Seemingly not, for a dispassionate reading of the history of those strategic weeks at the beginning of the Great War (e.g., Gilbert 1994; Tuchman 1962) can only produce the conclusion that Britain and France stopped the German invasion in the west by the narrowest of margins, that they held on by a hair's breadth. Had Germany been able to focus on a single front and hurl the full might of its formidable war machine against the Allies in the west, the war against Belgium and France would likely have ended with a German victory—a quick German victory. Whether England would have fought on is speculation, but a quick and complete German victory over France would have had major implications for subsequent pivotal events such as the Bolshevik revolution in Russia and Hitler's seizure of power in Germany—the former becoming uncertain, the latter almost certainly diminishing to a minor if not zero probability. And if those two events were changed or did not occur at all, the rest of the twentieth century becomes unreconstructible.

TIMES DIFFER

But what led to the reaffirmation of the original plan that afternoon rather than to its modification? Cultural forces, powerful cultural forces, seem to have played a major role, and those forces directly involve forms of human time. One force was the cultural preference for engaging tasks and events, a preference selected from a continuum of choices ranging from a strict one-thing-at-a-time attitude to a preference for being involved with many tasks and events simultaneously. This continuum of choices for sequencing activities is known as *polychronicity*, and it will be explored in depth in Chapter 3. But for now we simply need to know that German culture traditionally valued and preferred the one-thing-at-a-time option; indeed, it strongly preferred it (Hall and Hall 1990, p. 14).

However, there is more to time culturally than polychronicity, as fundamental as polychronicity may be. Another facet concerns the explicit emphasis given to organizing and coordinating action with schedules and plans. And the matter of flexibility once a plan is made or a schedule created is especially relevant to that August 1 decision. Some cultures emphasize flexibility as new information becomes available, whereas others believe plans and schedules should be inviolate. German culture traditionally tended toward the latter orientation, a point that could not be made any more clearly than Von Moltke did when he said "once settled, it [the plan] cannot be altered."

Ironically, the decision could have gone the other way, because the cultural values and beliefs influencing the Kaiser were not aligned; they were pulling him in opposite directions: one to change the plan, the other to keep it intact. The plans-are-inviolate value in German culture did push for keeping the plan unchanged, but the one-thing-at-a-time value pushed in the opposite direction and was likely a source of some of the Kaiser's anxiety as the German war plan was about to be implemented. Because of the one-thing-at-a-time value, a two-front war, anathematic to all generals, should have been particularly loathsome to German generals, so it is surprising that a plan for a two-front war was developed in the first place. Nevertheless, that was the plan, and in a culture such as turn-of-twentieth-century Germany's, once a plan was made, the preference was to leave it unchanged.

So the two values conflicted, and the plans-should-be-inviolate value prevailed. Had the culture differed on this point and taken a more flexible view toward plans and changing them, had the Kaiser been Romanian on this point rather than German, Romanian culture being relatively more flexible about plans (see Chapter 8), the decision—and the twentieth century—might have gone differently. And it is in this sense that time (i.e., German values about keeping plans unchanged) joined the Allies' cause that day, because the two-front war specified in the original German plan favored the Allies more than the Kaiser's single-front-in-the-west alternative would have.

The Kaiser's decision on August 1, 1914, to retain the original plan illustrates the importance of time in human affairs, and the discussion of it also illustrates that time and times vary; they are neither uniform nor the same from one moment to the next. Thus all times are not the same. There would be no point in writing this book if this were not true. Yet of its truth there can be no doubt, a truth that can be demonstrated easily because only one time need differ from all others to make it true. To demonstrate, consider Elias Canetti's penetrating question: "And what if you were told: One more hour?" (1989, p. 144). Who would argue that with such foreknowledge of one's final hour that any other hour would be its equal?

But Canetti addressed final hours. What of the first hours? The first hours also differed, and originally they did so from day to day. The first measured hours were called temporal hours, which sounds redundant because what else would an hour be? In this context, however, the *temporal* conveys the sense of "changing," since these first hours were defined as twelve equal parts of the day, and twelve equal parts of the night (Boorstin 1983, pp. 30–31; Dohrn-van Rossum 1996, p. 19). Hence within a day the hours were equal, but as the days passed and grew longer or shorter depending on the season of the year, so too would the comparable hours lengthen or shorten daily. For example, during the spring each new day is longer than the last, so each hour, defined as one-twelfth of the daylight period, would lengthen with each passing day too. In autumn the order reverses, and as the days grow shorter, likewise so do the hours. Although the concept of regular, absolutely equal temporal units was known, the best early technologies (e.g., sundials and sandglasses) could do was measure a few consecutive such hours, either before dusk came rendering sundials impotent or the hourglass turner fell asleep. And in civilizations such as those of Egypt, Greece, and Rome, the period of light (day) and the period of dark (night) were each divided into twelve hours, which meant that except near the equinoxes the length of daytime and nighttime hours differed (Gimpel 1976, pp. 167–68).[1] So temporal hours would dominate the measurement of time during the day until the fourteenth century.

Some people have always realized that times differed, and in the industrial era Henry Ford was one of them. But those differences bothered him. So Ford built a watch with two dials enabling it to tell two times. It told the sun time, which defined the hour for each local community, and it told the new railroad time (the developing standard time of time zones; see Chapter 6). It also illustrated the tendency to believe there is only one time, for Ford noted that the watch "was quite a curiosity in the neighbourhood" (Ford 1922, p. 24).

Ingenious as it was, this was not the first watch to tell time in two different time systems. Jo Ellen Barnett (1998, pp. 116–17) has noted that nearly a half century earlier similar watches were constructed in England to deal with the same problem. She also described how the victorious French revolutionaries attempted to completely overhaul the time-reckoning system used in France, an attempt that divided the day (one complete rotation of the earth) into ten hours rather than twenty-four, and that divided each hour into one hundred minutes, each minute into one hundred seconds. In their efforts to establish and institutionalize this change, the revolutionaries had watches built that displayed two sets of numbers: the familiar one through twelve, but presented twice around the outer circumference of the watch's face in a circle that surrounded an inner ring containing the numbers one through ten.[2] And in the spirit of proper revolutionary zeal, the inner ring for counting the revolutionary hours appears easier to read. But not for long. After about two years of trying to convert the nation, the effort was abandoned in 1795 (Barnett 1998, p. 56; Richards 1998, pp. 263–64). It seems reasonable to suggest that the revolutionaries' efforts were motivated not so much by a desire to improve time reckoning as by a disdain for anything associated with the ancien régime and the belief that just about anything that differed from its practices was better, or at least desirable. Hence their efforts were driven principally by an effort to differentiate themselves from the old order, a use of time in which they were not unique (see the discussion of the Sabbath later in this chapter).

But the skeptic would protest, saying human time is really "just" subjective experience and not real time anyway, that one's experience of that final hour might differ from the thousands of other hours experienced in a lifetime, but that hour, the passage of a single hour across the universe, is the same as any other of the nearly uncountable hours that have passed since the universe began. The skeptic would quote Isaac Newton: "Absolute, true, and mathematical time, in and of itself and of its own nature, without reference to anything external, flows uniformly and by another name is called duration" (Newton 1999, p. 408).

Newton was wrong. Even the concept of time used in contemporary physics discards the idea of a uniform temporal meter, Albert Einstein's work on relativity having "demolished" it (Coveney and Highfield 1990, p. 70). For Einstein's special theory of relativity describes a slowing of time for clocks moving with a constant velocity relative to a referent observer (Einstein 1961, pp. 36–37), and his general theory of relativity similarly describes clocks ticking at different rates if they are located in different positions within a gravitational field (pp. 80–81). These effects are known as time dilation and gravitational time dilation, respectively (Thorne 1994, pp. 37, 66–86, 100–104), and either extreme velocities or extreme gravitational forces are necessary to produce major temporal differences. For example, gravitational time dilation means that the closer things are to a source of gravity, the more slowly time flows relative to an external observer (Jack Burns, personal communication, 2001; Thorne 1994, p. 100), and on the surface of a neutron star, a star whose gravity is often a billion times stronger than the earth's, time flows about 20 percent more slowly relative to the earth (Davies 1995, p. 105).

But even without relativity theory, other theoretical developments in the physical sciences might have eventually overturned the concept of absolute time. For example, applying the second law of thermodynamics to the universe as a whole yielded the controversial conclusion that entropy—the degree of disorder in a system (Hawking 1988, p. 102)—increases continuously across the entire universe with each passing moment (Coveney and Highfield 1990, pp. 33–34). Although this does not mean time passes at different rates, it does illuminate the different nature of each passing moment, for *if true* (the point is debatable, see Chapter 2), it means that no two temporal intervals in the history of the universe are characterized by the same amount of entropy. So, if true, it means that all times are not the same, that none of them are, which means that all times are different.

One need not go that far, however, to realize that at least some times differ. And differing times mean variance among times, and that variance creates the potential to explain other phenomena because a constant explains no variance. It is the nature of these differences, especially among human times, that will be explored throughout this book. But to be concerned about such differences suggests that the differences matter, and matter they do, profoundly.

TIME AND THE HUMAN EXPERIENCE

The possibility that time can explain other phenomena, especially human behavior, is the scientific raison d'être for studying time and caring about it: If times differ, different times should produce different effects. And an important mechanism through which differing times affect human behavior is the definition of the situation.

Early in the twentieth century, William Thomas and Florian Znaniecki (1918, pp. 68–74) developed a fundamental explanation for human behavior, the definition of the situation, the implications of which would be stated most memorably a decade later: "If men [and women] define situations as real, they are real in their consequences" (Thomas and Thomas 1928, p. 572). Later, in his analysis of the self-fulfilling prophecy, Robert Merton would elevate this statement to the status of a basic theorem in the social sciences (1968, p. 475). It is so fundamental because human beings generally behave in ways consistent with their perceptions and interpretations of reality, most of which are based on social constructions developed through interactions with others. To understand these constructions is to understand and explain much of the behavior that follows from them. So if schedules and plans are seen as set once they are made, if they are thus perceived as immutable, the expected reaction to the proposal to change a plan is to reject the proposal—even if the proposal comes from the Emperor of Germany.

And temporal forms define the situation at the mundane level of everyday life just as they do at the strategic heights. For time has often served the social function of differentiating one group from another. Noting this in his cogent analysis of the week, Eviatar Zerubavel (1985) explains how the Sabbath was used this way to distinguish the three great monotheistic religions. Judaism developed first, so it had its choice of days, and the choice was Saturday. Then came Christianity and Sunday, followed by Islam and Friday. Further temporal differentiation was provided through rules dealing with prayer times. For example, prayer for Muslims came to be forbidden at sunrise, sunset, and midday so as to create a deliberate contrast with other religions (Dohrn-van Rossum 1996, p. 30). Thus the *time* of worship and devotions would distinguish one religion from another, just as would the *place* of worship.

Consistent with this interpretation was the practice of early Christians to celebrate both Saturday and Sunday (Zerubavel 1985, p. 21). This was a transitional era in which Christianity was not wholly distinct from Judaism, either theologically or temporally. So as Christianity developed as a theologically distinct religion, the observance of both days ultimately stopped, with only Sunday being observed. In this same vein, as Christianity became more organized because of events such as the Council of Nicaea held in 325, it also explicitly proscribed Easter from occurring at the beginning of Passover (Duncan 1998, p. 53). The emperor Constantine was even candid, if unecumenical (or worse), about why such a coincidence should be avoided: "We [Christians] ought not to have anything in common with the Jews" (quoted in Duncan 1998, p. 53).

But even after the two religions had become completely distinct, they still shared some things "in common," among them the general location of their respective Sabbaths within the week. Zerubavel noted that the two Sabbath days "touch," that they are temporal next-door neighbors (1985, p. 26). At first glance this would appear to be a poor strategic choice for a group attempting to distinguish itself from a well-established competitor. But that is the point: The new religions, first Christianity, then Islam, wanted to distinguish themselves. And as new religions they suffered from the liability of newness (Stinchcombe 1965, pp. 148–50), the set of disadvantages all new organizations face that make their survival uncertain at best because organizational mortality rates (i.e., ceasing to exist) decrease with organizational age (Hannan and Freeman 1989, pp. 244–70).

So the new religions likely attempted to reduce this uncertainty and enhance their survival potential by copying the temporal form of their more successful, more legitimate predecessor(s), a process Paul DiMaggio and Walter Powell (1983) describe generically as mimetic imitation. Rather than selecting Tuesday or Wednesday, the days farthest from Saturday, Christianity picked the temporally proximate Sunday. Similarly, when it was Islam's turn, it too selected a proximate day, Friday. By so doing, both Christianity and Islam tapped into an already familiar institution—a weekly holy day—and by juxtaposing their holy days with the holy days of their predecessors they created a temporal structure that communicated two messages. One message said, "We are different"; the other, "We are doing similar things." The use of the different day for weekly worship communicated the difference; the location next to ("touching") the other days of worship communicated the similarity and, it was perhaps hoped, some legitimacy too. The similarity would appeal to converts who, though they were now members of a different group, were still doing something that at a more general level was the same thing done by the other group (cf. Zerubavel 1985, p. 26)—even if in some centuries the other group might advocate burning them at the stake for doing so. For as DiMaggio and Powell have noted, "The modeled organization may be unaware of the modeling or may have no desire to be copied; it merely serves as a convenient source of practices that the borrowing organization may use" (1983, p. 151).

The developing religions' quests for legitimacy by positioning their Sabbath days adjacent to those of their predecessors also reinforced the practice of a week of seven days (Zerubavel 1985, p. 26). For if either of the new religions had instituted a week composed of a different number of days, the new Sabbaths would have been adjacent to those of their predecessors only occasionally. Moreover, unless a special exclusionary rule was included to prevent it, sometimes the new Sabbaths would have fallen on the same day as one of the other religions' Sabbaths, in such cases defeating the social functions of a new Sabbath. For the proper different-but-legitimate balance to be maintained, the new monotheistic religions had to use a seven-day week, and their weeks had to be aligned properly with the seven-day weeks of their predecessors. This alignment of weekly cycles is an example of *entrainment*, a phenomenon Chapter 6 examines in detail.

So time provides a tangible, observable way for groups to define who is and is not a member. In the case of religions, which day is the weekly holy day and which hours are and are not for prayers would clearly distinguish insiders from outsiders with relative ease. These temporal decisions and practices shaped the lives of their adherents, and in so doing led them to lead and experience different lives. But this is just a specific example of the more general principle: Different times produce different effects. Consequently, every chapter in this book includes major discussions of the differences produced by differing times and temporal practices. And one example of this principle, perhaps the most profound example of a differing time's effects, is useful to consider here.

The effect was produced by an ingenious piece of technology invented in the thirteenth century—by whom no one knows—that few residents of the twenty-first century have ever heard of, yet its effects were revolutionary. The invention was the escapement, a device that converted the power in a clock into gear movements of equal duration. The escapement made the mechanical clock possible, and the mechanical clock revolutionized time—and so very much more (Landes 1983).[3]

How big a revolution was it? David Landes (1983) ranked the mechanical clock among the great inventions: below fire and the wheel, but on a par with movable type for its impact on "cultural values, technological change, social and political organization, and personality" (Landes 1983, p. 6). Yet Landes's ranking notwithstanding, the clock did something that no invention has done before or since: It provided the archetype for the way Western civilization would see God and the universe. And by doing so the clock would become the greatest metaphor of the second millennium (as years had come to be reckoned in the West).

But the date on which this revolution began is unknown. The year, even the decade, in which the first shots were fired is uncertain. However, it seems likely that it began in either the 1270s or the early 1280s. Evidence for this is provided by J. D. North, who noted that a commentary about the most prominent medieval astronomical textbook discusses time and timekeepers but appears to be unaware of a mechanical escapement, and the commentary was written in 1271—as part of a course of lectures "at the university of either Paris or Montpellier" (North 1975, p. 396). Yet by 1283 records were made of a clock in Bedfordshire, England, that seemingly contained a mechanical escapement, the evidence for this conclusion being "persuasive" (1975, p. 384). No claim is made that this was the first mechanical clock to incorporate an escapement mechanism, just that it was described in the first records yet known about such a device, and those records date from 1283. If these bounds are accepted—no escapement-based clock is known in 1271 and the first records of such a clock appear in 1283—the escapement-based mechanical clock may have been invented sometime between 1271 and 1283, a conclusion consistent with a later statement by North, "in the 1270s, or thereabouts" (North 1994, p. 129). Perhaps the twelve-year interval from 1271 to 1283 is the best that can be done as far as determining the date of origin.

After 1283 but before 1300, records were made of several other escapement-based clocks in England (North 1975, pp. 384–85). Moreover, before 1300, references to mechanical clocks appeared in European literature (Crosby 1997, p. 79). And shortly thereafter Dante's *Paradiso*, begun in 1315 (Mazzotta 1993, p. 10) and completed in 1321 (Bergin 1965, p. 44), described the workings of a mechanical clock:

> And ev'n as wheels within the works of clocks
> so turn, for one who heeds them, that the first
> seems quiet, while the last appears to fly.
>
> (Dante 1921, p. 277)[4]

So the escapement-based mechanical clock is an invention of the latter thirteenth century, an invention that would be disseminated with amazing speed throughout the Western world—amazing given the difficulty of transportation at the time—along with its influence on "cultural values, technological change, social and political organization, and personality." It even influenced—sometimes dominated—the West's cosmic worldview, its weltanschauung.

In what may have been the centennial year of the mechanical clock's invention, 1377, at the behest of the king of France (Charles V), Nicole Oresme, the dean of the Cathedral of Rouen (later bishop of Lisieux), published a translation of important scientific works by Aristotle in *Livre du ciel et du monde* (*The Book of the Heavens and the World*) (Menut 1968, pp. 3–9). These translations incorporate Oresme's commentaries in the text, and it is these commentaries that present the Metaphor: "The situation [God creating the heavens and establishing their regular motions] is much like that of a man making a clock and letting it run and continue its own motion by itself" (Oresme 1968, p. 289). Oresme may or may not have been the first to put this metaphor into print, for he anticipated it himself in an earlier treatise (Mayr 1986, p. 38).[5] Oresme also cites an author named Tully as having written, "No one would say that the absolutely regular movement of a clockh appens [*sic*] casually without having been caused by some intellectual power; just so must the movement of the heavens depend to an even greater degree upon some intellectual power higher and greater . . . than human understanding" (Oresme 1968, p. 283). Regardless of who said it first, no one ever wrote it more dramatically than Daniel Boorstin, "a clockwork universe, God the perfect clockmaker!" (1983, p. 71) or with more poetic grace than Loren Eiseley, "God, who had set the clocks to ticking" (1960, p. 15).

This was an idea, an image of reality and God's relationship to it, that would shape the West's thinking to the present day. For after the invention of the mechanical clock a major argument for the existence of God would be presented in terms of the clock metaphor: "Clocks owe their existence to clockmakers; the world is a huge clock; therefore, the world, too, was made by a clockmaker—God" (Mayr 1986, pp. 39–40). As Otto Mayr noted, this became the successor to a similar argument based on the more general *machina mundi* (world machine) metaphor (p. 39).

Being seen as the quintessential machine, the mechanical clock became a template for scientists and mechanics alike. For example, early in the seventeenth century Johannes Kepler described his intent to a friend as "to show that the heavenly machine is . . . a kind of clockwork" (quoted in Koestler 1959, p. 340). And Kepler's intent was to develop an accurate description of the motions of the heavens, which eventually led to his laws of planetary motion. Descartes too used the clockwork and God-the-clockmaker analogies (Boorstin 1983, p. 71). Concerning the mechanics, Boorstin described the clock as the "mother of machines" (p. 64) because it led to the basic technology of machine tools. Clocks required precisely fabricated screws and gears, and these requirements led to improvements in lathes and other machines used to make them. These improvements in machine tools, in turn, led to improved, more precise and accurate clocks, a level of quality that would be captured in the phrase "like clockwork," used to describe any well-ordered, well-coordinated process.

But the mechanical clock and the Metaphor guided more than scientific thinking and the development of better machinery. As Gareth Morgan has made clear to students of the organization sciences, metaphor is a potentially powerful tool for understanding human beliefs and behavior, and the metaphors people hold about organizations (which encompass much of the way they define organizational reality) explain much about the decisions they make and the actions they take (see Morgan 1997, especially p. 4). Hence Morgan's analysis is consistent with the view presented here that the Metaphor offered a template, a tacit imperative for managing and organizing human life itself, especially in the workplace. A major example of this impact comes from the organizational achievements of the master mechanic who built the two-dialed watch mentioned previously, Henry Ford. And his biography may explain why, because Ford was immersed in this metaphor.

While growing up, Ford was fascinated with clocks and watches, and with seemingly intuitive acumen, he quickly developed the self-taught understanding and ability necessary to repair a large variety of timepieces (Nevins 1954, pp. 58–59). Word of his virtuosity spread quickly, and he often repaired the errant timepieces of many of the Fords' neighbors, to the displeasure of his father because Ford did not charge for the service (Simonds 1943, p. 28). After leaving home, Ford would work in the evenings for a jewelry store repairing clocks and watches to earn extra money (Simonds 1943, pp. 35–36). So Ford was well versed in clocks and clockworks by the time he turned to the organization of automobile production.

The way he organized production, the assembly line, was his greatest legacy, both bad and good. From the standpoint of the Metaphor, Ford's assembly line and all those that followed his example emphasized clocklike attributes, "the absolutely regular movements" of a clock. That was Ford's idea, a workflow that was regularly timed (the escapement) so as not to produce just the desired output (e.g., Model Ts), but to produce it at a steady, even pace— just as a mechanical clock produces not uneven temporal hours but a constant flow of hours of equal duration. Ford's assembly line was at least as much about when things were done as it was about what was done, so much so that Catherine Gourley wrote of his accomplishment, "Henry had created a giant moving timepiece" (1997, p. 30). Ford, a man who quickly taught himself how to repair clocks and watches, a man who loved the mechanisms of such devices and working on them, had developed a manufacturing process designed to run like clockwork. Nor was this necessarily an unconscious connection. Ford was aware of comparisons with clock mechanisms and coined one himself after the death of his mother, saying that the family home seemed "like a watch without a mainspring" (Simonds 1943, p. 34).

This interpretation of Ford's assembly line, an invention so strategically important because it became the archetype for manufacturing practice throughout the world, emphasized the regularity-of-movement aspect of the clock's mechanism. But the clockmaker component is at least as important a part of the Metaphor as is the regularity of the mechanism, and this aspect of the Metaphor can be found in managerial practice as well. Ford was obviously the assembly line's creator and designer, but he was anything but an absentee creator who just gave the assembly line a push the first time and then sat back and watched it "run and continue of its own motion by itself." In fact, his divine intervention included experiments such as a short-lived "Sociological Department," which employed one hundred investigators to visit workers' homes to ensure, among other things, that they used their leisure time properly (Wren 1994, p. 161).

But the image of God creating the universe, giving it a shove, and then never having to deal with its physical properties again is appealing. And it has particular appeal to managers and those who advise them. The managerial image is well illustrated in the television series *Star Trek: The Next Generation*. In this series starship captain Jean-Luc Picard is often given advice, and when he agrees with the advice, he issues the command "Make it so." Then, consistent with the view of management based on the Metaphor, whatever Captain Picard commands normally becomes "so" unerringly, and most important, without his subsequent intervention. This imagery of "good management" based on the Metaphor has serious implications for many managerial practices, including delegation, planning, and decision implementation. The implications involve expectations, including self-expectations, for managerial performance suggesting that if one plans or delegates well enough, the good manager will not have to intervene in the process thereafter—an impossibly high standard of both performance and omniscience for any mortal. Unfortunately, such expectations define managerial intervention as a sign of managerial imperfection, and even worse, of "bad management," leading to an unwarranted reluctance for managers to intervene once a decision is made, a plan is developed, or a task is delegated. There may be other sound reasons for managers not to intervene in a particular situation, but the idea that intervention represents bad management ipso facto should not be one of them.

CONSTRUCTING TIME

Time is a social construction, or more properly, times are socially constructed, which means the concepts and values we hold about various times are the products of human interaction (Lauer 1981, p. 44). These social products and beliefs are generated in groups large and small, but it is not that simple. For contrary to Émile Durkheim's assertion, not everyone in the group holds a common time, a time "such as it is objectively thought of by everybody in a single civilization" (1915, p. 10). This is so because in the perpetual structuration of social life (Giddens 1984) individuals bring their own interpretations to received social knowledge, and these interpretations add variance to the beliefs, perceptions, and values. Although there is usually sufficient similarity and agreement to justify the designation "shared," variation is inherent in the process. And when it comes to times, there is such variation that Elliott Jaques would write of time and people and say that no two people "living *at* the same time live *in* the same time" (Jaques's emphases; 1982, p. 3). Of course this implies that there are as many forms of time on the earth as there are people. Nevertheless, rather than the idiosyncratic forms, the shared forms, the socially constructed forms have by far the greatest impact on human life, both individually and collectively.

But how do the shared forms come to be? It is one thing to assert that they are socially constructed, another to explain how. In some cases the how is easily seen. For example, the U.S. government's practice of beginning its fiscal year on October 1 rather than January 1 draws attention to itself and makes the human agency (i.e., decision making and consensus building) in its construction more obvious. (How else could a year with exactly the same number of days, even in leap years, begin and end on different days than the calendar year if human choice were not involved?) Such agency was certainly apparent in the eighteenth century when firms began to prepare periodic accounting reports about their operations that used a fiscal year which ended at the low point in the firm's annual operations (Chatfield 1996, p. 457). The human agency is apparent, not just because the low point in annual operations might diverge from the end of the calendar year, but because it suggests a deliberate management strategy to locate the end of the firm's fiscal year at a time when there would be more time and resources available to perform the accounting work and to prepare the reports. (This is actually a form of the out-of-phase entrainment strategy that will be discussed in Chapter 6.)

In many cases, though, the social construction of times is much less apparent. So one time story in particular will prove most illuminating, the story of the A.D. (*anno Domini*) system of reckoning dates. Early in the sixth century, a Moldavian (née Scythian) monk labored on the dauntingly complex task of calculating the dates of future Easters, and his efforts produced a method for calculating such dates known as the *computus*, the method still in use today (Steel 2000, pp. 106–7). But as the monk performed these labors, he came to have his fill of the A.D. dating system, a system he felt was insulting to Christianity, especially if it appeared in a grouping with the day and month of Easter. This was because the A.D. in this system was not an abbreviation of *anno Domini*; rather, it stood for *anno Diocletianus*, the year of Diocletian, the devoutly anti-Christian Roman emperor (Duncan 1998, p. 74). So the monk, Dionysius Exiguus, decided to replace the old A.D. system, which followed the practice common in his time of dating events from the beginning of different emperors' reigns—the A.D. system based on Diocletian is still used today by Coptic Christians in Egypt (p. 75)—with a new one based on the year in which Jesus of Nazareth was born. An oddity of this system is that it locates the birth of Jesus in 1 B.C. (before Christ)! But this is a fortuitous oddity because it provides another opportunity to illustrate the socially constructed nature of time.

By the traditional tenets of Orthodox Judaism, a boy's life does not properly begin until two things happen: He is named and he is circumcised (Steel 2000, p. 110). And following Duncan Steel's insightful analysis (2000, pp. 110–11), this is relevant because (1) Genesis 17:12 prescribes circumcision when a boy is eight days old; (2) Luke 2:21 reports that Jesus was named and circumcised on his eighth day; and (3) December 25 was established as the date of Jesus' birth under the Roman emperor Constantine and was well established as such at least 175 years before Dionysius. If these three points are combined, they reveal that Jesus' life *properly* began, as defined by the social customs and beliefs of his time, on January 1 of the year following his birth. Thus December 25 is celebrated as Jesus' biological birth, but less well known is that his sociological life began on his eighth day, which is January 1. But not January 1 of year 0. Although he was physically born in 1 B.C., he was circumcised and named on the eighth day: in A.D. 1. Something appears to be amiss here, and it is to that missing something we now turn.

The problem was that the numbering system used in the West lacked one vital number, a number whose absence in Dionysius's era would result in all kinds of mischief roughly 1,570 years after Dionysius developed a replacement for the *anno Diocletianus* system. The missing number was zero, and because zero did not exist in the number system, it was impossible to designate a year 0. Moreover, it even might have been impossible for anyone to *think* of a year 0. This conclusion follows from the twentieth century's most provocative linguistic claim: "We dissect nature along lines laid down by our native languages. The categories and types that we isolate from the world of phenomena we do not find there because they stare every observer in the face; on the contrary, the world is presented in a kaleidoscopic flux of impressions which has to be organized by our minds—and this means largely by the linguistic systems in our minds" (Whorf 1956, p. 213).

Later known as the Sapir-Whorf hypothesis (Trask 1999, pp. 169–70), the principle that language is necessary to interpret reality suggests that neither a year 0 nor even the need for it ever occurred to Dionysius. Nor is it likely that he thought much about the labels for the years preceding Jesus' birth either. For Dionysius was not trying to develop a system of year reckoning for the world to use. Instead, his purpose was to develop a system for calculating future dates of Easter, and his disdain for the *anno Diocletianus* system led him to replace it with a numbering system that designated the first year of Jesus' life as year 1—and this was mainly for his personal reference and use by other clerics (Steel 2000, p. 108).

About two centuries later, the Venerable Bede briefly described Dionysius's system for calculating Easter in his *Ecclesiastical History of the English People* (1969), which was written and published in the eighth century. In this history he used Dionysius's method of designating years, using the phrase *anno ab incarnatione Domini* several times (literally meaning, "in the year of the incarnation of our Lord," but many times translated as "in the year of our Lord."[6] Although the Venerable Bede did use the phrase *ante uero incarnationis Dominicae tempus anno sexagesimo* (Bede 1969, p. 20) once (translated as "in the year 60 before our Lord" (p. 21), albeit "in the year 60 before the incarnation of our Lord" would be more literal, the use of the B.C. designation for the years before Christ's birth would not be used much until the seventeenth century (Steel 2000, p. 114). Even so, the problem is that missing year 0, not the years with negative numbers, a problem that has never been corrected by adding a year 0 to the chronicle of years. For by the time the concept and symbol of zero had migrated from India to Europe (see Kaplan 1999, pp. 90–115), Dionysius's system of designating the years had gained wide currency, being more or less accepted in Western Europe by the beginning of the second millennium (Richards 1998, p. 217). By then it was already too late to make the correction, and to do so today would be prohibitively chaotic because of the requisite correction of either every B.C. or every A.D. date recorded using Dionysius's original system. (Whether the A.D. or B.C. dates would require correction would depend upon whether the year A.D. 1 or the year 1 B.C. was converted to the year 0.) This even suggests that the A.D. or C.E. (Common Era) designations would add at least one more letter, C, for *corrected*, so a reader would know whether the date was a Dionysian date or a corrected one (e.g., 2 B.C. or 1 B.C.C.).[7] And if the A.M./P.M. system at times results in people arriving twelve hours early or late, one can only imagine the confusion and chaos a one-year correction would create, especially if the correction made A.D. 1 the year 0.

So even considering a correction is now unthinkable, but from time to time that missing year 0 leads to other problems, albeit often silly ones. The most recent manifestation of these is the millennium debate that reached its high point—one is tempted to say nadir—in 1999. This argument took the form of much smoke and fury about which year—2000 or 2001—was really the first year of the third millennium. Those advocating the year 2000 usually did so assuming a number line that begins with zero because they were unaware of zero's absence sixteen centuries earlier. Without that zero, though, no year designation tells the number of whole years that have passed since year 1, the beginning of such a year-reckoning system. Instead, the year designations tell that $N - 1$ whole years have passed (N being the year designation). This means that years ending in zero, even though they are evenly divisible by ten, cannot be the first year of a decade, century, or millennium. The proper first years of such time spans would be the respective years (i.e., the years evenly divisible by ten, one hundred, or one thousand, as appropriate) ending in zero *plus 1*. This would seem to resolve the debate in favor of those arguing for 2001 as the first year of the third millennium. (See Figure 1.1 for a comparison of year counts between time lines beginning with year 0 and year 1.)

However, those favoring 2000 cannot be dismissed so easily, and for at least two reasons. First, what is a millennium? A millennium is defined as "a period of one thousand years," and as "a thousandth anniversary" (see the primary definition of millennium given in the second edition of the authoritative *Oxford English Dictionary*). And this ambiguity supports those who argue for the

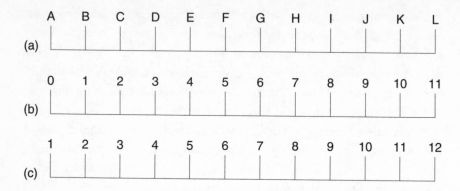

Each of the three time lines represents the same number of years. Counting the number of intervals gives the same number of years for each line: 11 years. But each time line gives a different answer to this question: On the first day of which year does the eleventh year *begin*? For (*a*), the answer is, on the first day of year K. For (*b*), the answer is, on the first day of year 10. And for (*c*), the answer is, on the first day of year 11. Line (*c*) represents the Dionysian year-reckoning system, which begins with the first day of year 1, so new decades begin on the first day of years xi, new centuries begin on the first day of years x01, and new millennia begin on first day of years x001.

FIGURE 1.1.   The millennium controversy

year 2000. Because if *millennium* means each thousandth anniversary of Jesus' birth, the second point supporting the year 2000 enters the debate: No one knows in which year Jesus was born. Several years seem plausible, ranging from 7 B.C. to A.D. 7 (Duncan 1998, p. 75), with 4 B.C. being the most commonly accepted date (e.g., Richards 1998, p. 218). Yet if this year is uncertain, so too must be its thousandth anniversary years, the millennia, and this uncertainty makes the designation of new millennia a matter of social constructions, several of which are involved in this story.

First, Dionysius's system was socially constructed. It was invented, not discovered (although, following Whorf 1956, all discoveries involve elements of social construction too). Moreover, it took several centuries to move from the status of a monk's proposal to the church hierarchy to a generally accepted social fact in Western Europe, only coming into widespread use in the eleventh century (Richards 1998, p. 217). And of special import to this explanation is the missing year 0 in this system.

The missing year 0 is the second social construction in this story, for zero is a social construction, as is the entire decimal (based on ten) number system of which it is a part, as are all number systems. Although the decimal number system has achieved worldwide acceptance, it is as much a social construction as are the Mayan vigesimal (based on twenty) or the Mesopotamian sexagesimal (based on sixty) systems (see Barnett 1998, p. 54). It just seems more real, more natural to twenty-first-century humanity because it is so well institutionalized that it is taken for granted as the True Number System, a status more easily maintained by the absence of encounters with alternative systems in everyday life.

Being a product of human interaction, the system for counting and designating the years is clearly a social construction, as are the number system and its elements, on which Dionysius's system is based. So then, what of the debate? Just as this problem was socially constructed, so too can it be resolved by developing a social consensus about its solution, a consensus that seems to have already occurred. It seems to have occurred because much of humanity already decided this issue late in the twentieth century by simply defining the year 2000 as the first year of the third millennium in this system, which is what it would be if the years were counted from a missing year 0 (see Figure 1.1). And as already noted, a millennium is properly regarded as a thousandth anniversary, in this case the thousandth anniversary of an event whose date will likely always be unknowable with complete certainty, so this socially constructed solution is as reasonable a solution to this dispute as any.

Indeed, the principal value of this debate is that it provides a good example of the socially constructed nature of time, in this case the temporal reckoning system used to designate the years. However, this system is relatively visible, and despite its nearly planetwide use for secular matters, the continued existence and parallel use of other calendars and year-reckoning systems such as the Jewish and Islamic calendars, whose year designations are very different from the Dionysian, occasionally remind humanity of the socially constructed nature of *all* calendars. Similarly, the Gregorian adjustment to what was then the Julian (for Julius Caesar) calendar reemphasizes this point, skipping as it did ten days in 1582 so October 4 was followed immediately by October 15, and also changing the system for designating leap years (Richards 1998, pp. 247–52). Thus humanity still receives occasional reminders about the socially constructed nature of calendars and year-reckoning systems, producing at least a semiconscious recognition of this point. Similar reminders are much less frequent, almost nonexistent, regarding that other major time reckoner in everyday life, the clock, so its system of reckoning the hours tends to be even more reified, even though it is equally a social construction.

THE PROFOUND IMPORTANCE OF TIME

In the seventeenth century both Cervantes and Newton wrote about time. Yet they reached fundamentally different conclusions about this abstruse phenomenon. To Newton, time was abstract and external to events, something that flowed "uniformly." Newtonian minutes were completely homogenous; one was the same as any other. Cervantes saw time differently. Although he might not have believed that all times were different, clearly he believed that not all of them were the same. Hence he wrote, "Que no son todos los tiempos unos (For all times are not the same)," the epigraph introducing this chapter.[8] As already noted, Cervantes' insight forms the basic premise upon which this book is based. Were it false, were Newton to prevail—as he did for several centuries —time would be reduced to a constant flow of banal, dreary, sterile moments, because the Newtonian concept of time was separate from events. Thus devoid of content, it could be characterized only by amount, for being reversible (Whitrow 1980, p. 3), it even lacked direction.

Although fungible Newtonian time has been fruitfully applied in many domains, its variability, being solely in terms of quantity, renders it not unimportant but extremely limiting, an "intellectual straitjacket" (Davies 1995, p. 17). To break out of that straitjacket, the strongest assumption underlying this entire book is that times differ, and they differ in many ways other than quantity, in ways that give time and times much greater potential for variance than Newtonian time. And the variance in times is a most profound sort of variance, so profound that Ilya Prigogine concluded that "time is the fundamental dimension of our existence" (1997, p. 1). Thus we strive to know time, not just to understand it, but to understand ourselves. And then not just to understand who we are or how we came to be, but to recognize the possibilities of who we might become. Because the most important findings of any investigation, empirical or theoretical, are not the discoveries of what is. The most important findings are the possibilities, the intimations of what yet may be. So ultimately this book is about possibilities—profound possibilities.[9]

2

Temporal Realities

Pythagoras also, when he was asked what time was,
answered, it was the soul of this world.

—Plutarch, *Morals* (in Platonic Questions)

As I wrote this book, I purchased a new watch. This would be unremarkable
except that one noteworthy feature of the watch led to its purchase: It will
never need to be wound nor have its battery changed, for it is solar powered,
and according to promotional material, it will run "forever." Forever being a
hyperbolically long time in this case, the word represents more the promoter's
use of poetic license than a realistic estimate of the watch's likely longevity.
More plausible would be a claim that the watch will operate properly without
winding or battery changes for the rest of my lifetime. (The warranty was for
a much shorter period than "forever.")

This is all well and good, but this watch is worth mentioning because it
seems especially infused with human temporality. Its face presents the millennia-
old template for reckoning the hours (discussed later in the chapter), a template
that is, of course, a social construction. Its solar-powered system directly links
the watch to sources of light, especially that fundamental light source, the sun,
and by doing so continues a linkage between human time and solar behavior
spanning several million years (discussed later in the chapter). And shared with
its time-reckoning contemporaries and forebears is the belief that it is measuring something, something real called time. But what is this something, this
time? This is the ancient question, a question this chapter addresses. And as if

that question is not challenging enough, still another even more esoteric question will also be examined: Why is there time?

Saint Augustine framed the first question as a paradox in the most famous quotation in all of temporal scholarship: "What is time then? If nobody asks me, I know: but if I were desirous to explain it to one that should ask me, plainly I know not" (1912, p. 239). Things had not gotten much better a millennium-and-a-half later: "It is impossible to meditate on time and the mystery of the creative passage of nature without an overwhelming emotion at the limitations of human intelligence" (Whitehead 1964, p. 73). Perhaps it is for this reason that so few books *focused on time* include a listing for "time, definition" in their indexes—the index to Elliott Jaques's *The Form of Time* (1982, p. 237) being a rare exception that proves the rule. This does not mean that the issue goes un-addressed, far from it, but definitions taking the form "Time is _____" have been avoided. This may be because the problem of time's ontology, of its fundamental nature, seems so intractable that its conceptualization in a simple declarative sentence proves elusive, to say the least. For, as Edward Hall has noted, "It is possible to philosophize endlessly on the 'nature' of time" (1983, p. 13). Fear not. This consideration will not endure "endlessly," just for part of this chapter. And perhaps one reason "time is" statements have been so rare is that time is a collective noun.

IMMODEST SUGGESTIONS OR COSMIC VERITIES?

I have written elsewhere that time "is a collective noun" (Bluedorn 2000e, p. 118). That pithy statement summed up the belief that there is more than one kind of time. For example, Paul Davies thought long and hard about time, especially as it is conceptualized in the physical sciences. Yet despite those labors, he felt time's mystery still: "It is easy to conclude that something vital remains missing, some extra quality to time left out of the equations, or that there is more than one *sort* of time" (Davies' emphasis; 1995, p. 17). So in the physical sciences just as in the social, the possibility is explicitly recognized that there may be more than one kind of time.

Interestingly, many of the categorizations of multiple types of time have been binary, and several category pairs illustrate this point. Isaac Newton's *absolute* time contrasts with Albert Einstein's *relative*; the Greeks distinguished *chronos* from *kairos* (Jaques 1982); Henri Bergson (1959) saw abstract and vital times; Paul Fraisse (1984), succession and duration; Stephen Gould (1987), John Hassard (1996), and Michael Young (1988), among others, linear and cyclical; McTaggart (1927), A- and B-series; and many (e.g., Clark 1985; Gersick 1994; Orlikowski and Yates 1999), clock-based time and event-based times. This exercise could continue to tedium, but the point is that several binary classification schemes have been developed that propose two forms of time.

A danger inherent in such systems is the tendency to argue that one or the other type is the *real* time, or at least the preferred time, and to thereafter see all phenomena as representatives of the preferred type or as a distortion of it. So when Lewis Mumford wrote that "each culture believes that every other kind of space and time is an approximation to or a perversion of the real space and time in which *it* lives" (Mumford's emphasis; 1963, p. 18), he was arguing that people tend to see their own views of time as the real time. And I argue that a binary classification system exacerbates this tendency, with one choice receiving the imprimatur of "real time" and the alternative being condemned as a "perversion" of it, if it is perceived at all.

Barbara Adam rejected such thinking in favor of dualities (1990, pp. 16–19). So did Wanda Orlikowski and JoAnne Yates, who argued against such false dichotomous thinking by suggesting that these distinctions be properly seen as "dualities, as both/and distinctions" (1999, p. 17). Another way of saying this is that there is no imperative to see such categories as mutually exclusive. Neither partner is the true, real, or even preferred time; instead, they may coexist, intermingle, and even be tightly integrated in specific social systems. This point is illustrated well by Peter Clark's research (1978, 1985), which revealed that both clock-based and event-based times coexisted in English organizations.

But not all analysts employ a binary classification system. Some identify more than two times (e.g., Richard Butler [1995] distinguished four types). Those employing phenomenological and ethnographic approaches have found many examples of many times, ranging from Eviatar Zerubavel's *duty period* (1979, pp. 32–34), to Frank Dubinskas's *developmental and planning times* (1988), to what is probably the most famous such time of them all, Donald Roy's *banana time* (1959–60, p. 162), which was a daily work-group ritual focused on the consumption of a single banana. But these times, colorful and insightful as they may be, are fully nominal-level distinctions, and as such they extend the binary approach of classifying times to systems permitting more than two types.

Some theorists have taken the multiple-types approach further and proposed multiple types of time that are arranged in hierarchies. It is interesting to note that these approaches all seem to rely on a hierarchical view of reality itself. This hierarchical view was evident early in the origins of modern social science, first in August Comte's concept of a hierarchy of the sciences (1970), then in Herbert Spencer's discussion of inorganic, organic, and superorganic evolution (1899), and again in A. L. Kroeber's discussion of the superorganic (1917). Further, in one of the few early in-depth sociological treatises on time, Pitirim Sorokin distinguished the categories of physicomathematical, biological, psychological, and sociocultural time (1943). (See Gurvitch 1964; Moore 1963; and Sorokin and Merton 1937 for three other relatively early sociological works that focus on time.)

An even more elaborate temporal hierarchy was developed by J. T. Fraser (1975, 1999), who proposed a six-level hierarchy that reflects the development of reality, the history of existence. The first three levels in Fraser's hierarchical theory of time are atemporality, prototemporality, and eotemporality, all of which deal with levels of physical reality; to wit, the absolute chaos of electromagnetic radiation at the instant of the Big Bang (the birth of the universe), the realm of particle-waves, and massive objects such as planets and stars, respectively. The fourth level is biotemporality, which is the time associated with living organisms, and among the characteristics of which are short-term time horizons. Following biotemporality in this hierarchy is nootemporality, which is the time of the human mind, with longer, open-ended time horizons. Atop the hierarchy is sociotemporality, the time of a society produced by a social consensus. This theoretical model is described in a set of eight propositions (Fraser 1999, pp. 26–43), the elaboration of which provide many of the model's details, including the points that the hierarchy is a nested hierarchy and that the hierarchy is open-ended, meaning that there is no necessary logic that indicates the time of human societies is the final temporal form that will evolve in the universe. Although not well known in the social science literature on time, this model of time as a hierarchy of nested temporalities is the most complex of the collective noun strategies. Whether it will also be the most successful remains to be seen.[1]

But what about an "is" statement? My assertion that time is a collective noun does not narrow the field greatly, for it describes approaches to defining time rather than time itself. The hierarchical theory of time provides such an "is" statement, defining time as "a hierarchy of distinct temporalities corresponding to certain semiautonomous integrative levels of nature" (Fraser 1975, p. 435). However, this definition requires the subsequent definition of each temporality for a complete understanding, which eliminates some of the simplicity desired in an "is" statement. Other "is" statements have been provided too, such as Whitehead's (1925b, p. 183), which will be discussed later in this chapter.

So "time is" statements do occur, even if they are rare, and they are likely to be debated as well. And not only the definitions of time are a subject of disagreement; so too is the question of time's direction. For if the second law of thermodynamics holds—and no less an authority than Albert Einstein felt that classical thermodynamics would "never be overthrown" (1949, p. 33)—along with its implication that all energy transformations are irreversible (Coveney and Highfield 1990, p. 150), then time, or at least some times, would have a flow, and that flow would have a preferred direction giving *objective* meaning to the concepts of past and future. Arthur Eddington described this preferred direction as "time's arrow" (1928, p. 69), and the debate about it has raged ever since (e.g., Coveney and Highfield 1990; Denbigh 1994; Fraser 1999; Harrison 1988; Hawking 1988; Novikov 1998; Savitt 1995).

A great deal of this debate focuses on the issue of entropy, which is the level of disorder in a system (Hawking 1988, p. 102; Davies 1995, p. 34), which is held to increase as irreversible processes occur in closed or isolated systems (Whitrow 1980, p. 5). A key point in the debate about time's arrow is whether or not the universe is a closed system. Because the issues of entropy's direction and level in the universe are closely linked to the question of whether the universe is a closed system, and because it would be fairest to say that no one really knows whether the universe is a closed system, a definitive resolution to this debate based on direct empirical evidence seems unlikely in the foreseeable future. This debate is made even more difficult because many different formulations of the second law of thermodynamics can be developed, perhaps twenty or more (Bunge 1986, p. 306). And the subject of the debate is formidable enough already, because it takes us, to use Emily Dickinson's penetrating phrase, "Into deep Eternity!" (1890, p. 116).[2] But who knows? The debate may be resolved satisfactorily sooner than anyone can foresee.

Of course, this debate is about the direction of time's flow, which is a profoundly fundamental attribute of time, whatever position one takes about it.

And if the time's arrow advocates prevail, the direction of time would assume a justified place in the definition of time itself, albeit both issues are likely candidates for continuing debate.

And the word *debate* provides an important insight into our views about time, for whatever side of the debate about time or time's arrow seems to prevail at any moment, that view about time or its direction is a social construction. This is because the various positions held by the debaters would not have occurred if they had lived in isolation for their entire lives. Their views and beliefs occurred only because of the debaters' direct and indirect interaction with other human beings—including their debates over these issues, debates being but one form of social interaction. So the vital point is that all conceptions of time are and always will be social constructions, which is, in Barbara Adam's words, "the idea that all time is social time" (1990, p. 42). After all, all human knowledge, including scientific knowledge, is socially constructed knowledge. But this point does not ipso facto invalidate any or all concepts of time. Their validity rests, instead, on their utility for various purposes, such as prediction and understanding. And as societies and cultures evolve, it is likely, perhaps even incumbent, for their concepts of time to evolve as well. So it would be well to understand *how* concepts of time differ in order to understand them and their differences better. Toward this end, a model of temporal differences is presented next.

A CONTINUUM OF TIMES

Investigations of human time reveal a profound distinction, and although investigators would recognize the respective labels used as kindred concepts, no two analysts appear to have employed the same names for the components of this fundamental dichotomy. As shall be shown, this dichotomy represents the two end points of a single continuum, a continuum anchored by two temporal archetypes: epochal and fungible times.

Fungible Time

In the beginning all human times were epochal times. But for how long it is difficult to determine because human times are intrinsically linguistic phenomena, and the date at which the Hominidae (the taxonomic family of which modern humans are the only living species) developed language is so far unknown. Estimates for this cultural watershed range from thirty-five thousand or so years ago to perhaps 2 million years or more (Cartwright 2000, p. 202), so the length of the era in which epochal times were the only form of human time remains obscure. But clearly it was for the vast majority of hominid history. The date or at least the era in which epochal time's opposite, fungible time, began can be more confidently dated in thousands rather than millions of years, and in anything like its extreme modern form to within the last ten thousand years. This Janus-faced development deserves the label "Creativity," the capital "C" denoting the culture-changing form of creativity Mihaly Csikszentmihalyi reserved for creative acts that change an entire culture or important segments of it (1996, pp. 7–8, 27, 30). In the case of fungible time, the change revolutionized the way humanity would think about the universe and its place within it.

Fungible time is Newton's absolute time, which he described as "absolute, true, and mathematical time, in and of itself and of its own nature, without reference to anything external, flows uniformly and by another name is called duration" (1999, p. 408). The key parts of this statement for the concept of fungible time are "uniformly" and "without relation to anything external." Elaborating directly or indirectly upon these points, other authors have used varying labels to describe this form of time, including abstract time (Bergson 1959), chronos (Jaques 1982), even time (Clark 1978, 1985), and clock time (Lee and Liebenau 1999; Levine 1997).

Several of the distinctions drawn by Joseph McGrath and Nancy Rotchford (1983, pp. 60–62) to describe the dominant concept of time held by Western industrialized societies in the twentieth century seem to describe this type of time well. This temporal form is homogeneous, which means that one temporal unit is the same as any other unit of the same type, and this means that such units are conceptually interchangeable with each other. So one second is the same as any other second, one minute is the same as any other minute, one hour is the same as any other hour, and so forth—the term *fungible* referring to things that are substitutable for each other without restriction. It is linear in that it extends "forward and backward without limit" (p. 60), a belief whose coming revolutionized science in the eighteenth and nineteenth centuries. And it is objective and abstract, something that is seen as existing apart from events (i.e., Newton's "without reference to anything external") and as *real*, not just the right stuff, but the *real* stuff. Most readers will be familiar with these descriptors, even if they have never considered how the industrialized West thinks of time. Indeed, the set of beliefs in this form of time, fungible time, is so deeply held that most westerners accept it as real time—which is the reason I chose *fungible* time as its label.

Several previously cited labels could have been used for this concept, and in many contexts their ability to serve the narrative's lexical requirements would compel their use rather than provoke the use of yet a new label. In this case, however, they all share a common failing: In both scientific and lay usage they have all played a role in developing the West's cultural beliefs about this form of time and have promoted the institutionalization of these beliefs. The mere presence of these labels touches deep convictions within the reader, convictions that generate a reaction that "now we are dealing with *real* time and not that made-up human stuff." For this reason a new label was desirable, so I built the concept for this form of time around the term *fungible time*, which I believe is the first use of this term to conceptualize this entire category or form of time. However, novelty for its own sake was not the primary motivation. Nevertheless, novelty is a virtue in this case because a new term would lack a reifying historical presence. Further, the term also needed to describe the phenomenon aptly, and by association with its more traditional usage, convey the attributes of a temporal form that is dull, dreary, and sterile.

The legal term *fungible* was novel, it emphasized the homogeneity of the form's divisible temporal units, and it communicated a less authoritative posture than words like *absolute* or *universal*. It thus gives both the reader and the analyst a better chance to see this form as a human construction and will seem in no way inherently superior or more profound than the epochal forms. After all, Alfred North Whitehead did write, "In fact absolute time is just as much a metaphysical monstrosity as absolute space" (1925a, p. 8). So despite its traditional authority, fungible time shares at least one vital property with all other forms of human time: It was invented, not discovered.

The Development of Fungible Time. Fungible time did not spring fully grown from an epochal ancestor. Even though it is a recent development in the 7 to 8 million years of hominid phylogeny, it still required a developmental process, one that proceeded steadily at times but was punctuated in others by revolutionary events (see the punctuated equilibrium discussion in Chapter 4 for a general description of such processes). One such watershed event was the development of minutes, seconds, and hours. And we can recognize this development's importance because it provided a way to see reality. (See the discussion of the Sapir-Whorf hypothesis in Chapter 1.)

Minutes, seconds, and hours became part of many linguistic systems and thereafter structured the way human groups saw time—not all human groups, though, because not all human languages included these words. The development of these temporal units led to efforts to measure them, efforts that themselves reinforced the units by drawing people's attention to them and getting people to think in their terms—which likely led people to be concerned about their measurement even more. The culminating measurement effort produced the mechanical clock and its attendant concept of time, a form of time that came to prevail as the dominant concept of time in much of European civilization for the last half, or at least the last third, of the second millennium. Gradually the fungible view came to be articulated in science, as the earlier quotation from Newton illustrates. Indeed, Newton's concept would dominate the scientific view of time until the twentieth century.

By the end of the nineteenth century, fungible time had become the dominant temporality in geology and biology. Uniformitarianism, the doctrine that the forces slowly operating to change the earth today also operated throughout the past in the same way and at the same rate (Asimov 1972, p. 251), was first proposed by James Hutton (1959), then developed and systematized by Charles Lyell (1868). Not only did it become the temporal framework that permitted small short-term effects to produce monumental long-term geologic change; it also made Darwin's (1859) claims for organic evolution by means of natural selection (normally a small short-term effect) possible and then plausible (Eiseley 1958, pp. 246–47). Evolution required time, and not just any time. It required a fungible time that could operate over a span of then unprecedented length. Thus fungible time became dominant in the natural sciences and concomitantly so, if not always smoothly so, in much of Western civilization (see Thompson 1967).

An Illustration of Fungible Time. Benjamin Franklin's famous metaphor "Remember that Time is Money" makes sense only if the variety of time involved is fungible time.[3] Indeed, this statement helped promote the fungible temporal form because it implied that money, a clearly fungible commodity, was interchangeable with time, implying that time was also fungible. Franklin's aphorism lives on in modern financial management as the time value of money.

The time value of money is the idea that the value of a cash flow depends on when it will occur (Emery, Finnerty, and Stowe 1998, p. 117). For example, it is better to receive a dollar today than to receive it a year from now, because interest can be earned on the dollar once it is received. And with the use of two formulas, an analyst can travel back and forth in fungible time. Travel into the future is made possible by the formula $FV_n = PV(1 + r)^n$, where the future value (FV) of a quantity of money at present (PV) increases as a function of the discount rate (r) and the number of time periods (n) (Emery, Finnerty, and Stowe 1998, p. 118–19). For example, \$1,000 invested today in a bond paying 5 percent per year will be worth \$1,628.90 in ten years. Similarly, the formula $PV = FV_n [1/(1 + r)^n]$ allows travel from the future to the past by allowing the present value (PV) of a monetary sum to be calculated for any future sum, provided the discount rate and the number of time periods are specified (Emery, Finnerty, and Stowe 1998, p. 122).

Present value is really a special case of what could be generically called past value. Because financial managers have traditionally been interested in the value *today* of an amount specified at a future date, the value of that amount at times in the past has not concerned them, nor has it usually been calculated, hence the label *present value* rather than *past value*. However, nothing in the formula prevents it from calculating the future sum's value at points before the financial manager's present. So the tradition of stopping at the analyst's present would return the \$1,628.90 from ten years in the future to \$1,000 today by applying the present value formula. To calculate its worth two years ago (twelve years before its future location) at the same discount rate, twelve would be substituted for n in the present value formula rather than ten, by that producing a past value of \$907.03. The formula doesn't know when the analyst's present is, it simply calculates the future sum's value for any point *n* periods into the past.

For either the future value or the present value formulas to function properly, the units of time must be completely fungible: Each unit of time, a period, must be equivalent to and interchangeable with any of the other units. Indeed, these units may take on the values of any clock or calendar interval (second, minute, hour, day, week, month, year, decade, century, etc.), and are thus generically fungible. The time value of money as it has been developed in contemporary financial management procedures would be impossible to calculate if the time involved were anything but fungible time. The form of time in the time value of money is just as fungible as the classical physicist's *t*.

Epochal Time

As claimed earlier, originally all human times were epochal. And the absence of clocks early on is not the reason. The absence of minutes and seconds is more determinate, for they are a very recent development, appearing only a few thousand years ago in the multimillion years of hominid evolution. But this is what epochal times are not, so a more positive explanation is required in terms of what they are.

Epochal time is defined by events. The time is *in* the events; the events do not occur *in* time. Events occurring in an independent time is the fungible time concept that Newton described so influentially as absolute time and Whitehead described so critically as a "metaphysical monstrosity." When the time is in the event itself, the event defines the time. To take an everyday example, is it time for lunch or is it lunchtime? Time for lunch could be determined by hunger, making it somewhat epochal, but in much of the industrialized world the time for lunch is usually signaled by the clock, often the arrival of noon, and lunch is the activity that fills a fungible time interval (e.g., noon to 12:30 P.M.). The epochal time analogue, lunchtime, is more apt to be linked to the individual's internal rhythms (e.g., the onset of hunger), external social rhythms (e.g., the flow of work that day), or both, making the definition of lunchtime *whenever the individual or group decides to eat lunch.* The event (eating lunch) defines the time; the time is *in* the event and the social and psychological constructions of it.

Although the lunchtime example is mundane, the principle that time is in the events has been proposed as a universal concept. Whitehead did so: "Time, Space, and Material are adjuncts of events," and "Events (in a sense) are space and time, namely, space and time are abstractions from events" (1925a, pp. 26 and 63). Einstein seemed to do so too: "I wished to show that space-time is not necessarily something to which one can ascribe a separate existence, independently of the actual objects of physical reality. Physical objects are not *in space*, but these objects are *spatially extended.* In this way the concept 'empty space' loses its meaning" (Einstein's emphases; 1961, p. vi). Since in Einstein's view time was part of space-time, just as Whitehead's, his comment points to time being in the events (physical objects in this case). Thus there is nothing necessarily any more contrived or constructed about epochal time than there is about fungible time.

The concept of epochal time and even its label were used by Louise Heath

(1936) and McGrath and Rotchford (1983), its use by the latter leading to its use here. (I have since discovered that Robert Smith [1961, p. 85] also used the phrase.) However, it was several years after choosing this label before I encountered Alfred North Whitehead's "epochal theory of time" (Whitehead 1978, p. 68). Interestingly—and I must admit, reassuringly—the concepts of epochal time in the discussion presented here and in Whitehead's work appear very similar. In Whitehead's formulation, temporality and time's arrow develop because of the becoming and perishing of episodic events and occasions of discrete experience (Lucas 1994, p. 670). Moreover, within this context, Whitehead offered an "is" statement about time: "Time is sheer succession of epochal durations" (Whitehead 1925b, p. 183).

Whitehead's concept of time as a succession of becomings and perishings, then, easily accommodates the realities of development ("becoming") described at the physical, biological, and social levels in contemporary chaos and complexity theory (Marion 1999; Prigogine 1997; Waldrop 1992), and it also accommodates the existence of entropy and entropic processes ("perishing"). Although the "perishing" element of this concept of time seems to take its origin in John Locke's work relating time to "perpetual perishing parts of succession" (e.g., Locke 1959, p. 238), a view cited by Whitehead (1978, p. 29), Whitehead's concept includes the "becoming" side of the coin too. So rather than perpetual perishing alone, poet Delmore Schwartz captured the essence of Whitehead's view of time as a succession of becomings and perishings in the following metaphor: "Time is the fire in which we burn" (1959, p. 67). Fire provides energy (heat) for becoming but also consumes (perishing).

So some times are fungible and others are epochal. But the distinction between fungible and epochal times has gotten lost or at least blurred to contemporary observers. People who work in organizations, which is most people in the industrial and postindustrial worlds, tend to eat lunch at about the same time every day. The routine becomes so habitual that the period even comes to be called the lunch hour (even if it is only thirty minutes long in some organizations), which does attach a content or event meaning to a fungible time span: That hour (or thirty minutes) is qualitatively different from the hours that precede and follow it. But because the lunch hour has become so well institutionalized that it always occurs at the same time, its epochal nature has become intertwined with fungible time, and eating lunch now describes the activity that occurs at a fungible time as well as an event defining it.

This intermixing of fungible and epochal times has happened many times. Consider geology and archaeology. In geology, uniformitarianism placed fungible time at the core of historical geological processes. At about the same time (the nineteenth century), though, an epochal description of geological time began to develop as well. The result is the well-known classification of historical geological events as four broad eras (Precambrian, Paleozoic, Mesozoic, and Cenozoic), periods within the eras (e.g., the Triassic, Jurassic, and Cretaceous periods within the Mesozoic era and the Tertiary and Quaternary periods within the Cenozoic era), and epochs within some of the periods (e.g., Paleocene, Eocene, Oligocene, Miocene, and Pliocene within the Tertiary period and the Pleistocene and recent epochs within the Quaternary period) (Berry 1968, p. 9).

William Berry's description of the principle used to define these different historical intervals helps illustrate the general concept of epochal time itself: "A time unit should mean an interval of time extending from events (or an event) that are unique in time and are used to denote its beginning (these events are included in the interval) to events (or an event) used to denote the beginning of the next time interval" (1968, p. 10). Thus these geological intervals (note that they are not intervals of an equal length in a fungible time sense), are defined by events (e.g., the appearance, expansion, and disappearance of specific species). Even though absolute methods of dating the intervals with methods based on the radioactive decay of elements have provided precise historical dates for these intervals, they are not defined by such dates. This is sometimes confusing to people who are accustomed to thinking in terms of fungible-time historical dates, because the absolute dates do not define the eras, periods, or epochs; physical and biological events do. Indeed, in principle the geologic time intervals could vary around the world in the absolute historical dates of their beginnings and endings, a point more easily seen in archaeology.

Analogous archaeological intervals begin with the three famous ages: Stone, Bronze, and Iron, which, interestingly, were developed by a Danish businessman, Christian Jürgensen Thomsen, who was given the task of organizing the rapidly growing pile of artifacts being sent to the Royal Commission for the Preservation of Danish Antiquities (Boorstin 1983, pp. 605–6). He sorted the artifacts by applying warehousing techniques used in the early nineteenth century, and when his museum opened in 1819, the artifacts went on public display grouped into the now familiar categories of Stone Age, Bronze Age, and Iron

Age. Thomsen had inferred that objects made of the same material were about the same age, and he also reasoned that the stone objects were older than the bronze, the bronze objects older than the iron (Boorstin 1983, p. 606). Thus the ages were defined in event terms: the use of different raw materials to make tools. And just as the geological eras were subdivided into shorter intervals, the Stone Age was later divided into the Paleolithic, Mesolithic, and Neolithic intervals (Oakley 1964); but again the subdivisions were event-based even though radiometric techniques (e.g., carbon-14, potassium-argon) allowed precise calendar dates to be assigned to them.

Although a cursory glance at contemporary organizational life would lead to the conclusion that it is dominated by fungible time, as in geology and archaeology both fungible and epochal time forms coexist. Peter Clark's (1978) observations of the importance of event-based seasons and their use for a firm in the textile industry were interpreted as comparable to the use of events such as changes in cloud formations (i.e., cloudier skies) by the Nuer to determine the end of the dry season (Evans-Pritchard 1940, p. 95). Indeed, ethnographically based organizational research commonly reports epochal times. For example, in a study of a high-energy physics laboratory, Sharon Traweek (1988, pp. 73–74) described many forms of ephocal time, among them "up" time (when the accelerator beam is running) and the "lifetime of a detector" (its life from gestation to obsolescence). In everyday life terms such as work time, playtime, teatime, and prime time all suggest epochal times as well.

The Temporal Heterogeneity Continuum

Although the focus of this discussion has been on the extreme forms of fungible and epochal times, a useful way to consider these two extremes is as the end points of a continuum whose defining principle is the relative distinctness of each form of time along the continuum, its temporal heterogeneity. This is so because some temporal forms have units or periods that are more distinct than others. A quotation attributed to Mark Twain (Least Heat Moon 1982, p. 10) illustrates this point well: "Although the past may not repeat itself, it does rhyme."[4] The lack of repetition can be taken to mean the absence of clonelike similarity, but the reference to rhyming indicates some similarity. To illustrate the point about degree of similarity, two words that rhyme in a poem might share phonetic similarity only by sharing at least one of the same phonemes, but in another stanza, two words might rhyme and also have similar meanings.

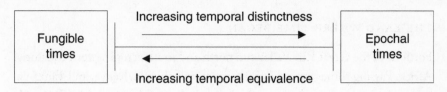

FIGURE 2.1. The temporal heterogeneity continuum

The latter case reveals more similarity than the former. But similar does not mean identical, and different does not require complete distinctness, as depicted in Figure 2.1. Thinking in terms of degrees of difference rather than just two extremes allows more precise statements to be made about the form of time under consideration than if one's conceptual portfolio contained only the two extreme forms.

As with the geological epochs and archeological ages, human times become more epochal as they become more homogeneous within themselves and more differentiated from other periods, units, or types. In analysis of variance (ANOVA) terms, the times become more epochal as the within-unit variance decreases and the between-type variance increases. Movement toward more epochal times is illustrated by phrases such as the "New York minute." This metaphor for the fast pace of life in New York City (see Levine 1997; also see Chapter 4) is so effective because it violates a tacit understanding about minutes: They should be equivalent and interchangeable because they are part of an extremely fungible time system. To indicate that some minutes are different from others violates a deep understanding by transforming a fungible time unit into a more epochal form—perhaps not completely epochal, though, because a "Boston minute" would be more similar to a "New York minute" than a "Los Angeles minute" would be to either (Levine 1997, pp. 148–49). These times do not repeat themselves, but they do rhyme.

So then, which type is the true time? Perhaps the best answer is to say they all are, a position Alan Lightman explained after he had described two different times, concluding, "Each time is true, but the truths are not the same" (1993, p. 27). And if the truths are not the same, once again we see that all times are not the same.

But why do times exist at all? Why do they differ? What produces these distinctions? And why have temporal differences persisted? To address such questions it is necessary to explore the origins of humanity itself.

THE LAND WHERE TIME BEGAN

Portions of the Great Rift Valley run north and south across eastern equatorial Africa. Passing through parts of contemporary Ethiopia, Kenya, and Tanzania, this geologic structure encompasses the birthplace of the hominids (the family of which modern humanity is the only living representative), and by some accounts it may even be responsible for their very existence. According to Yves Coppens's (1994) interpretation of the evidence, the valley formed about 8 million years ago, resulting in two very different ecological zones to its west and east. To the west, conditions remained humid and heavily forested. To the east, the climate dried, precipitation patterns became organized into seasons, and forest changed into savanna grasslands. As the climate and flora changed, so did the fauna, and humanity's ancestors were part of that faunal change. According to this interpretation, the lines leading to modern *Pan* (chimpanzees) and *Homo sapiens* (us) began to diverge after the valley was formed.

Our line likely began with a genus known as *Australopithecus*, beings who walked erect for at least 3 million years, who manufactured stone tools, and who may have had spoken languages (words fossilize poorly). The australopithecines branched into several species (e.g., *Australopithecus afarensis*, *A. africanus*, *A. robustus*, etc.) and eventually gave rise to a second genus, *Homo*. Modern humanity (*Homo sapiens*) is the only living representative of this genus, but it was preceded by *at least* two earlier species, *Homo erectus* and *Homo habilis*, the former seeming to have endured, albeit evolving, for well over 1 million years. Members of this subfamily also walked erect and manufactured stone tools, and its *H. erectus* and *H. sapiens* representatives used fire. Obviously, one species of *Homo* used spoken language, as may have all of its species.[5]

Against this historical backdrop, it is possible to see why the hominids developed a sense of time, temporal concepts, and the ability to perceive temporal aspects of the phenomena amid which they lived. These constructions and abilities provided survival value to help address the need, in Coppens's words, "for adaptation to the new habitat of the savanna, one that was drier and more bare than the preceding one" (1994, p. 92).

But how would developing temporal expertise increase the hominids' ability to adapt to their new environment on the savannas of East Africa? To increase adaptability, temporal expertise would have had to provide the hominids with capabilities they had not possessed before, but not just any capabilities.

They would have had to have been relevant capabilities, such that they would increase the hominids' probability of survival on the veld. As such, two capabilities seem to be conferred by temporal expertise, by forms of socially constructed time, which means all forms of time consciously and unconsciously used by any hominid group. These two capabilities are the abilities to coordinate and to provide meaning.

For example, Bronislaw Malinowski addressed the functions of time as follows: "A system of reckoning time is a practical, as well as a sentimental, necessity in every culture, however simple. Members of every human group have the need of coordinating various activities, of fixing dates for the future, of placing reminiscences in the past, of gauging the length of bygone periods and of those to come" (1990, p. 203). Sixty-three years later Barbara Adam would state it thus: "As ordering principle, social tool for co-ordination, orientation, and regulation, as a symbol for the conceptual organisation of natural and social events, social scientists view time as constituted by social activity" (1990, p. 42). The emphasis on the coordination function is more obvious in these statements, being mentioned explicitly in both of them, and this general function has also been noted in the organization science literature on time (e.g., Gulick 1987, p. 115). Less explicit and less emphasized in the literature is time's role in the creation of meaning. Malinowski and Adam hint at this capability in their statements: "sentimental necessity," "orientation," and "symbol for the conceptual organisation of natural and social events," the last of the three phrases most directly indicating time's role in generating meaning. So both capabilities increase with the development of greater temporal expertise. And in the remainder of this section the more familiar and intuitively plausible temporal function of enhancing the ability to coordinate will be discussed; in the next section, the ability to generate meaning.

Coordination could refer to the ability to coordinate personal activities, but because so much human activity occurs within a social context, much of the coordination function involves temporally ordered interaction with other human beings. Indeed, as Wilbert Moore concluded, "If activities have no temporal order, they have no order at all" (1963, p. 9). Although it is likely that the earliest hominids were social animals—many primates are social, including our closest living primate relatives, the chimpanzees and gorillas—the first forms of time to consciously emerge may have percolated up from prelinguistic knowledge that existed earlier in the lineage.

A basic dichotomy, one still tremendously important for organizing human affairs, may qualify as the first forms of hominid time. And these two forms were certainly epochal. They differed qualitatively in their feel and in their texture, in their meaning and in their purpose, in their importance and in their potential. Daytime was warm and bright, a time for maintaining life by hunting and gathering, the time for successful foraging to stave off hunger and death. Nighttime was different, a cooler, darker domain, a fearful time best spent resting, a time to avoid the things, much more powerful things, that *hunted* during the night. Thus what are likely the first two forms of hominid time, daytime and nighttime, differed not so much in their length—the equatorial periods of daylight and darkness being about the same in the Great Rift Valley and its environs—but in their attributes, and more important, in the expectations, reactions, and beliefs held about them by a nascent humanity. And although the lightness-darkness *cycle* may have been perceived, it is anyone's guess whether the two periods that constitute this cycle were perceived as a single unit *conceptualized* as a "day" this far back, or even when this idea would have developed, a point that reinforces the socially constructed nature of times.

Our hominid ancestors obtained survival value from being able to identify these two forms of time and by organizing their activities with respect to this planetary rhythm. Moreover, they learned to anticipate the onset of the relatively hostile nighttime environment so that they—slower, weaker, and with less visual acuity than their nocturnal predators—would not be caught in the open a long way from the group when the sun went down. There is a technical term for the australopithecine or early hominine in the Great Rift Valley area who was found alone in the open after dark: dinner. Indeed, in the company of a small band of modern hominids in Tanzania, I have heard the lion's roar after dark while camping on the Serengeti plains. Further, late one night everyone was awakened by the screaming of a baboon troop that was spending the night nearby on a large kopje (pronounced "copy," a huge rock formation). After dawn our guide explained that a leopard had passed through the area (camp!), a visitor neither species of social primate would have cared to experience a face-to-face encounter with, either individually or in a group.

Hominids must sleep to maintain their mental contact with the world and to avoid death (Coren 1996b, p. 59), and one wonders whether some hardwired requirement for sleep lies deep within the DNA, thereby conferring survival value to diurnal creatures like the hominids who are relatively helpless at night, by keeping them relatively inactive at night, hence less vulnerable to predators and accidents. And nothing has really changed in this regard after 7 or 8 million years of hominid evolution. The only way twenty-first-century hominids can function after nightfall is to turn the night into day by artificial means. Without artificial lighting or night vision apparatus, contemporary humans are just as disadvantaged after dark as were their forebears all those millions of years ago. So if being with the local hominid group after dark provided survival value, then the ability to judge distances, hence travel times, combined with the ability to estimate in travel-time terms the time to sundown would enhance survival potential by increasing the hunter's or gatherer's chances of returning to the larger group before the onset of night. Or if the entire group or parts of it tended to forage together during the day, the group's survival potential would similarly be enhanced by such estimation skills because such skills would allow the group to find or create relatively secure sanctuaries before night began. Either way, survival potential would be enhanced by such abilities.

Thus from the beginning of human time there is a link between time and space, making the concept of space-time in contemporary physics (Whitrow 1980, pp. 270–320) less a completely novel development than one that continues a tremendously ancient hominid synthesis of the two phenomena. Notably, at about the same time that Isaac Newton was declaring space and time absolute and distinct, John Locke was anticipating their formal twentieth-century synthesis: "To conclude: expansion and duration do mutually embrace and comprehend each other; every part of space being in every part of duration, and every part of duration in every part of expansion" (1959, p. 269).

The ability to anticipate nightfall, which was reinforced, indeed selected for every twenty-four hours, was likely an important survival adaptation that promoted, however slowly, the ability to anticipate future events over longer time frames. For instance, Donald Johanson and Blake Edgar (1996, p. 92) have suggested that the brains of fruit-eating species are often larger than those of their leaf-eating counterparts. Most directly relevant to the evolution of hominid temporal expertise is their suggestion that fruit eaters need larger brains "to process the more complex seasonal and geographic information about their environment" because "fruits are seasonal and more regionally distributed than leaves" (Johanson and Edgar 1996, p. 92). This interpretation complements

Coppens's (1994) explanation that the emergence of the Great Rift Valley organized precipitation into *seasonal* patterns, which would result in more distinctive seasonal variation in its flora. So at least indirectly the appearance of the Great Rift Valley led to the development of the first forms of human time such as daytime, nighttime, fruit season, dry season, and so forth.[6]

An important cognate question is, at what point in hominid evolution were the first words developed and spoken representing these forms of time? Which came first, the developing expertise or the words? This issue is wrapped up, of course, in the debate about the origins of language itself, something about which, because the issue involves spoken rather than written language, no *direct* evidence exists. So the principal evidence for this debate comes from an analysis of what appear to be the anatomical requirements for speech in modern humans (among them certain brain structures, position of the larynx and hyoid bone, and basicranium structure [Cartwright 2000, pp. 205–6; Johanson and Edgar 1996, p. 106]) and an examination of hominid fossils to see whether similar anatomical features are present. But this is tricky work because the fossil evidence is often frustratingly incomplete, so the debate extends from positions differing over a range of at least two orders of magnitude, from about thirty-five thousand years ago to 2 million years ago (Cartwright 2000, p. 202) —and perhaps longer. For although John Cartwright concluded that attributing language to the australopithecines seems an improbable conclusion, one notes his judgment that it "seems improbable," not that it was impossible (2000, p. 206). And he does note that some australopithecine cranial remains reveal brain asymmetries, such asymmetries believed to be associated with language capability and use.

If I were to choose, I would choose the long view rather than the more recent. For as Terrence Deacon concluded, "These data [evidence for the expansion of hominid brains] suggest that it is unlikely that speech suddenly burst on the scene at some point in our evolution. The ability to manipulate vocal sounds appears to have been in a process of continual development for over 1 million years" (1997, p. 252).

Further, I propose here that whenever spoken language began to emerge, temporal phenomena played an important role in that development—for several reasons involving the social nature of the hominids. My friend Carol Ward, a physical anthropologist, has noted that the only evolutionary function of language is for one hominid to influence other hominids (Ward, personal communication, 2001).[7] As has already been discussed, influencing others about temporal matters would have had important survival implications on the savanna, both long-term and short-term, individual and group. For example, if as seems likely, hunting and gathering were social rather than solitary activities, then a strategic matter would have been for the hunting and gathering parties to decide how far to search and when to begin the trek back to that night's base (or, in another image, when to look for the evening's base). On a longer scale, the ability to discuss when to shift general locations given the seasonality of food sources would have provided survival advantages to groups that could have articulated cues about seasonal shifts perceived by several members of the group. This would also have helped develop an ability to consciously engage longer time frames, from the twenty-four-hour cycle to cycles involving several months. Thus would have developed the concept of the future. How specific or well articulated such ideas and discussions would have been early on is nearly impossible to surmise. Indeed, the first records of even monthly cycles may be only about thirty thousand years old, if one accepts Alexander Marshack's (1964, 1972) interpretation of notches carved on antler and bone. So a not implausible conclusion is that temporal matters were important stimuli in the genesis of hominid language. (See Chapter 8 for a discussion of the strategic role linguistic abilities to conceive the future may have played in human evolution.)

And if I were to hazard a guess, I would expect that the ability to anticipate events arose first, and only later with greater language sophistication did the *concept* of a future arise. But long before the sapiens' era, forms of time developed, all epochal, forms that included daytime and perhaps parts of daytime such as sunrise and sunset, nighttime, past, present, and future.

Although *H. sapiens* inherited these forms of time from their hominid ancestors, they did not realize that they were *forms* of time. Just as with twenty-first-century humanity, the earliest of our ancestors did not think of these temporal forms as human creations and likely regarded them as givens, as a part of nature, and at some point probably added to them beliefs about the proper activities for the day and the night. These beliefs were so fundamental that they seldom entered these people's conscious awareness, and with some exceptions, these beliefs were also not taught consciously—though taught they were. They were tacit knowledge, knowledge learned and held unknowingly—though learned and held they were.

MESSAGES SOTTO VOCE

Much of temporal knowledge much of the time is part of the knowledge Michael Polanyi described when he wrote, *"We can know more than we can tell"* (Polanyi's emphasis; 1966, p. 4). It resides in the deepest level of culture, what Edward Hall (1983) called primary culture; Edgar Schein (1992), culture's basic underlying assumptions. And at this level, beliefs and values tend to be held unconsciously, are taken for granted, are treated as reality (Schein 1992, pp. 16–22).

This may explain why time has played such a minor role in the enterprise of social science. Barbara Adam said it well: "Much like people in their everyday lives, social scientists take time largely for granted. Time is such an obvious factor in social science that it is almost invisible" (1990, p. 3). As tacitly, unconsciously held knowledge, it tends to be in the background rather than the foreground (Backoff 1999), so everyone ignores it in the sense that they take it—whatever it may be—for granted, and it becomes a part of a very firmly defined reality. Yet as Schein concluded, "There is probably no more important category for cultural analysis than the study of how time is conceived and used in a group or organization" (1992, p. 114). And one important use is the generation of meaning.

In groups of all sizes time is used to generate meaning. According to Schein, the parts of culture found in the level of basic underlying assumptions define "what things mean" (1992, p. 22). So how does time generate meaning? One answer would be in temporal terms. Someone is early or late, or the pace of activity is fast or slow. Someone handles many things at once or only a few. Things are out of sync. All of these statements convey meaning in explicitly temporal terms. Interestingly, they are also all comparative examples in which one condition (e.g., fast or slow) is related to the other condition. And this leads to a more fundamental explanation of how time generates meaning.

Alfred North Whitehead asserted that "'significance' is the relatedness of things," that experience is too if experience is equated with significance, and that "it is thus out of the question to start with a knowledge of things antecedent to a knowledge of their relations" (1925a, p. 12). Thus for meaning (significance) to be attributed to events, behaviors, and objects, to things in general, they must be seen in their relationships with other things. They can have no meaning as isolated phenomena. An example of how time generates relatedness is seen in the role of the past as analyzed by Quy Nguyen Huy:

"Since one cannot distinguish a figure without a background, the present does not *meaningfully* exist without a past" (emphasis added; 2001, p. 608). The meaning of the present is impoverished without a connection to the past, without a relationship with it. Some of these connections and relationships are tacit but still passed on nonetheless. And several examples illustrate the point that relatedness confers meaning, and that meaning is transmitted and socially constructed.

The investigation of hominid evolution is by definition an effort to reconstruct the past, and by so doing provide a background for the figure of a modern humanity. What may be the most remarkable archaeological discovery in the study of hominid evolution did not involve a single fossilized bone. It was, instead, a trail left by modern humanity's forebears, almost literally, on the sands of time.[8] For in 1978 an expedition led by Mary Leakey discovered a trail of sixty-nine footprints left by at least two australopithecines, perhaps 3.6 million years ago. The prints indicate two individuals walked erect and side-by-side, and the difference in size of the pairs of footprints may reflect the species' sexual dimorphism, hence one walker may have been a male, the other a female (Gore 1997). To see the point about relatedness, one should immediately see the connection with hominids today: erect and bipedal, walking side-by-side, just like us.

The way contemporary hominids walk takes on greater meaning knowing that it is an ancient practice, a personal connection that Mary Leakey experienced herself. Writing of behaviors indicated by the footprints of the smaller of the two hominids, she provides a moving example of the potential meaning inherent in making such connections and seeing such relationships:

> Incidentally, following her path produces, at least for me, a kind of poignant time wrench. At one point, and you need not be an expert tracker to discern this, she stops, pauses, turns to the left to glance at some possible threat or irregularity, and then continues to the north. This motion, so intensely human, transcends time. Three million six hundred thousand years ago, a remote ancestor—just as you or I—experienced a moment of doubt. (Leakey 1979, p. 453)

But there is more, for there is evidence that a third individual walked with the other two: "The imprint of a second big toe in several of the larger prints suggests that another individual may have walked in the footsteps of the first, like children do in the snow" (Gore 1997, p. 80). Contemporary hominids are left to their own counsel to interpret what the possibility of that third individual might mean—both for the relationships among the three walkers and for the connections between the three walkers and twenty-first-century hominid social structures.

Although hominids have been genetically predisposed to bipedal locomotion for several million years, cultural variations seem to exist in how modern hominids walk (Hall 1983, pp. 184–85). Since it is cultural variation—such differences are not determined genetically—they are learned. But how are they learned? A large part of this type of learning would seem to occur through the semi- and unconscious observation and imitation of others in the group and in the subtle reactions of group members to the learners' behavior.

A concept in Anthony Giddens's structuration theory explains how patterns like these are maintained with such regularity and precision. The concept is "duality of structure," by which Giddens meant that "the structured properties of social systems are simultaneously the *medium and outcome of social acts*" (Giddens's emphasis; 1995, p. 19). He used language to illustrate duality of structure by noting that a speaker uses the syntactical rules of a language to form a sentence (outcome), and that by speaking the sentence according to the rules, the rules themselves are reproduced (p. 19). Applied to cultural distinctiveness in walking styles, a walker unconsciously follows the group's kinesthetic rules for walking and walks in an approximation of them. By so walking, the walker reproduces those rules, both for the walker and for others who can observe and imitate, albeit at the level of unconscious awareness.

To demonstrate that such a process exists with temporal matters, look at your wristwatch or at the nearest clock. Examine that piece of technology, because objects can incorporate the duality of structure as well as values, something Wanda Orlikowski so deftly demonstrated (1992). As mentioned earlier, the watch's dial has long historical roots. But there is something about the watch or clock that is obvious yet so subtle that it is taken for granted. That something is the movement of the hands. Why do they move the way they do? I knew, of course, what the term *clockwise* meant, and had known for a long time, but despite studying time for the better part of two decades this question had never occurred to me until my chance encounter with David Feldman's compendium of perplexing questions about everyday life's arcana, one of which is, Why do clocks run clockwise? And the answer to this question (Feldman 1987, p. 150) is not as arbitrary as one might think.

Before mechanical clocks there were sundials, and sundials were invented in the northern hemisphere. In the northern hemisphere the shadow on the sundial rotated in the direction the world now knows as clockwise. With the development of mechanical clocks and their dials—at first mechanical clocks marked the time audibly with bells rather than visually with a dial and moving hand (Crosby 1997, p. 80), most clocks having only one hand until the mid-seventeenth century (Barnett 1998, p. 78)—clockmakers simply followed the pattern established by the ancient pattern of the sundial and geared the clocks so as to drive the hand, an imitation of the sundial's shadow, in the manner known for generations now as "clockwise" (Feldman 1987, p. 150).

Following this pattern would have helped gain legitimacy for the new form of horologe via mimetic imitation, and Giddens' duality of structure process is also apparent. The fourteenth-century clockmakers followed the "rule" for the sundial's shadow and built clocks whose hands moved in the same way, thereby reproducing and reaffirming the rules. This successfully transferred the sundial shadow's rule, established by both the sundial's design and astronomical behavior, to a mechanical clock rule, a device whose design and behavior present more degrees of freedom to its human architects. And once the transfer of rules from sundial to mechanical clock was made successfully in the fourteenth century, the duality-of-structure cycle has delivered clocks that have run clockwise for over six centuries.[9]

Narrowing now the clocks-and-watches example to the single solar-powered watch on my wrist, the watch that was discussed at the beginning of this chapter, a final example is provided of how relatedness produces meaning. Of course this watch runs clockwise, thereby sharing the linkage to northern-hemisphere shadows on ancient sundials. But being solar-powered, it is an engineering design that returns to the sundial's dependence on the sun for its functioning. Just as this watch receives its energy directly from the sun, so did ancient sundials. And before sundials, the position of the sun in the sky or even its total presence or absence marked the time, the latter dichotomy certainly doing so for the human lineage over millions of years: "The recurrent round of the day was obvious to even the dimmest early hominid" (Barnett 1998, p. 174). Thus, compared with watches driven by batteries or springs, my solar-powered watch more directly shares the ancient hominid practices of involving the nearest star in time reckoning.

But in this case there is even more meaning attached to this watch, not just to the category of solar-powered watches, or even to this specific model, but to the specific watch on my wrist. Because after first being attracted to the watch for its low-maintenance potential (no batteries to change, no stem to wind), and then by its direct connection to the sun, hence its deep connection to traditional human timekeeping, the potential for adding yet one more complex of relatedness, hence of profound personal meaning, closed the sale.

As the watch was being discussed, the saleswoman happened to remark, "When I get my next watch, I'm going to get this one [pointing to one in the display case], the woman's version of the one you are considering." It was the same watch, just a bit smaller to comfortably fit a woman's wrist. And this was the year in which my wife and I would celebrate our twenty-fifth wedding anniversary. So an idea occurred to me—it has probably occurred to you too—and I suggested to Betty that we buy the pair of watches as part of our twenty-fifth anniversary celebration.

The two watches are now directly related to both our twenty-fifth anniversary and the original marriage ceremony to which that anniversary is linked, and by that link to each other as well. Betty and I socially constructed this relatedness in the watches within a much larger socially constructed temporal context (i.e., the practice of counting and celebrating wedding anniversaries and of attributing special significance as milestones to counted anniversary years evenly divisible by twenty-five), by that making the watches that much more significant, that much more meaningful. So when I put on my watch each day, not only do I don "a scientific instrument which has encoded within it a heritage extending from deepest antiquity to the recent past" (Barnett 1998, p. 162), I attach to my body a time-reckoning machine infused with extremely powerful socially constructed personal meaning as well. My watch is not for sale.

One of the key points emphasized in Chapter 1 is that all times are not the same, but not only do times differ, those differences make important differences in human experience and meaning. So in the specific case of the past and the present, different pasts lead to different relations with the present, hence different meanings, different experiences, different presents. And this is true for time in general. For example, Barbara Tedlock's (1992) research on Highland Maya (Quiché) time described forms of temporal organization very different from those of Anglo-European culture. Those differences led Edward Hall to summarize her work as indicating the organization of Quiché time produced a totally different "experience of living" (1983, p. 81) from that in Anglo-European culture. Elliott Jaques was right: "In the form of time is to be found the form of living" (1982, p. 129).

That the past generates meaning for the present through its socially constructed relationships with the present is one example of the principle underlying all the remaining chapters: different temporal realities, different human experiences. So having now described the general capabilities time provides, having considered the nature of time, and in Chapter 1 having developed important reasons for studying time, our attention will shift in all that follows to the overarching issue of the association between temporal differences and experiences. Thus specific ways in which times differ will be described and related to the different human meanings and experiences they produce. Different temporal realities, different human experiences.

3

Polychronicity

Nobody can do two things at once, you know.
—Lewis Carroll, *Through the Looking-Glass*

All human beings implement a fundamental strategy for engaging life. Not that they all employ the same strategy—far from it—but everyone develops a strategy for engaging the mixture of life's activities, transcendent and mundane, even if just by default. Elsewhere I have written of things fundamental and asked, "What is as fundamental as time?" (Bluedorn 2000e, p. 117); and within the domain of temporal matters, it is also reasonable to ask, What is a more fundamental process strategy than the choice of the pattern for one's activities, a pattern that becomes habitual? A process strategy is not about ends; it is about means. It is not about what; it is about how.

Strategy was defined by Henry Mintzberg as "a *pattern in a stream of decisions*" (Mintzberg's emphasis; 1978, p. 935). And although an infinite number of patterns are possible, all strategies for engaging life's activities fall along a continuum known as polychronicity, a continuum describing the extent to which people engage themselves in two or more activities simultaneously. That this choice is fundamental is revealed by the fact that most people most of the time are unaware that they are even making it. This is because the choice of strategy results from a combination of culture and personality, both of which store these choices and preferences at deep levels, very deep levels. Nevertheless, a choice or a decision made unconsciously is still a choice or a decision.

If something is fundamental, it should have important consequences, which means that if the polychronicity strategies of cultures and individuals are fundamental, they should have important consequences. And they do. But before exploring these consequences, it is necessary to understand the polychronicity concept more completely. So before considering polychronicity's relationships with other behaviors, and before evaluating the effectiveness of specific polychronicity strategies, the polychronicity concept itself will be described in greater depth.

THE CONCEPT OF POLYCHRONICITY

Polychronicity is about how many activities and events people engage at once (a more formal definition will be provided shortly). And though not using the polychronicity concept, research on managerial behavior reveals that managerial work seems to require more polychronic behavior than much nonmanagerial work (see the descriptions of managerial work in Guest 1956; Mintzberg 1973; and Stewart 1967). Indeed, Carol Kaufman-Scarborough and Jay Lindquist concluded that the average manager works polychronically (1999, p. 293). This point is illustrated in the behavior exhibited by one of the general managers John Kotter studied (1982).

In a detailed description of one general manager's day (Kotter 1982, pp. 81–85), the polychronic nature of managerial work becomes evident. The manager attended a regular morning meeting, and Kotter noted that during the meeting "Richardson [the general manager] reads during the meeting" (p. 82). Reading and involvement in the meeting at some level indicate Richardson's simultaneous engagement in at least two activities.

Further, Kotter's description reveals that Richardson's entire day was characterized by short episodic activities—the meeting where Richardson read lasted only fourteen minutes—many of which were interruptions of other activities, activities with which Richardson returned to active engagement after dealing with the interrupting activity. For example, at 10:50 A.M. on this same day Kotter described Richardson thus: "He gets a brief phone call, then goes back to the papers on his desk" (p. 83). The "going back" is indicative of the back-and-forth pattern of polychronic behavior, because it is another way of engaging several activities during the same time. Many of the day's interruptions result from people—subordinates, peers, bosses—simply walking into

Richardson's office to discuss something. Although Richardson's office was not attached to the large, plazalike reception area Edward Hall described as characteristic of office designs in polychronic cultures (1983, p. 47), having one's door open and promoting a work climate in which people feel free to enter and interrupt could be considered an American approximation of such space-time architectures.

But what if Richardson had behaved less polychronically? What if he had taken a more monochronic (i.e., one thing at a time) approach to his day? How would he have behaved differently? At the meeting he would not have read; instead, he would have focused his attention on the meeting and nothing else. The material he read during the meeting would have been deferred to later when he would have read it and done nothing else. Similarly, he would not have taken phone calls while he worked on other tasks, perhaps having his secretary screen the calls for him so he could return them at a time when he did nothing but make phone calls. And the open-door policy would have been changed to one where he would see people only during certain times of the day, and then only if they had appointments. So a monochronic Richardson would be very different from a more polychronic Richardson.

Henry Mintzberg has characterized much work as involving "specialization and concentration," but he concluded there is at least one major exception: managerial work, which allows "no such concentration of efforts" (1973, p. 31). Mintzberg's description of managerial work as highly varied and fragmented echoes the findings of other research on managers (e.g., Guest 1956; Stewart 1967). For example, Robert Guest described the foreman's job as consisting of constant "interruption, variety, discontinuity" (1956, p. 481), with the hourly employee's work being just the opposite. Even more telling is his finding that foremen had to "retain many problems in their minds simultaneously, and to juggle priorities for action" (p. 480). The retention of "many problems in their minds simultaneously" speaks directly to the definition of polychronicity, mental activity being a component of polychronicity as well as overt behavior (Persing 1999).

The point is that managerial work is polychronic work, at least compared with most nonmanagerial work. So to become a manager means to face a work context replete with polychronic demands. And consistent with Guest's (1956) study of foremen, nonmanagerial employees appear to cross the polychronicity Rubicon when they are promoted to supervisory positions; nevertheless, it may take time, in Mintzberg's words, for a manager to become "conditioned by his [her] workload," for the work to lead the manager "to develop a particular personality" (1973, p. 35). The supervisory level may be the socialization and selection ground for identifying individuals who are sufficiently polychronic for the demands of managerial work and for possibly developing polychronic behaviors within managerial candidates too. But polychronicity is about engaging life in general, not just work, so the concept should be examined in the widest possible context, which takes us to its origins.

Edward Hall (1981b) introduced the concept of polychronicity to describe fundamental differences in human behavior, and he continued to study and develop the concept thereafter (Hall 1981a, 1982, 1983; Hall and Hall 1987, 1990). Hall described cultural variance along the polychronicity continuum in the following way: "In the strictest sense, a polychronic culture is a culture in which people value, and hence practice, engaging in several activities and events at the same time. Monochronic cultures are more linear in that people prefer to be engaged in one thing at a time" (Bluedorn 1998, p. 110). But as polychronicity scholars have noted (Bluedorn et al. 1999; Palmer and Schoorman 1999), Hall implied in some of his work that polychronicity refers to a much larger set of phenomena (e.g., Hall 1981a, p. 17; 1983, p. 53; Hall and Hall 1990, pp. 13–15). Nevertheless, most polychronicity scholars employ the more focused version of the concept, which is how the concept will be defined here. Following Bluedorn et al. (1999, p. 207) and Hall (Bluedorn 1998, p. 110), polychronicity is the extent to which people (1) prefer to be engaged in two or more tasks or events simultaneously and are actually so engaged (the preference strongly implying the behavior and vice versa), and (2) believe their preference is the best way to do things.

This definition is preferable on two grounds. First, it is consistent with many of Hall's own definitions (e.g., Bluedorn 1998, p. 110; Hall 1983, p. 230). Second, it allows for the testing of empirical relationships between polychronicity and other variables rather than assuming they are all simply dimensions of a single phenomena, polychronicity. This assumption creates ambiguity for interpreting results when such an omnibus variable is related to other variables (Carver 1989). Indeed, David Palmer and David Schoorman (1999) analyzed a large complex of variables that might have constituted a broader definition of polychronicity. Their results showed that compared to models in which the preference-for-engaging-two-or-more-events-simultaneously dimension was combined on the same dimension with other variables, the model in which this was specified as a separate dimension revealed a much better fit to the data. Moreover, the correlations between the preference-for-engaging-two-or-more-events-simultaneously dimension and the other dimensions in the best-fitting model were very low, so Palmer and Schoorman concluded, "The three dimensions of time use preference [preference-for-engaging-two-or-more-events-simultaneously], time tangibility, and context do not represent highly correlated measures and should be considered separately" (1999, p. 336). Thus polychronicity is defined here in the narrowly focused preference-for-engaging-two-or-more-events-simultaneously sense, and research about its relationships with variables suggested by Hall and others will be presented later in this chapter.

Polychronicity is a continuum, and preferences exist for degrees of engagement. At one extreme is the pattern of focusing on one task at a time, interpreting other potential tasks and events as interruptions and attempting to shield one's chosen task from such interference. The other extreme is actually open-ended, and it involves engagement in several tasks simultaneously, sometimes literally simultaneously and sometimes in a frequent back-and-forth engagement pattern. And Kotter's description of how Richardson worked illustrates both ways in which multiple tasks can be engaged simultaneously (e.g., reading *while* participating in a meeting, and switching back-and-forth between phone calls and his other work).

The previous discussions of Richardson's actual behavior and its opposites lead one to think in terms of two types of behavior: the engagement-in-many-tasks extreme being the high polychronicity pole (though higher levels of polychronicity may be possible), and the engagement-with-a-single-task extreme being the low polychronicity pole, sometimes referred to as a monochronic orientation. Although it is easier to see this distinction in terms of a dichotomy—polychronic or monochronic—polychronicity is a *variable* that reflects an underlying *continuum* of engagement preferences and practices, and a potentially infinite set of gradations distinguish one individual's preferences from another's, as well as one culture's from another's.

Linear Versus Circular

Life strategies become more and more linear toward the low-polychronicity (monochronic) end of the continuum where one task follows neatly upon the completion of its antecedent, forming a temporal archipelago. Toward the other end of the continuum, tasks and events are engaged at the same time, or what would seem to be more common, by revisiting the same projects multiple times during a given interval. Though not a set of eternal returns (see Eliade 1954 about the eternal return concept), projects, activities, and events may be revisited many times. Two examples illustrate this difference.

Although naps are known in the United States, they are usually reserved for an individual's private time and are not a behavior approved of on the job. In Spain, however, a centuries-long tradition is observed for work to cease for a two-hour period during midafternoon. This allows for a siesta or other activities, after which people *return* to work. Contrast the American with the Spanish process strategies. In the United States the activities of sleep and work are kept separate; in Spain they intermingle. The American process strategy is less polychronic, whereas the Spanish is more so.

Another way to see this difference is with a demonstration involving your hands. Straighten the fingers of each hand, gather the fingers together so that adjacent fingers touch, keep each hand flat, and then place both hands in front of your face, palms toward you, so that each hand is viewed comfortably. In this position your two hands illustrate a monochronic process strategy (see Figure 3.1a). The left hand represents one task; the right hand, a different task. As each hand is distinct, so too are the tasks. They neither overlap nor intermingle. This process strategy illustrates a very low level of polychronicity, perhaps as low as is possible because there is no overlapping of the tasks whatsoever.

But your two hands, representing two different tasks, can also illustrate more polychronic strategies. To shift to a more polychronic strategy, begin with your hands in the monochronic position (as in Figure 3.1a). Then splay the fingers and interlace them as illustrated in Figure 3.1b. Because the two hands represent two different tasks, interlacing the fingers illustrates a pattern of moving back-and-forth between the two tasks, and if the fingers are completely interlaced, a process of revisiting each task several times.

Thus the monochronic pattern is linear, because one task (the right hand) is distinct from the other (the left hand) and follows after it. Polychronic strategies are cyclical, because they involve multiple visits—revisits—to the same tasks and events. Noting this distinction, Mary Waller, Robert Giambatista, and Mary Zellmer-Bruhn developed a measure of group polychronicity that included observations of how often groups "switched back to a previous phase"

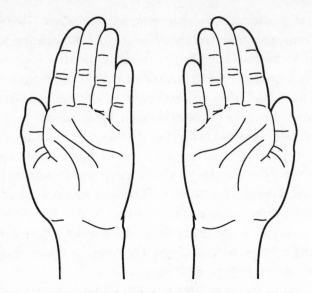

(a) Monochronic process strategy

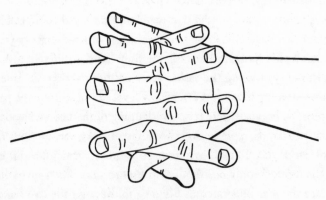

(b) Polychronic process strategy

FIGURE 3.1.   Monochronic and polychronic process strategies

of the problem-solving process (1999, p. 251). The two patterns—monochronic and polychronic—form a continuum, because polychronicity is the extent to which people prefer to engage in two or more tasks simultaneously, and the complete absence of any simultaneous involvements, engaging tasks one at a time, is the least polychronic position on the continuum.

Culture and Personality

From the beginning (Hall 1981b), polychronicity has been analyzed at both the group and individual levels. As such, it has been seen as both a cultural and an individual phenomenon. And values about the same phenomenon can and do occur in both cultures and personalities, but this does not mean that relationships involving them are the same across levels of analysis (e.g., Dansereau, Alutto, and Yammarino 1984; Robinson 1950). Relationships found at one level of analysis, individual or group, are suggestive of those relationships at another and are reasonable justifications for hypothesizing their existence as a prelude to their empirical investigation, such investigation clearly being necessary to establish the existence of relationships across multiple levels of analysis.

At the group level—*group* referring to all potential culture-carrying aggregations larger than a single individual (e.g., departments, organizations, societies, etc.)—polychronicity is a value and belief complex that manifests itself in overt process strategies. Although the strength with which it is held may vary, as a *fundamental* process strategy—it is fair to say *the* fundamental process strategy—whichever position along the polychronicity continuum is normative in a culture is apt to be held strongly. This is because such process strategies are mainly learned unintentionally, usually unconsciously. Such learned knowledge is retained at the level of culture Edgar Schein (1992) labeled basic underlying assumptions. This deepest of cultural levels normally contains beliefs and values prescribing behaviors that are so taken for granted and institutionalized that they seldom rise to the conscious level for extensive examination and discussion (Schein 1992, p. 22). Consequently, they are difficult to change, and in this sense they are strongly held.

When values and beliefs from this level do surface, a typical reaction is, "I never thought of that before." And while true of differences among cultures, this reaction is just as typical, perhaps more so, when people reflect on their own individual behaviors—which was the case for me when I learned about polychronicity for the first time while reading about it in Hall's *The Dance of Life* (1983)—I hope the "Aha" did not disturb many people on the plane. Although the strength with which elements of culture are held may vary, individuals, even within the same culture, also vary in their choices of fundamental process strategies. This is not surprising, because few would argue that cultures produce clonelike members who possess absolutely identical values and beliefs. If they did, a culture could be studied confidently by simply interviewing a single member and observing that member's behavior for a reasonable time. Though related, culture and personality do differ, making it necessary to study each in its own right.

At the level of individual beliefs and behavior, the nature of polychronicity as either a trait or a state becomes an important issue. If it is a trait, individuals will be much more consistent, even habitual, in the polychronicity process strategies they follow, more consistent than if polychronicity preferences are a state. But if polychronicity is a state, it will be affected much more by the contextual factors in an individual's environment, leading to much greater variability in patterns of polychronicity behavior. So the degree of stability or its converse, the amount of variability, would provide important clues about polychronicity's statelike or traitlike identity.

Such evidence has been provided in multiple forms. First, individuals have been asked about their preferences for being engaged in two or more activities simultaneously and have been able to complete psychometric instruments— questionnaires—designed to measure individual polychronicity (e.g., Kaufman, Lane, and Lindquist 1991a).[1] That people complete these instruments as readily as they complete other questionnaires indicates they are not baffled by them, which is at least modest evidence that people can recognize their own levels of polychronicity behaviors and values, which the questionnaires ask them to report.

A second and stronger piece of evidence comes from a study I conducted in which people in a large sample took the same polychronicity questionnaire nine days apart. The time interval allowed me to compare the similarity of their responses on the two questionnaires. Responses to the identical polychronicity scales on both questionnaires produced a substantial positive correlation: the higher an individual's score on the first questionnaire, the higher it tended to be on the second questionnaire.[2] So the people in this sample displayed a high level of stability in their perceptions of their own polychronicity.

Perhaps the strongest evidence comes from research that involved external observers. To study individual polychronicity, Jeffrey Conte, Tracey Rizzuto, and Dirk Steiner (1999) recruited one hundred pairs of friends from the student body at a large public university in the southern United States. The friends had known each other for at least one year, and part of the study required the pairs of friends (1) to complete a polychronicity questionnaire scale about themselves, and (2) to complete the same scale about each other. Conte, Rizzuto, and Steiner found a statistically significant positive correlation between the observers' ratings and the participants' self-ratings: The higher people rated themselves on the polychronicity scale, the higher their friends rated them on it too. It is noteworthy that the ratings concerned behaviors that the researchers had not briefed the subjects about beforehand, and that the subjects had likely not thought much about, if at all, before they took part in the research. Hence the results of this study indicate (1) that people are aware of such behavior patterns even if they do not usually attend to them consciously, and (2) that such patterns display sufficient stability for observers to detect them, albeit unknowingly. The apparent existence of stable patterns of polychronicity process strategies revealed by these several studies favors the interpretation of individual polychronicity as a traitlike property, as assumed in Slocombe and Bluedorn (1999, p. 95).

These studies provide important evidence that individuals' patterns of polychronicity display at least modest degrees of stability, meaning that individual polychronicity is unlikely to be purely a statelike individual characteristic. However, by itself this evidence does not indicate how traitlike polychronicity is, meaning this matter (cliché ahead) requires further research. This issue will be revisited toward the end of this chapter, but now the discussion will shift to a presentation of research about polychronicity's relationships with other variables.

THE CORRELATES OF POLYCHRONICITY

The fundamentally different strategies described by points along the polychronicity continuum have some importance as descriptions of the way human beings act (e.g., Richardson's behavior). However, polychronicity can achieve genuine importance as a variable in the social sciences only if it is systematically related to other variables. Such relationships will be discussed in this section at the group and individual levels, polychronicity being both a cultural and an individual variable. Further, as discussed earlier, polychronicity can be considered a description of basic life strategies, so the assessment of the strategic options along the polychronicity continuum—that is, their effectiveness—will be addressed in the section following this one ("The Effectiveness of Polychronicity Strategies"). But in this section, polychronicity's relationships will be considered with other variables that are not normally used as effectiveness criteria.

Group-Level Relationships

Consistent with the usage employed earlier in this chapter, *group* refers to all culture-carrying aggregations larger than a single individual. In this section polychronicity's association with other variables will be examined in two types of groups: organizations and nations.

Nations. Much of Hall's work on polychronicity involved cultural differences among nations. Based on his own observations, Hall concluded that cultures in the Mediterranean world—southern Europe, the Near East, and northern Africa—were more polychronic than the cultures of northwestern Europe —Germany and England (Hall 1983). In the New World, Latin America was more polychronic than the United States (Hall 1983). Similarly, Usunier's (1991) results indicated that Brazil was more polychronic than France or Germany. Writing later and focusing on business culture, Richard Gesteland used a three-category system and classified Nordic and Germanic Europe, North America, and Japan as monochronic; the Arab world, most of Africa, Latin America, and south and Southeast Asia as polychronic; with Russia, much of eastern and central Europe, and southern Europe, China, Singapore, Hong Kong, and South Korea as in between (Gesteland 1999, p. 55). These sets of conclusions converge well. Thus at a very macro level, a fairly complete polychronicity map of the world's general and business cultures has been created. One should note, though, Catherine Tinsley's (1998) findings that American managers were significantly more polychronic than their counterparts in Germany and Japan. This does not contradict Gesteland's classifications, but it does provide details about variation among the countries he classified as monochronic.

At least one association has been noted for polychronicity at the level of national culture, and that is with proxemics, people's use of space as prescribed by their culture (Hall 1982, p. 1). In a conversation I had with Edward Hall (Bluedorn 1998), I raised the question about whether personal space was related to polychronicity. Hall agreed that it was, and he indicated that the two were negatively correlated, that the more polychronic the culture, the smaller the personal space distances it prescribed. For example, in a more polychronic culture, people would stand closer to each other while talking. So time and space are related in the social as well as the physical world.

Organizations. After nations, the next largest groups in which polychronicity has been studied are organizations. Unlike the research on national cultures in which judgments about polychronicity were made based on observations and interviews, research on polychronicity as a component of organizational culture has mainly used psychometric scales, questionnaires, to measure polychronicity.[3] This approach often involves asking a sample of group members to each complete a polychronicity scale about the group as each respondent sees it and then averaging their scores. This average is the group's level of polychronicity. From a methodological perspective, using this approach requires the researcher to demonstrate sufficient within-group agreement to justify aggregating the individual perceptions as just described (Klein et al. 2000, pp. 513–14). Marina Onken (1999) has demonstrated such agreement in her sample of companies, and I have been able to do the same in a study of dental practices that Gregg Martin and I conducted.[4] These results demonstrate that such perceptions of polychronicity are a justifiable way of measuring polychronicity as a shared element of an organization's culture. Thus the following discussion of the relationship between organizational size and polychronicity is based on research that makes use of perceptions of the level of polychronicity in different organizations, albeit mainly by just one high-ranking corporate manager in each organization.

For about a two-decade-long period from the mid-1950s through the mid-1970s, organizational size stimulated a large amount of research in the organization sciences, even excitement (see Donaldson 2001; Scott 1975; and Slater 1985 for reviews). At a diminished level this interest persisted into the 1990s (Bluedorn 1993). But increasingly organizational size has come to be studied not as a variable in its own right but in the secondary role of a control variable. The following analysis of data from a project with which I have been involved goes against this trend.

In this project Stephen Ferris and I sent questionnaires to top executives (i.e., presidents/CEOs and COOs/executive vice presidents) of a large set of randomly selected, publicly traded companies in the United States. A representative sample of almost two hundred companies returned usable questionnaires, companies ranging in size from less than ten to over one hundred thousand employees (Bluedorn and Ferris 2000).

Although some results from this study have been reported elsewhere (Bluedorn and Ferris 2000), the results that follow are from original analyses performed for this chapter. The correlation between organizational size (number of employees) and polychronicity (the perception of the company's overall polychronicity by either its CEO or a senior vice president) was not statistically significant. However, the logarithm (base 10) of organizational size was significantly correlated with company polychronicity, and it was a positive correlation: Larger firms were more polychronic.[5]

This result is surprising because the one theoretical discussion of a possible size-polychronicity relationship indicated that monochronic cultures would be more appropriate (i.e., more effective) in large systems whereas polychronic cultures would be more appropriate in small ones (Schein 1992, p. 108). Although it does not predict it explicitly, Schein's analysis anticipates a negative correlation between size and polychronicity; indeed his analysis was the reason for examining the size-polychronicity relationship, although other findings (Lee 1999) did contradict part of the basis for Schein's original conclusion.

It is hard to explain this relationship as an idiosyncrasy of a small or unique sample because the sample is so representative of publicly traded companies in the United States (Bluedorn and Ferris 2000). So the question becomes, why are greater degrees of polychronicity associated with increasingly larger organizations?

Perhaps at least part of the reason can be found in the results of the size research conducted over the last half century. It is known that as organizational size increases, so does the division of labor as manifested in the differentiation of both work units and individual positions (Donaldson 2001). Perhaps the scope of organizational tasks and functions increases as organizations become larger, but the increasingly differentiated structures organizations develop to deal with their growing scope of activities may not be proportionate to the new activities they add. The scope of work may increase faster than the division of labor. Thus departments and even individual positions may not be able to focus on a constant set of activities as the organization grows. New activities may be added that must be accomplished in the same amount of time as well as additional work if integration is to be maintained through a variety of integrating mechanisms (see Lawrence and Lorsch 1967 about integration and integrating mechanisms).

If this explanation of the polychronicity-size relationship is plausible, then polychronicity should be positively correlated with an emphasis on speed as organizational units and their members attempt to accomplish more within the same time period. To accomplish more in the same time frame, work will have to be performed faster, which may engender an increased emphasis on speed.

At the level of organizational culture, two tests have been conducted on the speed-polychronicity relationship. Onken (1999) studied organizational polychronicity in a sample of twenty firms from the telecommunications and publishing industries. Using the mean of completed polychronicity scales from each firm to measure organizational polychronicity, she found a statistically significant positive correlation: The more polychronic the company, the more it valued doing things fast. Similarly, I found a significant positive correlation between polychronicity and speed values in the sample of publicly traded companies described earlier in this section (see the organizational size-polychronicity discussion).[6] Both studies reveal a positive correlation between polychronicity and speed values: The more polychronic the organization, the more doing things rapidly is valued in its culture. Although these consistent findings about the speed-polychronicity relationship support the explanation of the size-polychronicity relationship developed in this discussion, they are not a direct test of this explanation, which is, admittedly, speculative. More direct tests must await studies deliberately designed to investigate this explanation.

A small amount of research has been conducted on polychronicity at the department and small-group levels. But those studies involve polychronicity's relationships with variables whose story is the focus of Chapter 4, so those studies will be discussed in that chapter.

Individual Polychronicity

Individuals display polychronicity differences, even within national or organizational cultures. Just as cultures vary from one another in their polychronicity, so do individuals vary within each culture, albeit potentially around different averages. As such, polychronicity is no different from any other group value or belief. Few would argue that cultures can transmit their values to their members so unfailingly that, except for measurement error, within-group variance would become zero. Such a view is contrary to everyday experience and to empirical research. Indeed, Usunier (1991) found significant differences within nations for several temporal aspects of culture, including polychronicity. So this discussion will now examine such differences and variables with which they are associated. The discussion will be divided into two types of variables associated with polychronicity: demographic and psychological variables.

Demographic Characteristics. Three demographic variables have been investigated as correlates of polychronicity: gender, age, and educational level. Of these three, education (Slocombe 1999) and gender (Hall 1983; Manrai and Manrai 1995, p. 119) have received at least a small amount of theoretical attention, with Hall (1983, p. 52) having concluded that more monochronic time was male time and more polychronic time was female time. Thus it would seem to follow—if Hall was correct—that on average women would be more polychronic than men, a conclusion that will now be examined.

Hall may be correct, but the results from thirteen studies provide mixed findings about the relationship between gender and polychronicity. All of the studies used questionnaire scales to measure polychronicity, and in five of them women were more polychronic than men (Bluedorn 2000c; original analysis of a large student sample for this chapter). However, two studies found men to be more polychronic (Conte, Rizzuto, and Steiner 1999; original analysis of a sample of food service managers for this chapter), and six studies found no statistically significant differences between men and women in their respective samples (Conte 2000; Conte, Rizutto, and Steiner 1999; Kaufman, Lane, and Lindquist 1991a; Palmer and Schoorman 1999; and original analyses of data sets reported here for the first time: the dental practice personnel described earlier and a sample of college students).[7]

Although the bulk of the statistically significant relationships indicate that women are more polychronic than men, the overall results are very mixed, with the modal finding being no significant association between polychronicity and gender. If such predispositions were as fundamental as Hall believed they were —he labeled them "preconscious" (1983, p. 52)—one would anticipate a series of consistent correlations across a variety of populations. Instead, nearly half the correlations were not statistically different from zero (i.e., no relationship), and the significant correlations were not all in the same direction (i.e., in some samples women were more polychronic, whereas in others men were more polychronic). At a minimum these results indicate that gender is not consistently related to polychronicity; they even question whether there is a gender-based *predisposition* to polychronicity at all. And a similar albeit stronger conclusion can be reached about a possible relationship between polychronicity and age.

To the best of my knowledge, no one has suggested a relationship between individual age and polychronicity. It may be that no one has suggested such a relationship because it may not exist. Three studies have examined age over a large enough range of ages to reasonably test for an age-polychronicity relationship, and all three studies failed to produce statistically significant relationships (Kaufman, Lane, and Lindquist 1991a; the sample of dental practice personal mentioned earlier in this chapter; and the sample of food service managers also mentioned before). Admittedly, all three of these samples involve people over a wide range of the *adult* years, people in their twenties and much older, so if there is a relationship between age and polychronicity, it would have to involve changes that occur between youth and adulthood, a possibility that none of the three samples allows to be tested because all three were limited to adults. (None contained anyone over sixty-five either.) However, for the pre-retirement adult years, there appears to be no relationship between age and polychronicity.[8]

To this point things do not look promising for relationships between demographic variables and polychronicity. Nothing consistent seems to be happening between gender and polychronicity, and between polychronicity and age nothing seems to be happening at all—at least for the adult years. Both gender and age are, of course, demographic variables based on biological differences, so things may be more promising when demographic variables based on social rather than biological factors are examined. Biology is not destiny, at least as far as polychronicity is concerned. But sociology may be.

In their sample of 310 randomly selected adult residents of a residential neighborhood adjacent to Philadelphia, Carol Kaufman, Paul Lane, and Jay Lindquist (1991a) found that polychronicity was positively correlated with respondents' levels of formal education: the more formal education, the more polychronic the respondent. Kaufman, Lane, and Lindquist noted that respondents who reported having college or professional degrees scored highest on the Polychronic Attitude Index, the study's measure of polychronicity (1991a, p. 397). Because of the broad-based nature of the sample, and the well-designed method of ensuring its representativeness, reasonable confidence can be placed in the generalizability of these results—at least for the United States—despite their being based on a single sample.

But why are higher levels of education apparently associated with higher levels of polychronicity? If polychronicity is indeed a traitlike personality variable, does this imply that people who are more polychronic are apt to seek out more formal education? Or if personality continues to develop into early adulthood, might not the higher levels of formal education develop an individual's preference for engaging more tasks at the same time as well as the individual's ability to do so? Or perhaps, as Thomas Slocombe proposed (1999, pp. 318–19), people who acquire greater amounts of formal education may tend to have jobs that are more likely to require them to behave polychronically, which leads them to develop more polychronic patterns of behavior. Or might *all* of these processes be operating? And does this relationship hold in other countries, especially in countries that differ significantly from the United States in the polychronicity of their cultures? For example, in countries such as Brazil or Mexico, whose cultures are more polychronic than that of the United States, would greater individual polychronicity still be associated with higher levels of educational attainment? At present these questions remain unanswered.

Psychological Variables. The preceding discussion indicates that polychronicity is related systematically to amount of formal education. In this section similar relationships will be explored with psychological characteristics, including personality attributes, observable behaviors, and perceptions of organizational attractiveness.

An important contemporary trend in personality theory has been the development of a multidimensional model of human personality, the "Big-Five" model (Digman 1990). And if polychronicity is a traitlike variable, an important question to examine is how polychronicity is related to the big five dimensions—if it is related to any of them at all. Jeffrey Conte (2000) addressed this issue directly in his study of 181 train operators who worked for a large metropolitan transit authority. Conte found that polychronicity was not significantly related to emotional stability, agreeableness, or intellectance (openness to experience), but it was negatively correlated with conscientiousness and positively correlated with extraversion (*extroversion* in everyday language) (Haase, Lee, and Banks [1979, p. 273] report a positive correlation with extraversion too), both correlations being statistically significant. Thus the more polychronic the individual, the less conscientious and more extraverted the person is.

Conte also found that individual polychronicity was significantly related to two very important organizational behaviors: lateness and absenteeism. And these relationships persisted after controlling for respondents' gender, work experience, cognitive ability, extraversion, and conscientiousness. So the greater the individual's polychronicity, the more frequently the person was late and absent.

Another approach to personality is the well-known Type A–Type B distinction (Friedman and Rosenman 1974). Jeffrey Conte has also investigated polychronicity's relationship with this aspect of personality, this time with colleagues Tracey Rizzuto and Dirk Steiner (1999). Focusing their work on two health-related dimensions of the Type A–Type B distinction, these researchers found that polychronicity was positively correlated with both dimensions: The more polychronic the individual, the greater the striving for achievement and the greater the individual's general impatience and irritability, both correlations being statistically significant.

The Type A personality construct has received a great deal of attention because of its apparent relationship with health-related factors such as stress and heart disease (Friedman and Rosenman 1974). Nevertheless, despite polychronicity's positive relationships with two key Type A dimensions, Conte, Rizzuto, and Steiner (1999) found no significant correlation with stress. As they noted, the absence of a correlation with stress is consistent with Kaufman, Lane, and Lindquist's finding (1991a) of a negative correlation between polychronicity and role overload—the greater the polychronicity, the less the individual feels overloaded by work tasks—and Kaufman, Lane, and Lindquist's interpretation of this relationship that polychronicity may be an adaptive response to busy schedules. It should be noted that this interpretation is consistent with Mintzberg's idea cited earlier that the realities of work lead managers "to develop a particular personality" (1973, p. 35).

To further examine the possibility that higher levels of polychronicity may be a way of coping with busy schedules, I conducted new analyses for this chapter using data from the dental practice study discussed earlier. In this study dental practice employees, including the dentists, completed a questionnaire scale about their individual polychronicity as well as a measure of stress. As in Conte, Rizzuto, and Steiner's study (1999), the correlation between individual polychronicity and stress in these data was not statistically significant. However, the picture changed when I divided the sample into the two categories of (1) dentists and (2) all other practice employees. The correlation within the all-other-practice-employees category remained nonsignificant, but in the subsample of dentists, a very interesting and statistically significant positive correlation emerged. There was no relationship between polychronicity and stress among nondentists in the practices, but among dentists, the more polychronic the individual dentist, the less the dentist experienced stress (see Figure 3.2). And after controlling for age, gender, and number of years worked in the current dental practice, this difference between the two categories not only persisted but became more extreme.[9]

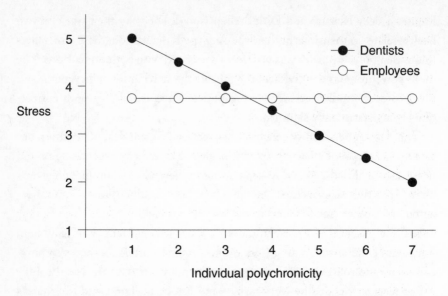

FIGURE 3.2. Relationship between polychronicity and stress for dentists and employees in a sample of dental practices (regression lines)

Why is polychronicity unrelated to stress for everyone in these practices except the dentists themselves? Perhaps the jobs performed in the two categories differ in fundamental ways. Nondentists' jobs may not require constant shifting to and fro among a variety of tasks in the same way that the dentists must move back and forth among several patients undergoing a variety of procedures. In other words, the way the workflow is structured in most American dental practices may require the dentist to work much more polychronically than the other practice employees. If so, by analogy with the positive outcomes associated with congruence between individual and work-unit polychronicity (Slocombe and Bluedorn 1999), it follows that if a job must be performed in a very polychronic manner, the more polychronic the role holder is, the more readily and comfortably, and the less stressfully, it can be performed.

The issue of congruence between individuals and groups will be dealt with in depth in Chapter 6, so it will not be explored further here, but Edgar Schein's example of polychronic time patterns in the relatively monochronic United States is suggestive of the explanation just presented for polychronicity's negative correlation with stress among dentists. Schein noted, "A doctor or dentist, for example, may simultaneously see several patients in adjacent offices"

(1992, p. 108), an indication that the dentist's job, at least as typically practiced, is quite polychronic relative to much work in the United States.

This interpretation of the relationship between polychronicity and stress among dentists is also consistent with Benjamin Schneider's (1987) Attraction-Selection-Attrition (ASA) theory. According to this theory, people and organizations with similar values will seek each other out and then be more likely to remain associated than people and organizations who are less similar. The dental findings suggest that polychronicity might be a value-based behavior pattern that leads individuals to seek out certain occupations and avoid others. But is polychronicity a complex of values, beliefs, and behaviors that makes some organizations more attractive than others? After all, similarity of values or beliefs that hold little importance to the individual or organization seems unlikely to result in significant mutual attraction.

To address this question I collected data to investigate the possible attraction between individual and organizational polychronicity, data that also revealed relationships between polychronicity and several other personality variables (Bluedorn 2000c). Using scenarios presented in Bluedorn et al. (1999) to represent high- and low-polychronicity organizations, respectively, respondents in two samples were asked to rate each organization according to how attractive it was to them as a potential employer. Respondents also completed a polychronicity questionnaire about themselves.

As anticipated, if polychronicity were to matter enough to generate an attracting force, respondents' individual polychronicity scores were positively correlated with the attractiveness of the high-polychronicity scenario—the more polychronic the respondent, the more attractive the high-polychronicity organization appeared—and negatively correlated with the low-polychronicity scenario—the more polychronic the respondent, the less attractive the low-polychronicity scenario appeared. And these relationships persisted in both studies (one sample comprised over two hundred college students, the other over three hundred) after controlling for the effects of age, gender, grade point average, orientation to change, propensity to creativity, locus of control, and tolerance for ambiguity (Bluedorn 2000c). Thus the polychronicity of a potential employer, if perceivable, appears to be a significant attribute to potential employees.

While investigating polychronicity as a potential attractor, I also examined the relationships between individual polychronicity and several of the control variables (Bluedorn 2000c). The results produced important findings because two of the personality variables revealed consistent relationships across multiple samples, relationships that persisted after the effects of several other variables were statistically controlled.

First, I found that after controlling for age, gender, grade point average, locus of control, propensity to creativity, and tolerance for ambiguity, a positive relationship persisted between polychronicity and orientation to change. The more polychronic the respondent, the more favorable the respondent was toward change in general.

The second important finding concerned polychronicity and tolerance for ambiguity, which is "a range, from rejection to attraction, of reaction to stimuli perceived as unfamiliar, complex, dynamically uncertain, or subject to multiple conflicting interpretations" (McLain 1993, p. 184). Tolerance for ambiguity was positively correlated with polychronicity, and these positive relationships persisted after controlling for the other variables (Bluedorn 2000c). These results replicate an earlier finding reported by Haase, Lee, and Banks (1979, p. 272) and indicate that the more polychronic people are, the more tolerant they are of ambiguity.

So individual polychronicity is related to several individual variables. Relative to less polychronic people, more polychronic people appear to have more of the following:

- Extraversion (extroversion)
- Favorable inclination toward change
- Tolerance of ambiguity
- Formal education
- Striving for achievement
- Impatience and irritability
- Frequency of lateness and absenteeism

Those same people appear to have less of the following:

- Conscientiousness
- Stress (only in some jobs)

But, as will be revealed in the following section, these are not the only individual variables to which individual polychronicity is related.

THE EFFECTIVENESS OF POLYCHRONICITY STRATEGIES

Polychronicity was described earlier as a continuum of life process strategies, which raises a number of important questions. For example, is a single strategy from the polychronicity continuum optimal for individuals and groups? This seems unlikely, as nearly a half century of research in the organization sciences indicates that a strategy's success is highly contingent on a variety of factors (Bluedorn and Lundgren 1993; Chandler 1962; Donaldson 2001, pp. 11–16 and 221–25). So the question should be restated: Which polychronicity strategies work best for individuals in which situations? And which polychronicity strategies are associated with the highest levels of group effectiveness?

Effectiveness has traditionally been defined as "the degree to which a social system achieves its goals" (Price 1972, p. 101). Yet there is no reason to limit the effectiveness concept to social systems such as organizations and work groups, because individuals have goals, and the extent to which they achieve them can be assessed as well. Thus this discussion will first consider polychronicity strategies and individual effectiveness, then group-level polychronicity and effectiveness.

Individual Effectiveness

As I have noted elsewhere (Bluedorn 1980), a problem with conceptualizing effectiveness, individual or group, in terms of goals is that almost anything can be a goal. Nevertheless, many individual outcomes are frequently considered desirable, such as having good health and succeeding in and enjoying one's work; hence such outcomes are likely to become individual goals, and polychronicity's relationship with several of these outcomes will be considered in this discussion.

Early in the twentieth century, Frank Gilbreth, the motion study pioneer, made an important discovery about polychronicity and health. In his constant quest to do things efficiently, Gilbreth attempted to shave with two razors simultaneously, two *straight* razors. But he abandoned the attempt because he lost more time applying bandages to the cuts this technique produced than he saved by shaving with the two razors (Gilbreth and Carey 1948, pp. 3–5). So, as far as shaving is concerned, even a moderately polychronic strategy is ineffective as judged by the criteria of total time taken and physical safety. But the self-inflicted damage Gilbreth incurred was minor by most standards, and especially so compared with the results of a form of polychronic behavior that developed nearly a century after Gilbreth concluded his ill-fated experiment with personal hygiene.

The results are traffic accidents, which are a major cause of human injury and death, so behaviors that appear to be related to accident frequency have major public policy implications. And one such behavior is a new form of polychronicity: using a cellular phone while driving. Several countries have made this behavior illegal, and as an important public health study shows, with very good reason.

Donald Redelmeier and Robert Tibshirani (1997) studied 699 drivers who owned cellular telephones and who had motor vehicle accidents that resulted in substantial property damage (but no personal injury). They found that using a cellular telephone while driving a motor vehicle was associated with a *quadrupling* of the risk of a collision during the period of the call—with no statistically significant difference between drivers who used handheld or hands-free cellular phones. As Redelmeier and Tibshirani noted, "This relative risk is similar to the hazard associated with driving with a blood alcohol level at the legal limit" (1997, p. 456). Even the modest degree of polychronicity involved in performing these two tasks simultaneously has important negative consequences, so driving, like shaving, is a task best performed monochronically.

The preceding two examples could easily lead to the erroneous conclusion that even modestly polychronic life strategies generally lead to negative health outcomes. But a series of studies linking the density of social networks to a large set of health outcomes would lead to exactly the opposite conclusion—density of social networks basically being the number of different *types* of social relationships (e.g., relationships with spouses, children, fellow employees, friends, etc.) in which a person is actively engaged. Starting with fewer instances of the common cold (Cohen et al. 1997), a greater diversity of social networks has been associated with lower levels of depression (Cohen and Wills 1985), heart disease and cancer (Vogt et al. 1992), and mortality (Berkman and Syme 1979). And if the difference in density of these social networks is great enough, the mortality risk becomes comparable to the differential between smokers and nonsmokers (House, Landis, and Umberson 1988). Because active engagement in multiple forms of social relationships indicates varying degrees of polychronicity—the more forms actively engaged, the greater the polychronicity owing to a greater amount of moving back and forth among them over

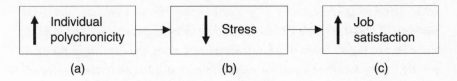

FIGURE 3.3. Model of the relationships among polychronicity, stress, and job satisfaction for dentists in a sample of dental practices

time—these findings reveal a salutary effect on health of polychronic life strategies regarding social relationships.

Overall, these health-related findings reveal a mixed set of results. Sometimes monochronic behavior patterns are best, sometimes polychronic. This mix of findings implies that no one level of polychronicity will produce the best results for all outcome variables. Consequently, each potential relationship must be investigated individually, so polychronicity's relationships with several other individual outcome variables will be examined.

Earlier in this chapter the relationship between polychronicity and stress was described for a sample of dentists. Revisiting that relationship is worthwhile here because it is part of a simple three-variable chain relevant to this discussion of polychronicity and individual effectiveness. Since adult Americans spend more waking time at work than involved in any other activity (Robinson and Godbey 1997, pp. 321–23), how much a person enjoys or receives gratification from that time (i.e., job satisfaction) is an important individual effectiveness criterion. And in the subsample of dentists, although polychronicity was positively correlated with job satisfaction, the correlation was not statistically significant. But this relationship is not the main point of interest. As presented earlier, the correlation between the dentists' polychronicity and stress was positive and statistically significant: The more polychronic the dentist, the lower the dentist's stress. Polychronicity, stress, and job satisfaction are three variables, but only two of the relationships have been reported. The third is the correlation between stress and job satisfaction, which is negative and statistically significant: the lower the stress, the higher the job satisfaction. And the diagram in Figure 3.3 presents a very reasonable albeit simple model of the relationships among these three variables.

Noting that this process is limited to just the dentists among dental practice employees, according to the diagram in Figure 3.3, polychronicity has a direct effect on stress, and stress has a direct effect on job satisfaction, but there is no direct effect of polychronicity on job satisfaction. So as a process, increasing amounts of polychronicity lead to reduced levels of stress, and reduced levels of stress lead to higher levels of job satisfaction—among the dentists in the sample. By having an effect on stress, polychronicity still has an impact on the individual effectiveness criterion of job satisfaction, albeit indirectly.[10]

Another index of individual effectiveness is job performance, which is important from both the individual's and the organization's perspective. Researchers have examined polychronicity's relationship with job performance for two occupational categories: college students and college professors. For a college student, the typical index of performance is the grade point average (GPA), and polychronicity's relationship with GPA has been examined in student populations at two universities. In a sample of 161 undergraduates at a large public university in the southern United States, Conte, Rizzuto, and Steiner (1999) reported a nonsignificant correlation. And I reported the results of five tests of this relationship on large samples of undergraduate students at the University of Missouri-Columbia (Bluedorn 2000c). One of these five tests produced a statistically significant positive correlation; the other four tests revealed nonsignificant correlations. I also tested this relationship in two samples that were not reported in Bluedorn (2000c), but neither correlation was statistically significant.[11] Overall, out of eight tests of the possible relationship, only one produced a significant correlation, a very modest one; the other seven correlations were all nonsignificant. These results indicate there is no consistent relationship between polychronicity and undergraduate college student GPA, perhaps no relationship at all, so everyone can breathe a sigh of relief!

As for the college professors, fewer tests have been conducted, but both that have been show the same significant relationship. In the first test, which was not conceptualized in terms of polychronicity, Taylor et al. found a significant positive correlation between the extent to which college faculty "engaged in multiple, concurrent, research and writing projects" (1984, p. 408) and their productivity (an index composed of number of research books, chapters for edited books, and articles in professional journals). Being "engaged in multiple, concurrent" projects is a way to describe someone behaving polychronically. Similarly Richard Frei, Bernadette Racicot, and Angela Travagline (1999) reported an even larger positive correlation between the same two variables, albeit their productivity index included a broader range of research activities and outlets.[12] In both studies the two concepts and their measures were fortunately very similar, which allows their results to be compared directly. Thus behavioral polychronicity—number of concurrent research and writing projects—is positively related to college faculty research productivity: the more behaviorally polychronic the faculty member, the higher the research productivity.

Group Effectiveness

Several studies have examined the relationship between group polychronicity and performance, most of them at the organizational level. And although several scholars have conceptualized their research explicitly as polychronicity research, others have not. For example, James McCollum and J. Daniel Sherman (1991) studied sixty-four companies with matrix structures and found that the highest levels of effectiveness occurred in companies where the highest percentage of research and development personnel were assigned to two projects rather than either to one project or to three or more projects, which suggests that a moderate level of polychronicity may be better than either high or low levels in companies with matrix structures.

Another example is provided by Kathleen Eisenhardt's intensive case studies of eight microcomputer firms, studies that led her to conclude, "The greater the speed of the strategic decision process, the greater the performance [of organizations] in high-velocity environments" (1989, p. 567). This is relevant to polychronicity's impact on organizational performance because Eisenhardt's data also led her to conclude, "The greater the number of alternatives considered *simultaneously*, the greater the speed of the strategic decision process" (emphasis added; 1989, p. 556). Thus dealing with decision alternatives polychronically leads to a faster pace in the process of strategic decision making—which given the relationship Eisenhardt found between speed of decision making and organizational performance should at least indirectly lead to better organizational performance.

This relationship is supported by experimental research on small-group decision making (Weingart, Bennett, and Brett 1993), which found a relationship between outcome quality and considering issues simultaneously (polychronically) rather than sequentially (monochronically). Moreover, William Judge and Alex Miller replicated Eisenhardt's work and found "strong support for the proposition that the number of alternatives simultaneously considered is a critical determinant of decision speed regardless of environmental context" (1991, p. 457). They also replicated the relationship between speed of decision making and organizational performance, but they discovered it was limited to organizations in high-velocity environments. These two studies suggest an indirect relationship between polychronicity and performance through polychronicity's impact on speed of decision making, although this relationship would be limited to high-velocity environments. (See Chapter 4 for more on the polychronicity-speed relationship).

Marina Onken (1999) conducted the first research to investigate the relationship between organizational performance and polychronicity with polychronicity explicitly conceptualized as such. She found statistically significant positive correlations between a company's polychronicity and both return on assets and return on sales: the more polychronic the company's culture, the better its performance as measured by these two performance indicators. But she did not find a statistically significant relationship between return on equity and polychronicity, although the correlation was in the predicted (positive) direction. Nevertheless, because her sample was so small (twenty companies), these results might change in larger, more broadly drawn samples.

And they did. Working with a much larger, more representative sample of American companies (the national sample of organizations I used to investigate the size-polychronicity relationship discussed earlier in this chapter), I found that a firm's polychronicity was not significantly related to either its return on assets or its return on sales. But a company's return on equity was positively correlated with its polychronicity, meaning the more polychronic the firm, the better its return on equity. Thus, as interpreted with customary levels of statistical significance, the results from the national sample data contradicted all three of Onken's findings: She found significant positive correlations between polychronicity and returns on assets and sales, whereas I did not in the national sample data, and although I found a statistically significant correlation between polychronicity and return on equity, Onken did not.[13]

Taken together, Onken's (1999) findings and mine from the national sample of organizations do not reveal a consistent pattern of relationships between polychronicity and organizational financial performance. However, they do suggest that such relationships may exist and deserve additional investigation.

Distinctive Competencies

Across the group and individual levels, various polychronicity strategies have been associated with desirable outcomes. Sometimes a high level of polychronicity was associated with the best outcomes, but in other circumstances a low level of polychronicity seemed to be most effective. Some evidence even indicated that a moderate level of polychronicity was best for producing some outcomes. Overall, these findings suggest that high and low levels of polychronicity each have their virtues, their own distinctive competencies.

The potential benefits of the monochronic strategy seem clearer, but that may be because I and many readers grew up in the monochronic United States. Nevertheless, a monochronic strategy confers the advantages of *focus*, including efficiency. As such, a monochronic strategy should generate a greater depth of involvement with a decision or activity, hence a more thorough knowledge of it, at least in the short term. So when substantial focus is required, such as in the task of driving a car, or when many details are involved, a monochronic strategy may be best.

But a major disadvantage of this strategy is that the events or tasks engaged monochronically are less likely to be well integrated with other tasks and activities. Moving back and forth among several tasks and activities might result in some cross-fertilization as well as greater integration among them, but at a minimum the more polychronic strategy should keep people more aware of the status and implications of all activities engaged. And this appears to be a key effectiveness factor for the managers of the successful project portfolios studied by Shona Brown and Kathleen Eisenhardt (1997).

Brown and Eisenhardt studied change and project management in computer firms and found that firms with less successful project portfolios demonstrated very low amounts of communication across projects. This was part of the context in which projects were planned, divided into small tasks, and then executed in a "structured sequence of steps" (1997, p. 14). A structured sequence is, of course, a monochronic strategy, and the low amount of communication is consistent with the proposition that monochronic strategies generate less awareness of other activities and tasks. One of the managers in their study remarked, "Most people only look at their part" (p. 14); another, "The work of everyone else doesn't really affect my work" (p. 14). These responses contrasted with the pattern of work in the companies that managed their portfolios of projects more successfully, which Brown and Eisenhardt characterized as "iterative" (p. 14). Iterative (repetitive) patterns are suggestive of the back and forth flow of polychronic strategies.

Another disadvantage of a monochronic pattern is its failure to provide either the timely feedback or the flexibility so that a flawed or problematic part of a project or activity can be corrected before the entire project is completed. Eisenhardt (1989, p. 558) interpreted Barry Staw's (1981) work on escalation of commitment to mean that, at least in decision making, considering multiple alternatives simultaneously reduces the escalation of commitment to any single alternative, whereas considering options sequentially does the opposite. Escalating commitment, of course, reduces the motivation to even look for trouble, and it also reduces flexibility about options, an interpretation consistent with descriptions of monochronic strategies. For the monochronic approach has been described as being associated with "strict planning" (Kaufman-Scarborough and Lindquist 1999, p. 289) and with a tendency to "adhere religiously to plans" (Hall and Hall 1990, p. 15). And in the case of the computer companies Brown and Eisenhardt studied, the sequential structure of project work added to the psychological mechanisms that produce escalating commitment.

Thus Brown and Eisenhardt noted that at the less successful companies it was difficult to adjust projects in changing conditions because "once started, the process took over. It was hard to backtrack or reshape product specifications as circumstances changed" (1997, p. 14). (The similarity of this statement to Von Moltke's rebuff of the Kaiser is almost eerie: "once settled, it [the plan] cannot be altered"; see Chapter 1). Conversely a more polychronic strategy, which by definition allows for the ebb and flow of an iterative pattern, increases the chances that people will more readily become aware of a problem and thus would not have as far back to go in order to correct it. But with a monochronic strategy, the distance back may be too great to undertake any modifications. Indeed, a monochronic strategy may lead to a greater degree of satisficing—picking a decision that simply meets a satisfactory level on one or more criteria (March and Simon 1958, p. 169)—than would a polychronic approach to decision making. Simultaneous consideration of alternatives not only reduces the escalation of commitment to a single option but also increases the speed of decision making (Eisenhardt 1989). As such there is less time pressure to make a decision, hence less reason to satisfice at lower and lower levels. Thus a polychronic decision-making strategy seems likely to produce more optimal decisions—or at least to seek them.

If the preceding analyses seem to overwhelmingly favor polychronic strategies, it is well to note that polychronic strategies have their downside too. For example, an unlimited flexibility could well lead to "unproductive dithering," a potential problem of polychronic strategies (Bluedorn, Kaufman, and Lane 1992, p. 23). The point is that both strategies have their distinctive strengths and weaknesses and that it is best not to let one's personal and cultural biases lead to the conclusion that one strategy or the other is *always* best. Both have their virtues and vices, and the best strategy is to recognize them.

SUGGESTIONS FOR THE THIRD GENERATION

Empirical studies of polychronicity are poised to enter their third generation. Edward Hall's work (e.g., Hall 1981b, 1983; Hall and Hall 1990) constitutes the first generation, and most of the other research cited in this chapter constitutes the second. Identifying a third generation implies a qualitative difference from what has gone before, and several possibilities suggest themselves for differentiating future polychronicity research from that conducted by the first two generations.

The core of polychronicity's formal definition is the extent to which people prefer to be engaged in two or more tasks or events simultaneously (Bluedorn et al. 1999). Although none of the conceptual work on polychronicity addresses the types of tasks engaged simultaneously, the work-design distinction between job enlargement and job enrichment (George and Jones 1999, p. 221) suggests that the types of tasks involved in a job matter in many ways (enlargement mainly involving similar tasks; enrichment, dissimilar tasks). So when considering behaviors along the polychronicity continuum, does it matter whether the tasks are similar, or whether they vary along one or more dimensions? Put another way, is a person who engages simultaneously in several different tasks more polychronic than someone who engages simultaneously in the same number of similar tasks? Or are both people equally polychronic? These questions suggest the model presented in Figure 3.4.

The typology presented in Figure 3.4 results from dichotomizing two continua—number of tasks engaged simultaneously and degree of difference among the tasks engaged—and cross-classifying them. The result is four types of behavior patterns: quantitative polychronicity and monochronicity, and qualitative polychronicity and monochronicity. A quantitatively polychronic pattern involves engaging several *similar* tasks simultaneously, whereas a quantitatively monochronic pattern involves engaging a task and completing it and then moving on to another *similar* task. Conversely, the qualitative polychronicity pat-

Differences among tasks

Low   High

		Low	High
Number of tasks engaged simultaneously	High	Quantitative polychronicity	Qualitative polychronicity
	Low	Quantitative monochronicity	Qualitative monochronicity

FIGURE 3.4. Forms of polychronic and monochronic patterns of behavior

tern involves engaging multiple *dissimilar* tasks simultaneously; the qualitative monochronic pattern, engaging a single task and completing it before engaging another but *dissimilar* task. And the question is, does the addition of the quantitative-qualitative task distinction add any explanatory power to theoretical statements about polychronicity's relationships with other variables?

Evidence that this distinction does make such a difference is provided in at least one set of studies already discussed in this chapter. In fact, it was these studies that led me to propose this typology of polychronicity patterns. The findings linking polychronicity in social relationships with several important health outcomes (i.e., colds, heart disease, cancer, mortality), all found differences related to the number of *types* of social relationships people engaged in *regularly*. The more types of social relationships engaged in regularly, the more favorable the health outcomes (i.e., less probability of contracting a disease or dying).

Of particular importance for the polychronicity typology just introduced is Cohen et al.'s (1997) remarkable—remarkable because participants allowed the researchers to deliberately expose them to a cold virus via nasal drops—experimental investigation of both the absolute number of social relationships and the number of *types* of social relationships and their relationship to colds. The key finding in support of the quantitative-qualitative distinction made in the typology deserves to be presented in the authors' own words: "In contrast to the diversity of the network, the total number of network members was not associated with colds. . . . Moreover, entering the number of network members into the first step of the regression equation along with standard controls did not reduce the association between diversity and colds" (Cohen et al. 1997, pp. 1942–43). Diversity of the network is a reference to the number of types of relationships engaged in regularly and repetitively (at least once every two weeks). Thus these results suggest that qualitative polychronicity was negatively related to the probability of contracting a cold: the more different *types* of social relationships people were engaged in, the lower their chances of contracting a cold. But simply the more social relationships people engaged in had no impact on the probability of contracting a cold. Qualitative polychronicity mattered, quantitative polychronicity did not.

Examining qualitative and quantitative forms of polychronicity requires a way to determine how similar tasks and events are, otherwise one would be unable to say whether multiple tasks differ from each other qualitatively. Unfortunately, no general method for classifying tasks and events has been developed that seems adequate for examining these questions about polychronicity, although some concepts and models might provide elements for such a classification scheme.

For example, the well-known job characteristics model (Hackman and Oldham 1976) offers several dimensions that might be useful for describing a task or event. Among the possible characteristics suggested by this model, skill variety seems the most promising because it seems most directly related to polychronicity. Skill variety is the degree to which a job requires a variety of different activities in performing work, and the activities themselves involve the use of a number of the individual's different skills and talents (Hackman and Oldham 1976, p. 257). The reference to "variety of different activities" suggests qualitative differences in work or tasks, as does the phrase "different skills." So skill variety, if it can be extended to cover life's tasks and events in general, might be a way to classify types of tasks and events as qualitatively similar or different.

Perhaps such a modified skill-variety approach could be augmented by or combined with part of Carol Kaufman and Paul Lane's (1996, pp. 139–41) approach to describing consumer product use. They described product use according to whether the use was monochronic or polychronic, and whether the use involved a single sense (e.g., vision) or multiple senses (e.g., vision and hearing). Thus how many senses are involved in a task or event might be combined with elements of the skill-variety dimension (e.g., number of different skills used and the variety of specific activities) as an additional attribute to determine how similar or different tasks and events are. But until a general method is developed for evaluating the degree of differences among life's tasks and events, results such as Cohen et al.'s (1997), though suggestive, will remain limited to each study's unique domain and methods.

But regardless of whether polychronicity's relationships with other variables is contingent on one or more task and event dimensions, other important questions remain about its fundamental nature, especially its psychological foundations. And three of these questions seem especially salient.

The first of these questions involves an assumption made implicitly almost from the beginning of polychronicity research: Polychronicity is more likely to be a trait, or at least traitlike, than it is to be statelike (e.g., Slocombe and Bluedorn 1999, p. 76). But how traitlike is it? How much flexibility can people display before experiencing stress from a too uncomfortable polychronicity pattern? As mentioned in the discussion of the polychronicity-education findings, do more polychronic people seek out more education, or does exposure to more education somehow lead to more polychronic preferences and behaviors, perhaps by leading people to jobs requiring more polychronic behavior as proposed by Slocombe (1999)? Of course, a third possibility is that both processes could be operating for each relationship.

But even if something is a trait, this does not imply that an individual's behavior patterns will never vary from the mean, that the standard deviation is zero. Instead, variability will be observed, perhaps even variability in the preference components as well as in overt behavior, and such variability deserves to be studied. For example, can some people vary along the polychronicity continuum more than others? June Cotte and S. Ratneshwar (1999) have certainly documented the ability of some people to vary their behavior radically along the polychronicity continuum as they moved between work and leisure activities. Hall suggested the facility to make such shifts may be related to what he called a "high adaptive factor" (Bluedorn 1998, p. 114), such people being more flexible along the polychronicity continuum than others. In a life context of varying polychronicity demands, perhaps an individual whose own polychronicity lies near the average of the varying environmental demands might be able to cope most readily with them because the largest adjustment required would be smaller, hence less potentially uncomfortable or stressing than from any other position on the polychronicity continuum. (See Chapter 6 for more

about the issue of congruence between individual and group polychronicity.)
But again, this suggestion supposes a traitlike nature of polychronicity.

A second, related question asks, is polychronicity temporally scalable? The
issue here is whether an individual's polychronicity maintains itself across differing amounts of time. For example, if a person behaves very monochronically
within a period of two or three hours, will the same pattern reveal itself at
higher orders of magnitude such as weeks, months, or years? That is, however
the tasks or events are defined, and they may need to be defined in terms of
larger magnitudes as time frames increase, would the same one-thing-at-a-time
pattern be invariant and change little across time frames? This question has not
really been investigated, for regardless of whether it is based on observation,
phenomenological interviews, or questionnaires containing psychometric scales,
the existing research seems likely to have dealt with time frames of a single day
or less, likely frames of just a few hours. So does polychronicity scale? Or is it a
nested phenomenon whereby someone might be monochronic within hour-
long intervals but polychronic when the frame enlarges to a month? And if so,
what might be the consequences of different nesting combinations?

Such questions not only point the direction for expanding our knowledge of
polychronicity but also suggest the likelihood of various social and psychological determinants of it. One such determinant seems especially intriguing, and
it is the individual's breadth of attention. Breadth of attention is "the number
and range of stimuli attended to at any one time" (Kasof 1997, p. 303). This
concept is used to describe screeners, people who focus on a small range of
stimuli and filter or "screen out" other stimuli. Conversely, nonscreeners attend
to a large range of stimuli and are aware of a much larger range of potentially
unrelated stimuli (Kasof 1997). Breadth of attention is basically a phenomenon
that describes differences in how individuals perceive the world from moment
to moment, and its definition makes it seem likely to be related to polychronicity. Hence Joseph Kasof communicated the following thoughts to me about
this potential relationship:

> Polychronicity must be low among people who have dispositionally narrow
> breadth of attention, because if one's attention capacity to simultaneously
> maintain multiple streams of thought is very low, it would be practically im-
> possible to simultaneously engage in multiple activities at the same moment.
> Over time, difficulties in doing multiple tasks simultaneously would cause
> individuals who are dispositionally low in breadth of attention to hold less

favorable attitudes toward engaging in multiple tasks simultaneously. (Joseph Kasof, personal communication, 2000)

Although untested empirically, the propositions suggested in Kasof's insights may, if supported, reveal at least some of the major psychological bases of polychronicity and help account for individual polychronicity variation within cultures.

Clearly other important questions about polychronicity can be framed, but these three—how traitlike, scalability, and relationship with breadth of attention—will have to be answered before any claims can be made that we truly understand this so fundamental of behavior patterns. And certainly other questions come readily to mind. For example, until now the term *multitasking* has not been used in this chapter—and for good reason. That reason is the multitasking concept combines both speed and activity-pattern dimensions rather than simply focusing on activity patterns (i.e., polychronicity). As such it is only partially synonymous with polychronicity and will be dealt with in Chapter 4, where speed is a principal focus.

So perhaps it is fairest to conclude by describing the status of our knowledge of polychronicity as Winston Churchill once did other matters: "Now this is not the end. It is not even the beginning of the end. But it is, perhaps, the end of the beginning" (Churchill 1943, p. 266).

Seldom Early, Never Late

Time goes, you say? Ah no!
Alas, Time stays, *we* go.
—Austin Dobson, *The Paradox of Time*

How much is being on time worth? How much value do some societies place on punctuality, on the temporal precision of their machines as well as their people? Nearly three hundred years ago, in 1714, the British Parliament provided a precise answer to this question. Parliament set its value at twenty thousand pounds, which is equivalent to over 5 million contemporary U.S. dollars (Landes 1983, p. 112). This fortune was to be paid to whoever the "Constituted Commissioners for the Discovery of the Longitude at Sea" determined had been able to "Discover a proper Method for Finding the said Longitude," if the Commissioners declared the method "Practicable" (Act of Queen Anne, 12, cap. 15, as reproduced in Sobel and Andrewes 1998, p. 65).[1] The promise of this reward led to the solution, a punctual clock known as the "marine chronometer," which within a narrow range was never early, never late. And therein lay the solution to the longitude problem; for if you have a sufficiently punctual clock and set its time to that of a place whose longitude you know, you have the basis for later determining your ship's longitude accurately throughout the voyage.

The process works like this. There are 360 degrees of longitude, and the earth rotates on its axis once in twenty-four hours. Thus in one hour the earth rotates 15 degrees of longitude, which results from dividing 360 degrees by twenty-four hours (see Brown 1949, p. 210; Sobel and Andrewes 1998, p. 7). Because of this relationship between longitude and time, in 1530 Flemish astronomer Gemma Frisius proposed the idea of using a portable timekeeper to identify a ship's longitude (Andrewes 1994b, p. 346). The method requires the time at a place of known longitude (e.g., Greenwich, England) to be kept on a clock aboard the ship, and for this clock to keep this time accurately throughout the ship's voyage. This is so the time at the reference location (e.g., Greenwich) will be known at any moment during the voyage. To determine the ship's longitude, the local time is determined and compared with the time at the reference location. Because of the relationship between time and longitude (15 degrees per hour), the difference between the local time and the accurate time of the reference longitude location converts easily to the ship's longitude at its present location (Andrewes 1994b, p. 346). So if Greenwich was the location of known longitude, and the marine chronometer said it was 2:00 P.M. in Greenwich when the local time was *noon*, there would be a two-hour difference in time, indicating a 30-degree difference in longitude (15 degrees per hour multiplied by two hours equals 30 degrees), so the ship would be at a longitude of 30 degrees west of Greenwich.[2] (The north-south position, latitude, can be determined by direct observations of the sun, a somewhat simpler problem to solve.)

Of course, Frisius's method will give the longitude with the desired accuracy only if the on-board clock providing the time at the reference location can do so with sufficient precision. And such a clock or watch would not be made for another two centuries after Frisius made his proposal, at least not one that could function accurately under the challenging conditions of sea travel (i.e., the constantly pitching ship, widely fluctuating temperatures, etc.).

But how accurate—how punctual—did such a clock need to be? To win the full twenty-thousand-pound prize, the longitude had to be determined accurately to within one-half of a degree, which is about thirty nautical miles (Andrewes 1994b, p. 346). This amounts to a margin of error (gain or loss) of no more than three seconds each twenty-four hours (Sobel and Andrewes 1998, p. 72). With 86,400 seconds in twenty-four hours, this meant the clock had to average an accuracy level of at least 99.9965 percent every day during a six-week voyage from England to the West Indies—the test specified in the Longitude Act of 1714. But why was such accuracy needed? The answer lies in the catastrophes that could befall ships whose captains did not know where they were with sufficient precision.

By being partially lost all of the time prior to the solution to the longitude problem, sea voyages lasted longer than necessary, which increased the likelihood of voyagers contracting an especially nasty disease, scurvy. Given that scurvy was caused by a lack of vitamin C, the longer the voyage, the greater the chances of contracting it, because during this era food sources that provided vitamin C were not part of the seafarer's diet (Sobel and Andrewes 1998, p. 19). But as appalling a fate as scurvy was to mariners, the more dramatic and terrifying threat was shipwreck, a catastrophe often caused by not knowing a ship's location or being mistaken about it. For example, Rupert Gould (1960, pp. 2–3) described several such results of faulty navigation. One of the most infamous is the sinking of four ships in the fleet of Admiral Sir Cloudisley Shovel, including the admiral's ship. The fleet was returning to England from Gibraltar in October 1707 when in bad weather it crashed into the Scilly Islands off the coast of Cornwall in southwest England. Four ships went down, resulting in the deaths of 1,647 sailors, including Sir Cloudisley (Sobel and Andrewes 1998, pp. 15–17).[3] This tragedy was one in a series of such events that led to the Longitude Act of 1714, which offered the twenty-thousand-pound prize for a practical solution to the longitude problem. That it took nearly seven years for the British Parliament to respond officially may reflect the pace of life in the early eighteenth century, a pace of life that differs radically from that experienced by most readers in industrialized nations in the late twentieth and early twenty-first centuries. (The pace, or speed, of life will be discussed later in this chapter.)

An ironic temporal asymmetry surrounds the rest of this story. Although the punctuality of the marine chronometers was at the heart of concern in this entire matter, it being required, as noted, to be 99.9965 percent accurate, the Board of Longitude was not nearly as concerned with comparable punctuality in awarding the prize for the longitude problem's solution. The Board of Longitude was the body of experts—astronomers, mathematicians, navy officers— and government officials selected to judge solutions for "the discovery of the longitude," a phrase that came to mean something of "practical impossibility" (Gould 1960, pp. 16–17). And after more than half a century of effort the celebrated horological wizard (i.e., virtuoso clockmaker) John Harrison produced "No. 4" (a.k.a. "H-4" and "H.4"), a kind of giant pocket watch 5.2 inches in diameter (see Barnett 1998, p. 112; Gould 1960, pp. 53–54; Sobel and Andrewes 1998, pp. 129–31), which won the prize. When tested on a voyage from Portsmouth, England, to Bridgetown, Barbados, in the West Indies during the spring and summer of 1764, H-4 proved to be even more accurate than the prize required, as it allowed the calculation of the longitude to within ten miles (Sobel and Andrewes 1998, pp. 148–52). This meant the watch had been accurate to within about one second per day, or about 99.9988 percent accurate.

But after the success of this test, "the Board of Longitude allowed months to pass without saying a word" (Sobel and Andrewes 1998, pp. 152), perhaps anticipating contemporary bureaucratic hubris. Indeed, not until October 28, 1765, did John Harrison receive a certificate from the Board of Longitude authorizing him to receive *a portion* of the prize, seventy-five hundred pounds plus twenty-five hundred he had received from the board earlier (Gould 1960, p. 62), which one could argue was over a year late. Compared to the three seconds per day margin of error allowed Harrison, the Board of Longitude was permitted an absolute margin of error over *10 million times greater*. Seven years later, in 1772, and only after obtaining the support of King George III, Harrison received an additional £8,750 when Parliament passed a special money bill for this purpose (Gould 1960, pp. 66–67). This sum combines with other amounts given to Harrison over the years to constitute the remaining portion of the full twenty-thousand-pound prize, the delay in awarding this portion of the prize constituting an expansion of the absolute margin of error to something over *70 million times greater* than Harrison's watch was allowed. This example illustrates with quantitative fungible time precision the principle that the more powerful are allowed greater discretion to be late and keep others waiting (i.e., margins of error) than the less powerful (see Levine 1997, pp. 109–14), and as such, of the more general principle that time and the control of it are important political matters.

The politics of punctuality will be one of the matters discussed in the following section on punctuality, and that section will be followed by a discussion of the closely related phenomenon of speed in human life. As will be seen, speed is a matter closely related to punctuality, so the relationships between speed and punctuality will form part of that discussion.

PUNCTUALITY

As with so many temporal matters, what it means to be punctual, to be "on time," has varied widely throughout history and across cultures. It even varies within organizations, for as Deborah Ancona and Chee-Leong Chong noted, if a product cycle lasts eight years, being three months late is being on time; but in a nine-month product cycle, three months late is late indeed (1999, p. 44). This variation is a reminder that punctuality is socially defined, hence a human construction that produces a wide range of variation across both time and space in the definition of what it means to be "on time." And when it comes to punctuality, that variation has had a direction over the last two or three millennia, a direction toward greater precision and more demanding targets. Despite wide-ranging contemporary variation, this general trend becomes evident by examining observations of earlier eras and lifestyles.

For example, punctuality was very different in the Rome of two thousand years ago. The Romans used sundials and water clocks to reckon the hours, but the imprecision of these mechanisms, along with the constantly changing temporal hours (see Chapter 1), made it difficult to measure and determine the hour precisely. This led Daniel Boorstin to conclude, "Since no one in Rome could know the exact hour, promptness was an uncertain, and uncelebrated, virtue" (1983, p. 31). Uncertain and uncelebrated it may have been, but that would all change with the invention of the mechanical clock (Landes 1983, p. 7; Levine 1997, p. 60). For before that event, in the late thirteenth century (see Chapter 1), the inability to reckon the hours precisely during the day or night vitiated attempts to coordinate activities, except through the use of relatively unambiguous temporal markers such as dawn, noon, and dusk, albeit noon in the hands of nonspecialists was itself problematic (Levine 1997, pp. 60–61). Robert Levine, a scholar whose work plays a prominent role in this chapter, described the era before mechanical clocks well when it came to punctuality: "Before the invention of the first mechanical clocks, the idea of coordinating people's activities was nearly impossible. Any appointments that had to be made usually took place at dawn. It is no coincidence that, historically, so many important events occurred at sunrise—duels, battles, meetings" (1997, p. 60).

Thus, with the exception of a few *relatively* unambiguous moments such as dawn, noon, and dusk, twenty-first-century-style appointments—"Let's meet for a cup of coffee at 10:15"—were literally inconceivable because times with that precision were unmeasurable in everyday life. But that would change radically after the invention and rapid diffusion of mechanical clocks in Europe during the late thirteenth and early fourteenth centuries.

To illustrate what was coming, E. E. Evans-Pritchard's famous ethnography of the Nuer as they lived in the first part of the twentieth century provides an important description of time and punctuality in an agricultural/pastoral society, a society based on domesticated plants and animals, a society without a technology for reckoning the hours.

> The Nuer have no expression equivalent to "time" in our language [English], and they cannot, therefore, as we can, speak of time as though it were something actual, which passes, can be wasted, can be saved, and so forth. I do not think that they ever experience the same feeling of fighting against time or of having to co-ordinate activities with an abstract passage of time, because their points of reference are mainly the activities themselves, which are generally of a leisurely character. Events follow a logical order, but they are not controlled by an abstract system, there being no autonomous points of reference to which activities have to conform with precision. Nuer are fortunate. (Evans-Pritchard 1940, p. 103)

In terms of the continuum presented in Chapter 2, Nuer times were extremely epochal. Moreover, the references to being "controlled by an abstract system" and to "autonomous points of reference to which activities have to conform with precision" obviously point to comparisons drawn with the system of time reckoning and temporal values existing in the West during the twentieth century. This set of values had been anticipated nearly fifteen hundred years before by Christianity's development of monastic life during the first millennium, a life that emphasized reciting prayers at set hours known as the canonical hours (Landes 1983, pp. 58–66). Hand in hand with this development, the monastic orders also emphasized and enforced a regimen of performing each prayer precisely at its prescribed time. David Landes has suggested that the emphasis on punctuality surrounding these prayers was due to a desire to avoid giving offense by missing a prayer at its appointed hour or by being late and having to shorten or rush through the prayer. More speculatively, he suggested that "simultaneity was thought to enhance the potency of prayer" (1983, pp. 62–63).[4]

Prayers prescribed at regular times were not unique to Christianity. Such practices exist in both Judaism and Islam, for example, but the difference lies in the precision attached to the times prescribed for the prayers. In Judaism and Islam "the times of prayer are bands rather than points" and as such, timepieces were not required to identify the times for the prayers (Landes 1983, p. 59). Thus Christianity's monastic practices developed a concern with time, its measurement, and with punctuality well over a thousand years before manufacturing developed in the growing European towns and cities of the late Middle Ages, and certainly before the industrial revolution. This concern provided a cultural foundation for the increased emphasis on punctuality that would accompany the invention of the mechanical clock, a device that gave impetus to increasingly precise demands for being on time. And not being innate in the human condition, these demands required instruction to develop an increasingly precise and strict temporal discipline.

Learning to Be On Time

Before the new temporal discipline could become part of the tacit knowledge passed on by role modeling and subtle cues, it had to be taught explicitly—a task that took centuries. Yet even the ancient sundial provided lessons in temporal discipline long before the escapement transformed time reckoning, a point lamented by those who felt subject to its dominion as revealed in the following lines from a play written twenty-two hundred years ago:

> May the Gods confound that man who first disclosed the hours, and who first, in fact, erected a sun-dial here; who, for wretched me, minced the day up into pieces. For when I was a boy, this stomach was the sun-dial, *one* much better and truer than all of these; when that used to warn me to eat, except when there was nothing *to eat*. Now, even when there is something *to eat*, it's not eaten, unless the sun chooses; and to such a degree now, in fact, is the city filled with sun-dials, *that* the greater part of the people are creeping along the streets shrunk up with famine. [Plautus's emphases] (Plautus 1902, pp. 517–18)[5]

Fallible though the sundial may have been, the increased precision and temporal discipline wrought by the concept of hours and the technology of the sundial to measure them was obviously seen as a mixed blessing by this playwright, if a blessing at all (e.g., the reference to the sundial determining when to eat, culminating in the people "creeping along the streets shrunk up with famine"). As later events would demonstrate, it seems that each major advance in technology's ability to measure time precisely was attended concomitantly by the resentment and resistance of many, and especially by those upon whom the new time discipline was being imposed. Reasons for such resistance are manifold, including the possibility of a general human tendency to resist fundamental change, a tendency described memorably by James Baldwin: "Most of us are about as eager to be changed as we were to be born, and go though our changes in a similar state of shock" (1985, p. 643). And "shock" certainly describes the speaker's reaction to hours and sundials in the lines from the Roman play.

But beyond whatever general resistance people exhibit toward change, there were also sound political and economic reasons to resist the change toward more precise standards of punctuality and time discipline. For example, in 1335 King Philip VI of France authorized the city government of Amiens to use a bell to signal when people were to eat and when they were to begin and finish working (Crosby 1997, p. 86). This put the control of time, hence control of much of life, into the hands of other human beings rather than the natural markers of dawn, noon, and dusk. Not that dawn did not sometimes come too soon or dusk not soon enough, but with the timing of life now coming under the control of other people rather than nature, both the possibility and the reality of manipulation in the rhythms of time signals for personal benefit became objects of dispute. Public clocks, especially those with dials, made the verification of time a continuous possibility for all of the population (Landes 1983, p. 75), thereby reducing the possibility of abuse. But governments are never completely neutral, and who would say, in fourteenth-century Amiens for example, what was the right time to begin work? One doubts that the workforce was consulted extensively about this matter before a decision was made; one doubts anyone consulted the workforce at all.

Such matters are the subject of what may credibly be described as the most famous scholarly article ever written about time: E. P. Thompson's "Time, Work-Discipline, and Industrial Capitalism" (1967; see Glennie and Thrift 1996 for a discussion of this article's influence). Focusing on England during the seventeenth, eighteenth, and nineteenth centuries, Thompson chronicled the deliberate efforts made to instill in the entire population, but especially in the workforce, a time discipline based on obedience to the clock and to the appointments made at times specified on it (e.g., the time to begin work). And illustrating the premise that control of time reckoning was anything but a politically neutral issue, Thompson quoted records left by two nineteenth-century workmen, both of whom were employed in factories of the day and both of whom testified that workers were not allowed to have their own clocks or watches on company grounds.[6] Because this prohibition contrasts so strikingly with contemporary practices where watches are ubiquitous, one of the statements will be repeated here:

In reality there were no regular hours: masters and managers did with us as they liked. The clocks at the factories were often put forward in the morning and back at night, and instead of being instruments for the measurement of time, they were used as cloaks for cheatery and oppression. Though this was known amongst the hands, all were afraid to speak, and a workman then was afraid to carry a watch, as it was no uncommon event to dismiss any one who presumed to know too much about the science of horology [clock and watch making]. (quoted in Thompson 1967, p. 86)

Although the contemporary reader will find the prohibition on watches novel and likely disconcertingly oppressive, a contemporary parallel would be the practice in many organizations of keeping salaries secret and attempting to reinforce the secrecy with norms that proscribe telling one's salary to others in the organization, norms that define such telling as "unethical." In both cases an obvious reason for the information blackout is the cover it provides for "cheatery and oppression," something that Thompson documented well in the case of time and watches.

But cheating workers by manipulated time reckoning would not seem to be the primary motivation for management's concern about time discipline in the workplace. That concern stems from the organization of work itself: "Attention to time in labour depends in large degree upon the need for the synchronization of labour" (Thompson 1967, p. 70). And the need for such synchronization became especially salient when employees were brought together daily in factories and other enterprises. This need for synchronization, hence for time discipline or punctuality, eventually produced a new set of values and attitudes toward time, new values and attitudes that were taught through a variety of devices.

Thus by 1700 some English enterprises can be described as possessing "the familiar landscape of disciplined industrial capitalism, with the time-sheet, the time-keeper, the informers and the fines" (Thompson 1967, p. 82), several devices and practices that rewarded the desired time discipline and punished its violators. But these were devices that altered and reinforced the habits of adults, adults who needed to be converted to the new temporal practices from such older patterns as the weekly Saint Monday, the habit of taking Monday off each week to relax, to socialize, and most salient from the standpoint of the new time discipline, to opt not to show up for work. And if people could be taught the new time discipline early in life, they would be better prepared to meet the growing synchronization demands of the workplace—or at least to make the manager's ability to achieve it easier.

So a variety of forces, religious as well as economic, led to an increased emphasis on teaching punctuality in the schools, a life skill that was taught in both England and the United States (Thompson 1967; O'Malley 1990). For example, Michael O'Malley cited the following warning from a nineteenth-century *McGuffey's Reader*: "Little girl, never be a moment too late. It will soon end in trouble or crime" (1990, p. 20). This emphasis extended to the teachers as well as to the students, a point Thompson noted: "At the Methodist Sunday Schools in York the teachers were fined for unpunctuality" (1967, p. 84).

The twentieth century continued the nineteenth's emphasis on a strict punctuality as shown in the certificate awarded to my maternal grandmother when she was thirteen years old in 1903 (see Figure 4.1). A noteworthy feature revealed in this certificate is how fine-grained the school's system was for monitoring attendance and tardiness. The telling phrase on the certificate is "having been neither absent nor tardy during the month ending." These ten words speak volumes. First, they tell us that the school was concerned that my grandmother appeared both daily ("neither absent") and at the appointed times during the day ("nor tardy"), albeit the certificate does not specify the temporal point of no return a student needed to cross to be declared "tardy." But there is more, because the ten words also tell us that this laudatory behavior occurred during "the month ending," a detail indicating these awards were given on a monthly schedule throughout the school year. The certificate does not say whether it was presented in any kind of ceremony, but one easily envisions a modest monthly ceremony for bestowing these honors, not only to reinforce the recipient's behavior but also to remind the temporally deficient of the school's expectations in this regard.

The existence of a certificate at all is, of course, evidence for the importance placed on habits of punctuality. And to make sure the recipient understood the fundamental reason she was being rewarded, for why punctuality was so important, the certificate's first words present an aphorism attributed to Franklin: "Lost time is never found again." So in 1903 the Limerick School in Cedar County, Iowa, taught punctuality to its students well, perhaps too well, if punctuality implies an emphasis on speed. For a too great emphasis on speed is a matter addressed later in this chapter as well as in Chapters 7 and 9.

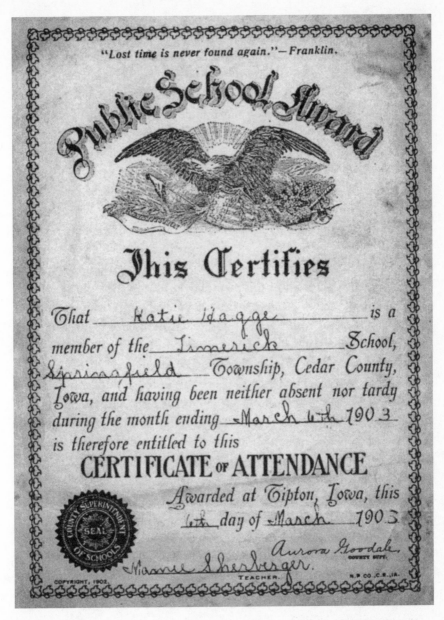

FIGURE 4.1. Certificate awarded in 1903 for an exemplary attendance and punctuality record in an American public school. The recipient was the author's maternal grandmother who was thirteen years old when she received the certificate.

Deadlines

Deadlines are the temporal markers defining punctuality; they are the standards used to determine whether something or someone is late, early, or on
time. Being on time defines the virtuous; being late, the villain; and being early,
usually just a mild irritant. In a fungible time sense, being one hour early is
arithmetically just as unpunctual as being an hour late, but the stronger, and
negative, reaction goes to the person or thing that is an hour late. And the origins of the deadline concept may explain why.

A deadline once referred to a physical line around a military prison, a line
beyond which any prisoner who ventured would be shot (*Oxford English Dictionary*, 2nd ed.). Translating this spatial phenomenon into its temporal counterpart, being late, corresponds to the unfortunate prisoner who ventured beyond the deadline. In both cases we may shoot the transgressor—or want to,
at least in some cultures. But prisoners should stay on the right side of the
prison's deadline, and someone arriving early, though potentially an irritation,
can be simply asked to leave and return later. Early arrivals present fewer
problems and can usually be dealt with more easily than late arrivals or no-
shows, and fewer people want to shoot them. Indeed, in the formalized rules
of punctuality discussed by Thompson, one rule instructed students to arrive a
few minutes before the deadline, "a few minutes before half-past nine o'clock"
(1967, p. 84). So being a few minutes early was not considered unpunctual; it
was considered part of being on time. But how many minutes are a few? And
can one also be a few minutes late without being considered late? Unless a social consensus exists about such matters, it is hard to determine whether someone is early, late, or even on time.

And the consensus itself may change. In an assessment originally written
seventy-six years after the Limerick School honored my grandmother for a
month of perfect punctuality, Ellen Goodman would lament:

> I am a member of a small, nearly extinct minority group, a kind of urban
> lost tribe who insist, in the face of all evidence to the contrary, on the
> sanctity of being on time.
>
> Which is to say that we On-timers are compulsively, unfashionably
> prompt, that there are only handfuls of us in any given city and, unfortu-
> nately, we never seem to have appointments with each other. (Goodman
> 1979, p. 106)

Times, they were a-changing in the twentieth century. For example, the practice of awarding certificates for perfect punctuality was unknown in the public schools my children attended late in the century. So across three or four generations the importance attached to being precisely on time was relaxed, not through a formal act of Congress, but in the less formal process of general social change. And in this case it is difficult to assess the change as either good or bad. Robert Heinlein (1973, p. 242) once quipped, "Roman matrons used to say to their sons: 'Come back with your shield, or on it.' Later on, this custom declined. So did Rome." Shall we likewise say, "Americans used to instruct their children: 'Be on time.' But then this custom declined, and so did America"? Before reaching such a pessimistic conclusion, this matter can be informed by examining punctuality from a cross-cultural perspective.

Many visitors to other cultures report that after language difficulties, temporal differences create the most problems, especially differences in punctuality (Spradley and Phillips 1972). And just such differences plagued a collaborative effort between banks from two countries—the United States and Mexico (DePalma 1994). The American bank, Banc One of Columbus, Ohio, signed on with Bancomer in Mexico to work with the Mexican bank in developing its credit card operations. A team from Banc One traveled to Mexico to work with their Bancomer counterparts on the credit card project, but the two groups experienced major difficulties stemming from punctuality issues. The Bancomer managers wanted to hold their meetings with the American bankers at 7:30 P.M. as part of their regular workday, which often extended from 9 A.M. to 9 P.M. This practice would intrude into time the Americans considered home time or recreation time, but certainly not work time (all times are not the same). But another problem arose, even after the U.S. bankers agreed to meet in the evenings: The Mexican bankers often arrived for the meetings sometime *after* 7:30 P.M.

To overcome this cultural impasse, the bankers from both countries agreed to a temporal compromise, with each group agreeing to behaviors contrary to their cultural traditions. For the Americans, this meant meeting during the evening. For the Mexicans, this meant agreeing to actually arrive at 7:30. And to ensure compliance with this now explicit agreement, the bankers developed a unique and highly visible enforcement mechanism. The bankers acquired a piggybank and placed it on the meeting table. The overall agreement specified that any late-arriving banker, American or Mexican, would drop a small number of pesos into the piggybank for each minute the banker was late (details about the banks and meetings from DePalma 1994, p. F5). Obviously the financial cost to the individual did not generate the motivation to be on time, for that motivation would have been social: To show up in the evening promptly at 7:30 demonstrated the commitment of both groups to the overall project, and either not to show up at all or to show up after 7:30 would demonstrate the opposite, an act that was then publicly sanctioned with deposits in the piggybank. A piggybank is, after all, one type of a bank, and it likely added important symbolic overtones to the mechanism for enforcing punctuality.

As this example illustrates, American culture and Mexican culture treat punctuality differently, and as a form of cultural variation, punctuality has been investigated comparatively across many countries. In the investigation of such differences, Robert Levine and his colleagues have taken the lead and have done so with consummate ingenuity. Levine discovered cross-cultural variation firsthand during a stay as a visiting professor at the federal university in Niteroi, Brazil (Levine and Wolff 1985). As he walked to class on his first day, he received a remarkable range of information about what time it was over a span of only a few minutes, information that included 9:55, 10:20, 9:45, 9:43, and 3:15 (Levine and Wolff 1985, p. 30). Welcome to Brazilian time, Professor Levine!

To a professor from California State University in Fresno, this initiation to temporal diversity certainly qualified as culture shock, but with his wits about him Levine was soon able to articulate what was happening: "Their timepieces are consistently inaccurate. And nobody minds" (Levine and Wolff 1985, p. 30). But more was happening than that, for Levine's unexpected immersion in a new temporal sea stimulated a stream of research on cross-cultural temporal differences that as of this writing has spanned more than two decades. The research has focused on punctuality differences and speed, or pace, of life differences, two interrelated temporal variables, both of which are the focus of this chapter—and to both of which Levine and his colleagues have made the leading contributions.

The first of the studies (Levine, West, and Reis 1980) examined punctuality differences between the United States and Brazil. To measure punctuality objectively, the Levine team checked the accuracy of fifteen randomly selected bank clocks in the downtown regions of Fresno, California, and Niteroi, Brazil (both cities had populations of about 350,000 at the time of the data collection). They used the time given by the local telephone company as the correct time and calculated to the nearest minute each bank clock's deviation from this standard. As they had hypothesized, the American bank clocks were significantly more accurate than their Brazilian counterparts, on average by almost a full minute (Levine, West, and Reis 1980, p. 543). That clocks in a city vary, hence their authority does also, was observed two millennia before Levine began to use such variation as an index of the importance a society places on punctuality. In the first century Seneca provided the evidence: "It was as impossible to find agreement among the clocks of Rome as to find agreement among Roman philosophers" (Seneca as quoted in Boorstin 1983, p. 31). So, as was demonstrated with the problem of the longitude, behavioral punctuality is limited to the accuracy of the technologies used to reckon time.

Levine's success with this index of punctuality led to its use for subsequent research in other countries. With his colleague Kathy Bartlett, Levine employed the same method in five more countries and repeated it in the United States (Levine and Bartlett 1984). To make more plausible claims about entire societies, they checked the accuracy—as defined by the time given by the local telephone companies—of fifteen randomly selected bank clocks in each of two cities in each country (thirty clocks per country). The researchers measured the accuracy of bank clocks in Japan, Taiwan, Indonesia, Italy, England, and the United States, and did so in one large city (population over 1 million) and one medium-sized city (population of about five hundred thousand) in each country. The results showed that the Japanese bank clocks were the most accurate, followed in order by those in the United States, Taiwan, England, Italy, and Indonesia. However, statistical analyses revealed that only Indonesian clock accuracy differed significantly from the others, and city size had no statistically significant effect on clock accuracy (Levine and Bartlett 1984, p. 238).

Then, over a span of several years in the 1990s, Levine and Ara Norenzayan conducted the most ambitious international study to use the bank clock method yet (Levine and Norenzayan 1999). Basically, they sampled the world with a total sample of thirty-one countries. The countries were arrayed as follows in terms of bank clock accuracy (from most to least accurate): Switzerland, Italy, Austria, Singapore, Romania, Japan, Sweden, Poland, Germany, France, Ireland, China, England, Hong Kong, Costa Rica, South Korea, Bulgaria, Hungary, Jordan, United States, Taiwan, Canada, Czech Republic, Kenya, Netherlands, Mexico, Syria, Brazil, Greece, Indonesia, and El Salvador (Levine and Norenzayan 1999, p. 190). The United States ranked twentieth among the thirty-one countries, which may not be surprising since punctuality may not be emphasized as much as it once was in American society. Although they probably have not disappeared entirely, certificates for punctual behavior do not seem as common today as they were in the America of 1903 (see the attendance and tardiness certificate in Figure 4.1).

But more important than any contest to win the punctuality championship are factors related to punctuality. Here too Levine and his colleagues have provided the leading work. For example, in the thirty-one-country study just discussed, Levine and Norenzayan (1999) reported analyses revealing that the accuracy of a country's bank clocks, hence its emphasis on punctuality, was greater in countries with colder climates, more productive economies, and more individualistic rather than collectivistic cultures. Indeed, they reported a substantial correlation between these three variables and punctuality (Levine and Norenzayan 1999, pp. 195–96), which indicates these three variables provided considerable predictive power for explaining the extent to which a country emphasizes punctuality. They also found that punctuality positively correlated with two measures of speed—the greater the emphasis on punctuality, the greater the speed with which things were done in a country. This finding replicated results from Levine and Bartlett's (1984, p. 244) earlier work, which produced even larger positive correlations between punctuality and the same two measures of speed.[7]

Because of these findings, I included measures of both punctuality and speed values in the national study of American companies discussed in Chapter 3. Doing so allowed me to test the speed-punctuality association that Levine and his colleagues found in their studies of national cultures, but at the level of organizational cultures. And I found a statistically significant positive correlation between the psychometric scales measuring the extent to which punctuality and speed were valued in this random sample of all publicly traded American companies.[8] So in organizations, at least American organizations, as in countries, the greater the emphasis on being on time, the greater the emphasis on doing things fast.

Thus it seems that cultures that try to do things fast also try to do things with greater temporal precision in terms of observing deadlines and being on time. Moreover, economic productivity was linked to a greater emphasis on punctuality, as was individualism, which is an orientation to the individual and the individual's nuclear family rather than to the welfare of one or more larger collectives (Levine and Norenzayan 1999, p. 182). As more researchers investigate these relationships and include more variables in their analyses, it will be interesting to see whether the physical climate retains its significant relationship with punctuality.

But other things are related to punctuality, several of which Levine, Laurie West, and Harry Reis (1980) discovered in their research on punctuality differences between the United States and Brazil. They conducted two studies, the first of which involved about four hundred randomly selected pedestrians evenly divided between Fresno, California, and Niteroi, Brazil. As with the bank clock results described previously, the watches worn by American pedestrians were significantly more accurate than those worn by their Brazilian counterparts. And indicating the degree of internalized concern with the time, Americans in the sample who were not wearing watches estimated the time of day significantly more accurately than did their watchless Brazilian counterparts. In fact, the errors in the Brazilian estimates were more than twice as large (an average error of 14.24 minutes) as those of the Americans (an average error of 6.93 minutes).[9]

However, both sets of estimates reflect the impacts of centuries of mechanical clocks and the form of time, clock time (i.e., fungible time), their influence promotes. That this influence has been deeply institutionalized is reflected in how we take the phrase "o'clock" for granted. This phrase is an abbreviation of the longer phrase "of the clock," which indicates this phrase originally identified a source of the time worth distinguishing from other sources, hence that more than one type or source was used or possible (Barnett 1998, p. 77). But even though punctuality is emphasized more in American culture than in Brazilian culture, there is no doubt that the least punctual American or Brazilian from either sample would consistently outperform the best pedestrian selected from either first-century Rome or tenth-century London to perform the same task, to answer the simple question: What time is it to the nearest minute? Actually, the concept of a "minute" would likely have been a mystery to the average citizen of either city during the first or tenth centuries.

So punctuality varies by eras, and within eras, by culture. What also varies by culture is the definition of punctuality itself. This was revealed when Levine, West, and Reis (1980) gathered a sample of 107 students from California State University in Fresno and 91 students from the Universidade Federal Fluminense in Niteroi, Brazil, a sample to which they then administered a questionnaire asking about several punctuality issues. Consistent with less concern about being on time, the Brazilian respondents reported using longer time intervals before defining someone as "early" or "late." For example, the American respondents considered someone *early* who arrived for a date about 19 minutes before the set time, whereas the Brazilian respondents defined being early for a date as arriving about 24 minutes before the set time, a difference of about five minutes. This contrasts with a much larger difference for the definition of arriving early for a lunch appointment with a friend: The American respondents defined early for this event as arriving about 24 minutes before the set time; the Brazilians, about 54 minutes early—a difference of half an hour. The American students were also stricter than their Brazilian counterparts in defining "late" for these events. For a date, arriving about $17\frac{1}{2}$ minutes after the set time was late for the Americans; for the Brazilians, arriving about 20 minutes after the set time was being late (a $2\frac{1}{2}$-minute difference). And for an appointment for lunch, arriving about 19 minutes after the set time was being late to the Americans, and about 34 minutes after the set time was late to the Brazilians, a difference of 15 minutes.[10] So the definition of punctuality differs by culture, and it also differs by event within cultures. The tolerances vary by both culture and event.

The event-contingent nature of punctuality tolerances shows once again that all times are not the same because what is considered late or early for one type of event is not considered so for another. Such socially constructed definitions provide the templates by which behaviors are informed, such as the decision for when to leave one's office to meet a friend for lunch, or whether to end an interaction with someone because of an appointment with someone else. The extent to which such choices constitute a dilemma will depend upon many factors, both personal and cultural, but if the definitions for early and late have greater tolerances, the frequency and intensity of such conflicts should be reduced, as would be the concern about being early or late. And Levine, West, and Reis (1980) found such differential concern, with American respondents reporting greater fear when they were late than the Brazilians. Consistent with this concern, the Brazilians responded more favorably (i.e., more likable, more relaxed, more happy, and more successful) about a person described as "always late for appointments" than the Americans did, with the opposite judgments made for a person described as "never late for appointments" (Levine, West, and Reis 1980, p. 548). Finally, and also consistent with the other results, the Americans rated punctuality as a more important trait in both a businessperson and a friend than the Brazilians did, again reflecting the greater emphasis on punctuality in American culture as well as its stricter definition of punctuality.

Perhaps a reason the Brazilians thought of the person described as "always late for appointments" as "more relaxed" than the Americans is the relationship between punctuality and the emphasis given to schedules and deadlines. Though *almost* a tautology, emphasizing being on time and giving priority to schedules over other considerations are conceptually close but empirically distinct. Jacquelyn Schriber and Barbara Gutek (1987) developed measures of both constructs, and the statistical analysis they conducted revealed two distinct dimensions rather than the single dimension that would be expected if the two constructs were the same thing.

Using Schriber and Gutek's scales, Bluedorn et al. (1999) reported a statistically significant positive correlation between the two variables for a sample of 199 departments in a large hospital system: The more punctuality was valued in a department, the more adhering to schedules and deadlines was valued too. At the individual level of analysis, Charles Benabou (1999) found a similar significant positive correlation in a sample of 301 graduating management students at the Université du Québec à Montréal. Also basing his work on Schriber and Gutek's scales, Benabou asked these students the extent to which they would "agree to work in an organization described by the following statements" (1999, p. 262), the statements being Schriber and Gutek's scales. Benabou's findings indicate that individuals who preferred an employer that emphasizes punctuality also preferred an employer that emphasizes schedules and deadlines.

And data from the national sample of American companies (discussed in Chapter 3 and earlier in this chapter) included Schriber and Gutek's punctuality and schedule-adherence scales, which permitted this relationship to be examined at the level of organizational culture. These data revealed a statistically significant positive correlation.[11] So punctuality and schedule-adherence values are positively correlated in organizational cultures, departmental cultures, and in the organizational attributes individuals prefer for jobs. But the correlations' magnitudes (see note 11), though consistently positive, argue that the two phenomena are not the same thing. Perhaps they are related reciprocally in the sense that creating schedules is a means for staying on time, and values of adhering to a schedule lead to valuing punctuality.

How these two values work together is illustrated in a story told by Richard

Gesteland (1999, pp. 58–59). A Malaysian businesswoman told him about an unpleasant experience she had in the United States. She had flown from Malaysia to Boston for a meeting with managers at an American company, arrived very late the night before, overslept the next morning, and then got lost driving her rental car around Boston trying to find the company, which made her four hours late for the meeting. Citing full calendars, the Americans told her they could schedule another meeting with her—nine days later. Unfortunately, she had to be back in Malaysia before then.

The story includes both punctuality and schedule-adherence values. The Malaysian woman was late for her appointment by American standards. Thus she violated the norms of American punctuality. The story also illustrates the use of schedules. The reference to calendars is an explicit reference to a formal schedule, and once the schedule was set, the Americans were not going to change it. This is reminiscent of General Von Moltke's conversations with the Kaiser, and Von Moltke's attitude that "once settled, it [the plan] cannot be altered" (see Chapter 1). The Americans were certainly not going to alter their schedules either. Somehow this response to the Malaysian woman's unintentional lack of punctuality seems extreme, especially for anyone who has ever experienced the ordeal of a flight that long (well over twenty hours) and then tried to navigate a major metropolis in a foreign country by themselves. It would have been unreasonable to expect the Americans to drop what they were doing as soon as she arrived, but perhaps a reasonable accommodation would have been for them to meet with her during "off hours," just like the Banc One group agreed to do in Mexico.

A true deadline is really a very precise appointment, and when appointments become deadlines, those appointments have a strong structuring effect on human behavior. In fact, an explicit deadline serves as the fundamental boundary condition for the punctuated equilibrium model of group development. Before it was applied to groups or other social phenomena, the concept of punctuated equilibria was originally developed by Niles Eldredge and Stephen Gould (1972) to describe patterns in the evolutionary (fossil) record, patterns in which long periods of stability would be "punctuated" by brief (by the scale of geologic time; see Gould 1987, p. 3) events that produced large, discontinuous change: "stasis punctuated by episodic events" (Eldredge and Gould 1972, p. 98). Working from this theoretical base, Connie Gersick proposed the punctuated equilibrium model as a paradigm for describing the pattern of change in several domains, including group development and other social processes, and in doing so indicated that some incremental change may occur during the periods of stability (Gersick 1991, p. 16).

For example, Thomas Kuhn's (1970) famous analysis of scientific paradigms fits the punctuated equilibrium pattern (Gersick 1991). What Kuhn called "normal science" would be the relatively long periods of stability characterized by incremental knowledge accumulation and change, all occurring within the boundaries of the existing scientific paradigm. These periods of normal science would then occasionally be punctuated by scientific revolutions, when a new theory or interpretation would appear and redefine the paradigm dramatically, perhaps totally replacing it, leading to a new period of stability and incremental knowledge growth.

But perhaps the most successful application of the punctuated equilibrium paradigm for explaining and interpreting social phenomena is Connie Gersick's (1988, 1989) own model of group development, a model of development for groups, that is, with *deadlines* for completing a project. According to Gersick's model, groups with a specific project and an explicit deadline for its completion develop as follows: During an initial phase they experiment with different approaches to the project and develop a direction for dealing with their project; approximately halfway to the deadline the group will significantly reorient itself and develop a new approach for completing the project and follow that approach to complete the project on time (by the deadline). After a direction is selected in the first phase, it becomes a period of stability, which is then punctuated by a major reorientation that becomes another stable period lasting until the end of the project. Although alternative interpretations have been proposed for what happens in such groups (Lim and Murnighan 1994; Seers and Woodruff 1997), these interpretations and Gersick's model all have one thing in common: The motive force in each interpretation is located in the deadline, an appointment taken seriously and seen as having a low far-side tolerance. The deadline has little if any slack for accepting projects completed late. So a concern for punctuality leads to structures beyond just being on time, and by that reaffirms once again Elliott Jaques's cogent observation: "In the form of time is to be found the form of living" (1982, p. 129).

But despite the appointments and deadlines and the structures they engender, what if one is running late, however late might be defined? To still come in on time, many people, groups, and societies have adopted the strategy of increasing the speed with which life, or at least part of life, is engaged. And this brings us to a consideration of speed itself.

THE QUEST FOR SPEED

Several terms apply to the referent of speed. Robert Levine and Ara Norenza-yan (1999, p. 178) used *pace*, and Robert Lauer (1981, p. 31) called it *tempo*. Although the choice of term may be largely a matter of personal taste, to me the term *speed* seems most fundamental, so as did Stephen Kern (1983), I will use the term *speed* in this discussion, and it will be defined as the frequency (number) of activities in some unit of social time (Lauer 1981, p. 31). As such, speed refers to "the frequency of activities" in general and is not restricted to a specific domain of activities such as the speed of change, a restricted usage Lauer allowed (1981, p. 31) but which will be disqualified here. And although the speed of life may be accelerating in general (Gleick 1999), the speed of change, contrary to received wisdom, may not be accelerating (Allen 2000). Lauer (1981, p. 32) also included *perceptions of speed* in his concept of tempo, but that phenomenon will be treated separately, in Chapter 7.

But if James Gleick (1999) and others (e.g., Robinson and Godbey 1997) have assessed things correctly, why did the speed of life accelerate in the twentieth century—and perhaps continues to do so? Several factors may be involved, but a major underlying mechanism seems to be efficiency and the desire for it.

Efficiency and productivity are basically the same thing: the ratio of output to input in an organization (Price 1972, p. 101; Price and Mueller 1986, p. 205). Output is whatever the organization produces (e.g., cars, college graduates, cured patients), and input is all of the resources used to produce the output. Although Price discussed efficiency in an organizational context, the concept is actually more universally applicable, since it can be applied to systems of all sorts, as well as to individuals. And speed is directly related to efficiency, for the faster output can be produced with the same amount of input, the more efficient the system becomes.

But efficiency is sometimes confused with a similar-sounding concept: effectiveness, which James Price defined as "the degree to which a social system achieves its goals" (1972, p. 101), and which Peter Drucker argued, while acknowledging the importance of efficiency, should be given primacy over efficiency (1974, pp. 45–46). Further, because individuals obviously have goals too, the effectiveness concept can be applied properly to individuals (as was done in Chapter 3), just as it can be applied to social systems. In general, speed is directly about efficiency and only indirectly about effectiveness. To illustrate this point, my experience with a high school speed-reading course is instructive. The course was a voluntary experience that lasted for a month or so. Despite being ultimately clocked at about two thousand words per minute (supposedly), I never enjoyed approaching written material that way, a point the speed-reading instructor highlighted one day. She explained that she used speed-reading frequently herself and had just speed-read a novel the night before. Even though she said that over thirty-five years ago, I can still remember my internal reaction as if it were yesterday: Why? What was the point of doing that? Isn't the point of reading a novel to *enjoy* the experience?

I could understand speed-reading materials such as bureaucratic rules and reports—to minimize the pain—but the idea of using speed-reading with material one is reading for pleasure seemed upside down. Yes, speed-reading would get one through a book faster. But if that comes at the cost of reduced gratification from the process, not to mention the loss of subtle and important insights, matters that often require pause for reflection, speed becomes a distortion, the ultimate end, and efficiency will trump effectiveness when it should be exactly the other way around. Efficiency is usually good, but effectiveness is always better. One is a means, the other an end, and to focus on efficiency entirely is in the truest sense to miss the point. Efficiency is about how; effectiveness is about why. Efficiency is to effectiveness what intelligence is to wisdom.

As mentioned earlier, until recently the matter of speed mainly concerned efficiency. But the emergence of speed-based competitive strategies (Blackburn 1991; Fine 1998; McKenna 1997; Meyer 1993; Stalk and Hout 1990; Vinton 1992) has linked speed much more directly to effectiveness. The strategy concept has traditionally been oriented toward the attainment of basic organizational goals (e.g., Chandler 1962, p. 13), so a speed-based strategy is much more directly linked to goal achievement (i.e., effectiveness) than is a strategy that focuses on increasing production speed simply to enhance internal efficiency, because the latter strategy routes the impact of speed on effectiveness *indirectly*, through efficiency. In some ways speed-based competition reverses the old scientific management sequence, which suggested that increased speed led to increased efficiency, which led to greater effectiveness. The whole point of time-and-motion studies was to identify and develop the proper procedures and worker move-

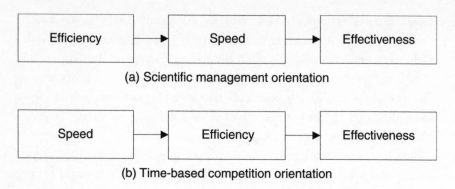

FIGURE 4.2. Different specifications for speed in two orientations to organizational effectiveness

ments to allow more work to be done, perhaps with less worker effort, during the same workday. The order is now increased efficiency leading to increased speed, which leads to increased effectiveness (see Figure 4.2).

The efficacy of speed for effectiveness has some empirical support too. As noted in Chapter 3, Kathleen Eisenhardt's research (Bourgeois and Eisenhardt 1988; Eisenhardt 1989) revealed that faster decision making by top executives was associated with higher organizational effectiveness. William Judge and Alex Miller (1991) then replicated these results but found that they seemed to apply only to high-velocity environments. And in a related piece of research, Marina Onken (1999) found a positive correlation between companies' speed values and both return on equity and return on sales. So both greater speed, at least the speed of executive decision making, and a greater emphasis on speed values have been related to higher levels of organizational performance, at least in some contexts—but not, perhaps, without cost.

THE COSTS OF SPEED

Some of the attempts to create greater efficiency actually lead to a sense of falling behind, of being late, that adds to the quest for speed. This can be seen in the movement to downsize organizations that was so prevalent in the United States during the 1990s. These efforts were seldom accompanied by plans to reduce the scale or scope of the organization's operations, so presumably the amount of work did not decline, only the number of people who had to do it.

Having more to do then meant an individual had to either work longer hours or work faster, or both, to complete the larger workloads. And the combination of more to do and doing each task faster produces the work pattern known as multitasking.

Multitasking shares some elements in common with polychronicity (see Chapter 3), because both involve the engagement of several tasks simultaneously. But a different orientation to speed distinguishes the two concepts. Polychronicity is purely about preferences for sequence: one thing at a time or moving back and forth among several tasks. It is not about getting more things done; it is not about doing things faster. Conversely, multitasking seems to be a combination of a relatively polychronic pattern with an overriding quest for speed, to get more things done. A familiar variety show act that Rhetta Standifer and I described illustrates the difference (Standifer and Bluedorn 2000).

The act is the well-known performance of the plate spinner, the individual who attempts to keep a dozen or more plates spinning simultaneously, with each plate spinning atop a stick that is fixed to the top of a table. We used this example to illustrate polychronicity, but we should have been more careful, because the image the plate spinner's act engenders would actually be closer to the polychronicity-plus-do-things-faster pattern just described as multitasking. That Aram Khachaturian's *Sabre Dance* is often the musical accompaniment to this act reinforces this point, which is of an almost-perfect visual and audio representation of the frenzied multitasker. But to isolate the polychronic component from plate spinning, the plate-spinning act needs to be moved to the moon. On the moon, with only one-sixth the gravity of the earth (Heiken, Vaniman, and French 1991, pp. 27–28), the performer can move more leisurely because the plates will fall less quickly should they stop spinning. Lunar plate spinning comes closer to capturing pure polychronicity than its terrestrial counterpart. On the earth, plate spinning is a multitasking activity.

So is some organizational work. For example, Deloitte Consulting listed "multi-tasking" as essential for success in its job description for systems analysts.[12] And as presented in Chapter 3, polychronicity and a speed emphasis were positively correlated as components of organizational culture in two different studies (Onken 1999; the national sample of publicly traded American companies). These positive correlations indicate that the greater the polychronicity, the greater the value placed on speed in organizational cultures, but because the correlations are no larger than they are (see the discussion of these correlations in Chapter 3 and its notes), they also indicate that polychronicity and speed values are two different variables, albeit positively correlated ones in these American samples.

But is multitasking good or bad? Since many of the good and bad effects associated with polychronicity have already been discussed in Chapter 3, the focus here will be on the speed component. Some of the most intriguing research on the relationship between speed and human well-being has been conducted by Robert Levine and his colleagues, and the results are startling: Speed kills.

In two studies Levine's research teams have objectively measured the speed of life (he likes to call it the pace of life) and found it related to rates of death from coronary heart disease. In a study of thirty-six American cities, Levine et al. (1989) measured the speed of life in each city. They did so by measuring (1) the walking speed of pedestrians in downtown areas, (2) the speed of bank clerks in responding to a standard request for change, (3) the talking speed of postal clerks responding to a standard question, and (4) the proportion of pedestrians wearing watches in downtown areas. They combined these four measures to form a pace-of-life index, which revealed that the northeastern United States had the fastest pace of life, followed in turn by the north central, southern, and western regions (Levine et al. 1989, p. 515). Los Angeles had the slowest pace of life; Boston, the fastest (Levine 1989, p. 45). So perhaps the reference should be to a Boston rather than a New York minute (New York had the third-fastest pace of life).

But the finding that dilates the pupils is the relationship between pace of life and coronary heart disease death rates. The correlation was statistically significant and positive, and it was large regardless of whether pace of life was adjusted for the age of each city's population or not. The faster the pace of life, the greater the rate of death from coronary heart disease.[13]

Levine and Norenzayan (1999) conducted an even more ambitious test of this relationship in their study of thirty-one countries discussed earlier. Using a similar pace-of-life index, they found a statistically significant positive correlation between speed of life and coronary heart disease death rates, a relationship that statistical controls revealed could not be explained by economic factors (Levine and Norenzayan 1999, pp. 191–94).[14] So the same relationship Levine et al. (1989) found in the study of American cities was replicated in the cross-cultural study of thirty-one countries: the faster the speed of life, the higher the death rate from coronary heart disease. But the relationship is not strong enough to declare a speed imperative. A city with a very fast pace of life in the American city study, Salt Lake City, with the fourth-fastest pace of life, just behind New York, had a very low coronary heart disease death rate. Similarly, in the study of thirty-one countries, Japan had the fourth-highest pace score but one of the lowest coronary heart disease death rates. Such cases suggest that the impacts of speed on human health are contingent on other factors, such as smoking and diet (Levine et al. 1989; Levine and Bartlett 1984).

Another case of the apparently contingent nature of speed's impact occurs in its relationship with a general willingness to help others. And again, Levine's work provides the findings. Investigating the hypothesis that pace of life would be negatively correlated with a willingness to help others in need, Levine et al. (1994) conducted additional research in the same thirty-six cities used for the coronary heart disease research. In each city they tested, behaviorally, for people's willingness to do such things as help a blind person, mail a lost letter, and help someone with an injured leg. Their results were going as predicted until the returns arrived from California. Although the speed of life in the eleven California cities included in the study was generally low, so was the average willingness to help people in these cities (Levine 1997, p. 163). Perhaps as a result of this set of cities in the sample, the researchers found no significant relationships between pace of life and any of the seven measures of willingness to help another included in the study (Levine et al. 1994, p. 77).

An interesting conceptual distinction grew out of these findings, the difference between helping and civility. For Levine noted, "In New York City, helping often appeared with a particularly sharp edge" (Levine 1997, p. 165). Thus help can be delivered with a variety of styles.

As has been discussed, speed is related to mortality from at least one major category of disease (coronary heart disease), but it can also have positive effects on such human endeavors as organizational effectiveness and efficiency. And efficiency has been an underlying motivation for much of the desire to increase speed. But there is a paradox to efficiency, a paradox James Gleick described as the effect of a web growing tighter and becoming more vulnerable to small disturbances that can "cascade through the system for days" (1999, p. 223). The web Gleick referred to is the web of service offered by airlines to the cities they serve. In terms more commonly used in organization science, the tightening web would be a system whose parts are becoming more tightly coupled (Weick 1979), which makes them more vulnerable to both small and large disruptions (see Weick 1995, p. 179, and Perrow 1984, pp. 62–100).

I believe this is a reason why speed and punctuality values are positively correlated. Being late or early is a disruption, especially in a tightly coupled system whose components are operating at close to maximum speeds. If schedules have been set to maximize efficiency, that is, with very small slack time tolerances allocated to deadlines such as takeoff and landing times, the efficiencies designed into the system will actually make the disruption a much bigger problem than if the system had not been designed to be quite so efficient in the first place (see the description of slack in Perrow 1984, pp. 89–90). In tightly coupled systems, the general association will be especially strong between valuing and practicing strict punctuality and doing things rapidly.

In a very real way this point returns us to the problem of longitude. The punctuality tolerances required for accurately determining the location of a ship at sea were extremely tight, allowing an error of less than one one-hundredth of 1 percent per day, the tolerance set by Parliament and enforced by the Board of Longitude. For errors greater than this, the threat of a disruption cascading through the system was horrific indeed, much to the sorrow of thousands of mariners. So despite problems they may cause, punctuality and speed play vital roles in promoting human well-being. The problem seems to be in striking the proper balance, in deciding how punctual? How fast? These are issues to which we will return in Chapters 7 and 9.

5

Eternal Horizons

> Does eternity only stretch one way?
> —Charlotte Perkins Gilman, *The Home*

Winston Churchill thought eternity stretched both ways, for he believed, "The longer you can look back, the farther you can look forward" (1974, p. 6897). Mary Austin reached the same conclusion even more strongly than did Churchill: "The arc of my mind has an equal swing in all directions. I should say the same of your mind if I thought you would believe it. But we are so saturated with the notion that Time is a dimension accessible from one direction only, that you will at first probably be shocked by my saying that I can see truly as far in front of me as I can see exactly behind me" (1970, p. 41).

These perspicacious observers of the human condition reached much the same conclusion about a connection between past and future. This chapter will examine this proposed connection between past and future, and even more important, examine why it exists and why it is important. But before examining these issues, we will begin with two simpler questions, questions whose answers will lead us to these weightier issues. One question asks, How far ahead do people look? (In Churchill's terms, how far forward do they look?) The other asks, How far back do people look? (In Churchill's phrase, how long back do people look?)

For example, how often do you think about things that might happen 250 years from now, that is, 250 years ahead? As we are about to learn, such behavior is rare, at least in the United States, but as we will see later in the chapter, at least one prominent CEO thought about matters two-and-one-half centuries ahead and took them very seriously.

To address the question of how far ahead people look, I asked a large sample of students at the University of Missouri-Columbia for information about how far ahead they looked when they *made plans or decisions.* (The 362 people—students are people—in the sample ranged from nineteen to forty-one years in age, with the average age being 20.83 years; women made up 46 percent of the sample.) They responded to the three questionnaire items that follow, and you are invited to answer them for yourself:

1. When I think about the short-term future, I usually think about things this far ahead. _____

2. When I think about the mid-term future, I usually think about things this far ahead. _____

3. When I think about the long-term future, I usually think about things this far ahead. _____

The respondents' answers are presented in Table 5.1 and are likely representative of this age group at colleges and universities in the United States. These responses also provide a distribution against which you can compare your answers.

Although Table 5.1 presents a wealth of information, several points are particularly notable. First, although over half of the respondents defined the short term as three months ahead or less (55.5 percent), over one-fourth of the sample (26.8 percent) thought of the short term as at least one year ahead—just under 3 percent defining it as five years ahead. These data reveal considerable variance in what sample members defined as short term, with a fair number defining it further ahead than stereotypes might have led us to anticipate. The variance certainly indicates that not all short terms are the same. As for the long term, close to half of the sample (45.1 percent) defined the long term as being ten or more years ahead, which again is likely further ahead than stereotypes would have led one to predict. And similar to the short term, the variance in this distribution indicates that not all long terms are the same. Nevertheless, and despite the variance, few if any of the respondents indicated they thought about things 250 years into the future—as did the CEO whom we shall encounter

TABLE 5.1

Distances respondents typically looked into three regions of the future.

Distance into the Future	Short-Term		Mid-Term		Long-Term	
	Region of the Future					
One day	17	(4.7%)				
One week	58	(16.0%)	4	(1.1%)		
Two weeks	41	(11.3%)	8	(2.2%)		
One month	47	(13.0%)	38	(10.5%)	5	(1.4%)
Three months	38	(10.5%)	33	(9.1%)	9	(2.5%)
Six months	59	(16.3%)	46	(12.7%)	11	(3.0%)
Nine months	5	(1.4%)	10	(2.8%)	3	(0.8%)
One year	72	(19.9%)	76	(21.0%)	46	(12.7%)
Three years	15	(4.1%)	59	(16.3%)	43	(11.9%)
Five years	10	(2.8%)	62	(17.1%)	82	(22.7%)
Ten years			23	(6.4%)	81	(22.4%)
Fifteen years			2	(0.6%)	22	(6.1%)
Twenty years			1	(0.3%)	39	(10.8%)
Twenty-five years					14	(3.9%)
More than twenty-five years					7	(1.9%)
Total	362	(100%)	362	(100%)	362	(100%)

later in the chapter. As would be expected, the answers about the mid-term future tended to be somewhere in between those for the short and long terms.

But do the differences presented in Table 5.1 make a difference? That is, are they related to other phenomena that someone somewhere, social scientist or layperson, considers important? If they are, the differences become much more important, because as Alfred North Whitehead (1925a, p. 12) argued, significance accrues through the relationship of one thing with another (see Chapters 2 and 7). Such is the case with the differences about the futures just described. As you have probably anticipated by now, these differences are related to other phenomena, important phenomena, phenomena that are themselves related to yet other phenomena, such indirect relationships making the kinds of temporal differences just presented even more meaningful, even more important. And one such phenomenon is the distance people look into the past, which will be considered in its own right in the next section, as well as its relationship with the distance people look ahead.

TEMPORAL DEPTH

If asked to summarize the findings presented in the preceding discussion, one would likely say something like "They were about time horizons." Yet the phrase "time horizons" never appeared in that discussion. The distances into the future that people look, individually and collectively, have traditionally been labeled "time horizons," both in general discourse and in organization science research (e.g., Ebert and Piehl 1973; Judge and Spitzfaden 1995; Mannix and Loewenstein 1993). But the distances into the *future* that people look are only part of a larger phenomenon that I have labeled temporal depth (Bluedorn 2000e).

I originally defined temporal depth as the temporal distances into the past and future that an individual typically considers when contemplating events that have happened, may have happened, or may happen (Bluedorn 2000e, p. 124). Although temporal depth certainly applies to individuals, it also applies to collectivities, especially as manifested in the cultures of groups (e.g., departments and organizations), so the phrase "individuals and collectivities" now replaces "an individual" in the definition, making the definition of temporal depth as follows: *the temporal distances into the past and future that individuals and collectivities typically consider when contemplating events that have happened, may have happened, or may happen.*

Thus temporal depth refers to both individual and cultural phenomena. It also deals with time in two directions, adding to the future a consideration of the past, the past generally being ignored in organization science (for exceptions, see March 1999; Thoms and Greenberger 1995; Webber 1972; and others cited later in the chapter), not that the rest of the social sciences are much less deficient in this regard (see Zimbardo and Boyd 1999, p. 1272). Because temporal depth also encompasses the past, I asked the 362 college students who answered the three questions about the future (see Table 5.1) for three parallel pieces of information about how they typically considered the past when they made plans or decisions. The three items they responded to follow, and as before, you may wish to respond to them yourself and compare your answers with those given by the respondents in this sample:

1. When I think about things that happened recently, I usually think about things that happened this long ago. _____

TABLE 5.2

Distances respondents typically looked into three regions of the past.

Distance into the Past*	Recent		Middling		Long Ago	
			Region of the Past			
One day	28	(7.7%)				
One week	123	(34.0%)	10	(2.8%)	1	(0.3%)
Two weeks	92	(25.4%)	30	(8.3%)	1	(0.3%)
One month	65	(18.0%)	100	(27.6%)	7	(1.9%)
Three months	28	(7.7%)	69	(19.1%)	21	(5.8%)
Six months	19	(5.2%)	57	(15.7%)	24	(6.6%)
Nine months			6	(1.7%)	13	(3.6%)
One year	5	(1.4%)	54	(14.9%)	67	(18.5%)
Three years			23	(6.4%)	74	(20.4%)
Five years	2	(0.6%)	11	(3.0%)	86	(23.8%)
Ten years			1	(0.3%)	51	(14.1%)
Fifteen years			1	(0.3%)	10	(2.8%)
Twenty years					5	(1.4%)
Twenty-five years					2	(0.6%)
Total	362	(100%)	362	(100%)	362	(100%)

* No respondents selected the category "More than twenty-five years," so it is not included.

2. When I think about things that happened a middling time ago, I usually think about things that happened this long ago. _____

3. When I think about things that happened a long time ago, I usually think about things that happened this long ago. _____

The respondents' answers appear in Table 5.2. As noted with their answers to the questions about the future, these responses are likely representative of this age group at colleges and universities in the United States. And as before, these responses also provide a distribution against which you can compare your answers.

So what do the answers in Table 5.2 reveal about these individuals' past temporal depths? Almost the entire sample defined the *recent past* as extending into the past no longer ago than six months, with slightly over two-thirds of the sample (67.1 percent) defining it as two weeks ago or less. At the other

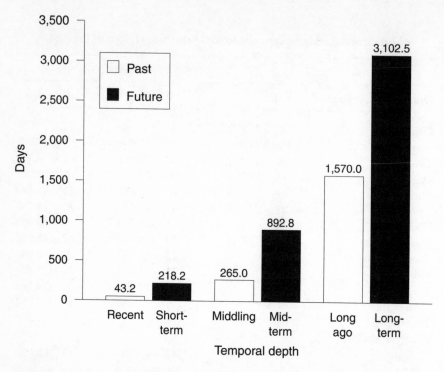

FIGURE 5.1. Average lengths of time that defined three past temporal depths and three future temporal depths for a sample of college students. *Note*: To calculate the average for the long-term future temporal depth, the seven respondents who selected "More than twenty-five years" as their responses were treated as if they had given "Thirty years" as their responses. No respondent chose "More than twenty-five years" for the other two future temporal depths or for any of the three past temporal depths.

extreme, 62.7 percent of the sample defined "a long time ago" as sometime from one year to five years ago, but no one defined it as beginning more than twenty-five years ago. And as with their responses about the future, the responses about "a middling time ago" fell somewhere in between those for the recent past and a long time ago.

The results about past and future temporal depths seem to parallel each other. To see how closely, I converted the fifteen temporal depth categories the respondents were given to choose from (presented in Tables 5.1 and 5.2; see the Appendix for how these categories were presented on the questionnaires as part of the Temporal Depth Index) and calculated the average number of days for short-term (recent), mid-term (middling), and long-term (long-ago) intervals for both the future and the past. The results are presented in Figure 5.1.

The key differences depicted in Figure 5.1 are all statistically significant, which is to say that (1) the differences among the future regions are all statistically significant, (2) the differences among the past regions are all statistically significant, and (3) the differences between the components of each parallel pair (e.g., short-term future and recent past) are all statistically significant.[1] And what are these differences? Perhaps the most noteworthy of the differences is that each of the future regions extends much further into the future than their past counterparts extend into the past. The short-term future extends about five times further than does the recent past; the mid-term future, about three-and-one-third times as far as the middling past; and the long-term future, about twice as far as the long-ago past. So although the steplike pattern is similar for both the past and future regions, the future depths extend over substantially larger amounts of time than do those in the past.

The Proposed Connection

The results presented so far present a great deal of information about the two "how far?" questions: How far ahead do people look? How far behind do people look? But they have not addressed the connection between future and past temporal depths proposed by Churchill and Austin. Are they right, that in Churchill's words, "The longer you can look back, the farther you can look forward"? The data from which the results have been presented so far allow the proposed connection to be tested by correlating the average of the three future temporal items with the average of the three past temporal depth items (see the Appendix about averaging these items), and the result is a statistically significant positive correlation (see the Appendix). The proposed connection is accurate: The longer the respondent's past temporal depth, the longer the respondent's future temporal depth. And this result was found not only in this sample but also in four other large student samples as well. (The details about these samples and correlations are given in the Appendix, which presents the Temporal Depth Index and a description of its development. The Temporal Depth Index is a questionnaire scale that combines a structured response format with the same six items about future and past temporal depths presented earlier in this chapter.)

Results from a group-level investigation of this connection found the same relationship. In the study of the national sample of publicly traded companies (see Chapters 3 and 4), Steve Ferris and I found a statistically significant positive correlation between past and future organizational temporal depths, a relationship that persisted after controlling for several organizational and environmental variables (Bluedorn and Ferris 2000). So at the organizational level as well as at the individual, past temporal depth and future temporal depth are positively correlated: The longer the depth of the past, the longer the depth of the future.

Such findings seems plausible, even intuitively plausible, but only so post hoc, for little research, almost no systematic research, has been conducted on this relationship at either the individual or the group level. This is not to say that there was a complete absence of clues about it, because there were clues, and they provide some of the rationale for examining this relationship in the first place. They form the basis, perhaps, along with the observations of Churchill and Austin, for the intuition that a positive correlation between past temporal depth and future temporal depth would be found. So an examination of these clues may provide greater insight into why this connection exists between past and future temporal depths.

The Temporal Depth Demonstration

For the better part of a decade I have presented versions of the following demonstration to groups of many kinds: "In Columbia, Missouri, a time capsule was buried in 1966, and the inscription on it specifies the date on which it is to be opened." I then ask each group, What date does the inscription specify for opening the time capsule? And regardless of the group I ask—undergraduate or graduate students, managers, college faculty, the general public—remarkably, the large majority of every audience, audiences that number over ten thousand people collectively, overwhelmingly responds with an estimate of sometime within one hundred years following the date on which the capsule was buried. I estimate that over 90 percent of these people have been able to answer this question with uncanny accuracy (within fifty years)—without ever having seen the instructions on the capsule. (The instructions are to open the capsule in 2066.) But the demonstration is not over. It proceeds with a description of a second time capsule, one in another country.

Half a world away, American entrepreneur Stephen Chubb visited the site of a then recently buried Japanese time capsule. In itself this was not noteworthy, but the date for opening it was—by American standards. For the instructions were to open the capsule *five thousand years* hence, a temporal depth Americans seldom consider, as is indicated in the dates for opening the American capsule (and Table 5.1). Referring to the United States, Chubb wondered, "How long would someone put a time capsule in the ground in this country?" (Murray and Lehner 1990, p. A16). This difference between Japanese and American practices is an order-of-magnitude difference between the two countries, between the two cultures, so the audiences at my time capsule demonstrations often gasp audibly when I say, "five thousand years."[2]

To reinforce the claim that one hundred years is a typical future temporal depth selected for opening American time capsules, that the Columbia, Missouri, time capsule is not idiosyncratic, one more piece of evidence will be presented. And that evidence, like Stephen Chubb's experience, comes from across an ocean, but this time the Atlantic rather than the Pacific.

Approximately two hundred yards inland from the French coast, the sidewalk from a parking lot makes an abrupt right turn and heads toward the sea. Where the sidewalk pivots, a time capsule was buried on July 6, 1969. At Omaha Beach. At the American Cemetery.

Beneath the cluster of five-pointed stars arranged in the pentagon-shaped insignia of a five-star general, the marker's inscription explains:

> In memory of GENERAL DWIGHT D.
> EISENHOWER and the forces under his
> command this sealed capsule containing
> news reports of the JUNE 6, 1944
> NORMANDY LANDINGS is placed
> here by the newsmen who were there
> June 6, 1969
> TO BE OPENED JUNE 6, 2044

The event the capsule commemorates occurred on June 6, 1944, which makes it the appropriate date, not the date of burial, from which to measure the length of time to opening. The interval spans, of course, one hundred years, as does its counterpart in Columbia, Missouri. But unlike its Missouri counterpart, indeed unlike most time capsules, this time capsule will likely be opened, and opened when specified. And there is more than one reason that it will be opened on schedule. First, the capsule is part of a cemetery holding tremendous symbolic significance to an extremely powerful and wealthy nation, which quite properly maintains the cemetery meticulously. The image of the nearly ten thousand gravestones arranged in geometric perfection is an image known worldwide, an image that is almost as memorable when seen in a photograph or on television as it is moving when seen in person. So the capsule is part of an administrative structure designed to attend to such matters.

The second reason the capsule will be opened as specified is less formal, but powerful in its own way. It is a spiritual reason. For to walk through this place is to be humbled; it quietly compels reverence, a reverence and gratitude that deepen as one learns that among the graves lie thirty-three pairs of brothers, side by side in death as they once were in life. Even a short visit produces a profound respect, both for the deceased and for the larger effort of which they were such a significant part. Those buried in this cemetery cannot be honored directly; instead, reverence is extended to their shrine, to their memory. And by honoring it, we connect with them, thereby adding meaning to our own lives. Such meaning makes us want that capsule to be opened on time. And it will be: on June 6, 2044.

Thus as revealed in their time capsules, even profoundly symbolic ones, future temporal depths vary from nation to nation. Moreover, a clue is hidden in the time capsule demonstration about the relationship between past and future temporal depths, a clue that several audience members have brought to my attention spontaneously over the years. Without prompting, one or two audience members will approach me after my presentation to discuss the demonstration. They will say something like "Isn't it to be expected that the Japanese would pick a date much further in the future to open their time capsule because Japan is so much older than the United States?" They will usually not provide the detailed logic for a relationship between age of the country and the depth of future for opening time capsules, but the intuitive logic is there: the older the country, the longer its future temporal depth.

So this demonstration helps tease out one clue that past and future temporal depths would be positively correlated, albeit that was not its original intent. I originally developed the demonstration to reveal how deeply temporal matters are embedded in cultures, in the core of culture known as basic underlying assumptions (Schein 1992, pp. 16–27). The demonstration works well to illustrate that point, as no one in the audience has ever been formally taught the proper time to open time capsules in America. It also demonstrates the major cultural differences about capsule-opening times, hence that such times, like all times, are socially constructed. (All times are not . . .) But the demonstration also suggests the relationship between past and future temporal depths, thereby offering one clue to that relationship.

Time's Arrow

Even before developing the time capsule demonstration, I had learned of the remarkable discovery made by Omar El Sawy (1983) and described it and its importance (Bluedorn and Denhardt 1988), a practice I have continued to the present (e.g., Bluedorn 2000e). This discovery provides the second clue that past and future temporal depths are positively correlated, and it does so explicitly—albeit El Sawy did not use the term *temporal depth*.

El Sawy conducted an experiment with CEOs from high-technology companies in Silicon Valley. He asked each CEO to think of ten events that happened in the past and when each of those events happened. Similarly, he asked each CEO to think of ten events that might happen in the future and when those events might occur. But he asked half the CEOs to answer the questions about the past events first; the other half, the questions about future events first. That was the experimental manipulation, a design that allowed inferences about cause and effect. To wit, if there is a relationship between past and future, which temporal direction affects the other?

Responses from thirty-three of the CEOs produced two sets of results that answered this question unambiguously. First, no statistically significant differences appeared between the two experimental groups for either the median age of the past events or the age of the oldest *past* event identified. Thus the experimental manipulation had no statistically significant impact on the CEOs' past temporal depths. El Sawy concluded, "It is safe to conclude that based on the available data, the CEO will always invoke the same past span whether he looks forward first or backwards first" (1983, p. 145). Looking forward first referred to listing ten events that might happen as the first task; looking backward first, to listing first ten events that had happened. Having the CEOs think about the future first had no statistical impact on how far into the past the CEOs would think about events.

El Sawy's second finding completes the answer to the question of which direction affects which. Having already determined that thinking about the future leaves past temporal depth unaffected, El Sawy found a major effect for thinking about the past first. The CEOs who were asked to think first about events that had happened in the past thought about events significantly further into the future than did the executives who thought about the future events first. (The distances into the future were measured by the median distance into the future of the ten future events as well as the single future event among the ten envisioned to occur the furthest distance into the future.) And as El Sawy noted, "Not only is there statistical significance, but there is also operational significance" (1983, p. 146). For example, the average (mean) temporal depth of the event thought of as occurring the furthest into the future was 5.11 years for the CEOs who listed the future events first. For the CEOs who listed past events before listing future events, the average (mean) temporal depth of the longest event was 9.18 years (El Sawy 1983, p. 147). This is a difference of 4.07 years, and as shown in Table 5.1, this is a difference large enough to move thinking from the short term to the long term for all but about 7 percent of the respondents who provided the results reported in that table. This is operational significance indeed.

El Sawy achieved these effects by simply asking half of the CEOs to think about events in the past before then asking them to think about events in the future. He did not ask the past-first group to think about events in the past that occurred a long time ago or before a specific year. Thus he provided no temporal structure for when in the past to think about events, just anytime in the past. This aspect of the design enhances the power of the results, because the results are more "natural" this way and less likely to be limited to the experimental situation. In this experiment, as in real life, when the CEOs thought about the past, they went wherever they wanted to go. So the results revealed another version of time's arrow, a version consistent with the position taken by those who argue for an arrow of time in the physical universe (e.g., Eddington 1928; Coveney and Highfield 1990; see Chapter 2). The past leads to and influences the future, but the future does not influence the past.

Thus El Sawy's research provided a second clue that past and future are related, and it even added a causal direction (i.e., "A connection to the past facilitates a connection to the future" [March 1999, p. 75]). This leads to findings about organizational age, which may provide a third clue and suggest at least part of the reason for the connection between past and future temporal depths.

The Organizational Age Connection

Steve Ferris and I found that organizational age was positively correlated with both past and future temporal depths, and that these relationships persisted after controlling for several organizational and environmental variables (Bluedorn and Ferris 2000). The older the organization, the further its members looked into both the past and the future, and the positive temporal depth correlations with the organization's age may suggest why.

With greater age comes a larger past—though *a potentially larger past* might be the more accurate phrase (see Butler 1995, p. 929)—a past that in the case of organizations apparently becomes received history. Not that these correlations speak to the truth status of the received history, such history being subject to social construction processes as much as any other human phenomena, including the future, as Paul Fraisse noted: "We construct our past as well as our future" (1963, p. 177). Indeed, George Orwell made this point well in *1984*: "And if all others accepted the lie which the Party imposed—if all records told the same tale—then the lie passed into history and became truth. 'Who controls the past,' ran the party slogan, 'controls the future: who controls the present controls the past'" (1961, p. 32).

Orwell was writing, of course, in opposition to the machinations of totalitarian regimes that consciously write histories without concern for fact or evidence, but write them only to facilitate their own ends. Yet his point exemplifies the larger issue that history is always a matter of interpretation, of construction, of points of view consciously or unconsciously held. And in the case of organizational age, greater age does not point to greater truth; it simply provides a longer timescape within which to search for material, a wider temporal loom upon which to weave the fabric of history.

And the determination of organizational age illustrates the constructed, enacted nature of the past, because what at first glance seems like a simple, even objective matter becomes ambiguous when mergers and acquisitions are involved. Is the founding date the date that the oldest of the merger partners began operations, or is it the date when the last partners merged? Families can face the same ambiguities when one or both spouses have been married previously and they and their children combine to form new families. As the definition of the situation principle teaches (see Chapter 1), the important issue is when the people in the organization or family *believe* it was founded. It is important because it places a temporal boundary on the past and by doing so limits the span to be searched when looking to and for the firm's (or family's) history.

So the history of a firm, nation, or family is socially constructed. But why is this important? And what about the past links it to the future and gives the past dominion over the future? El Sawy's findings empirically support the past's primacy, and his theoretical explanation was based on the work of Paul Fraisse and Karl Weick, work that must be examined before such explanations can be extended.

THE PAST AS METAPHOR

Fraisse saw the future as representations individuals draw from experience, as something that is "imagined as a repetition of the past" (1963, p. 172). And as individuals mature, they develop the ability to conceive a future "which is a creation in relation to our [their] own history" (p. 172). So to Fraisse the future was linked inextricably to the past, a connection Karl Weick (1979, 1995) asserted forcefully.

To Weick, human beings were sense-making creatures, and all sense-making is retrospective, all explanation relies on the past: "All understanding originates in reflection and looking backward" (1979, p. 194). So to understand the present and the future, one turns to the past—not just can, but must. Buttressed by the work of Alfred Schutz (e.g., 1967, p. 51), which reached much the same conclusion, Weick argued that the past was used to understand the present and the future, that neither could be understood without the past. And how the past can provide this understanding, this meaning, is a major insight.

Both Fraisse and Weick have essentially argued that people think about the future as if it will be like the past, albeit Weick does so in greater detail. So people will generally anticipate the future as if it will be like the past. Another way to say this is that people use the past as a metaphor for the future. As such, it is instructive to examine the way metaphors are generally used to develop understanding and to enhance meaning. Aristotle was one of the earliest writers to explain how this happens, and he did so in his analysis of literary forms when he wrote, "It is a great matter to observe propriety in these several modes of expression, as also in compound words, strange (or rare) words, and so forth. But the greatest thing by far is to have a command of metaphor. This alone cannot be imparted by another; it is the mark of genius, for to make good metaphors implies an eye for resemblances" (Aristotle 1911, p. 87).

Having a command of metaphor was the poet's greatest gift, and its essence was an "eye for resemblances," which means knowing both when things resemble each other and when they do not. This point is made even more explicitly in another translation of the same passage: "It is a great thing to make a fitting use of each of the forms mentioned as well as double and foreign words, but greatest is the use of metaphors. For this alone cannot be gained from others and is a sign of the naturally well-endowed poet, for to make good metaphors is to observe similarities among dissimilarities" (Aristotle 1961, p. 44).

The emphasis Aristotle placed on metaphor suggests he regarded it as an especially powerful way to produce insight, understanding, and meaning—the core competencies of a good poet. So if the conclusion just reached is correct, that the past is a metaphor for the future, its use as a metaphor should focus on observing similarities among dissimilarities. And this is basically what Robert Neustadt and Ernest May recommended, twenty-four hundred years after Aristotle, in their discussion of how to use history in decision making: "Comparing all those seemingly analogous situations with the present one, what are *Likenesses* and *Differences*? Compare 'now' with 'then' *before* turning to what should be done now" (Neustadt and May's emphases; 1986, p. 41).

Unfortunately, the similarities, the likenesses, may overwhelm the differences (see Morgan 1997, pp. 4–5). And according to Weick, "people who select interpretations for present enactments usually see in the present what they've seen before" (1979, p. 201). In terms of the past-as-metaphor perspective developed in this chapter, "what they've seen before" implies the use of "an eye for resemblances," the ability to see the similarities. But as Aristotle, Morgan, and Neustadt and May all noted, there is more to the mature use of metaphor than detecting the similarities between events and situations; the differences matter too. They matter, in part, because the ability to detect and deal with novelty may be a key to both organizational learning and performance (Butler 1995, pp. 944–46).

And if Weick has drawn the correct conclusion about how the past is used to enact the present, being able to note the differences may be even more important than being able to see the similarities. This is especially so in equivocal enactments, which Weick (1979, p. 201) described as involving a figure-ground construction, one in which the ground consists of the strange and unfamiliar (the differences); the figure, that which is known. Because the figure draws one's attention, the ground may grow, may change and become even less familiar and more different without one being aware of these changes. And if unnoticed—the natural tendency is to attend to the figure—these changes make the people involved susceptible to a "figure-ground reversal" wherein "nothing makes sense" (Weick 1979, p. 201).

Although Weick discussed metaphors and their use in organizational contexts (1979, pp. 47–51) and even referred to his own concept of retrospective sense-making as a "key metaphor" (p. 202), to my reading he did not explicitly frame his discussion of the past's use to enact the present in terms of metaphor. But as we have just seen, his ideas greatly inform the past-as-metaphor frame, which provides a cogent link between Aristotle's description of metaphor, Neustadt and May's advice, and the matter of how metaphor can be used to make sense of things in organizations.

Or how it can come to make less and less sense of things. When seen in this contrarian way, these principles of metaphor can explain what otherwise seem to be absurd decisions and behaviors. An example of such absurd behavior is the process Danny Miller (1990) described as the Icarus Paradox.

Put briefly, the Icarus Paradox describes how effectiveness leads to ineffectiveness, how success leads to failure. The gist of the process Miller described is that as organizations become successful, their members, especially the more powerful decision makers, begin to attribute the reasons for the organization's success to the way the organization does things, and they grow confident in these practices. If the organization continues to be successful, or if its success increases, the attributions continue and people's confidence in how the organization operates increases. But at some point, the confidence becomes over-confidence, even hubris, and the organization stops trying to adjust and adapt because the people in it, especially the powerful decision makers, believe they have found *the answers*, not just a good way to do things, but the *only right way* to do things. As a result the organization does not change when it needs to change, and its effectiveness diminishes. In extreme cases, the organization goes out of business.

Miller picked his label for this process aptly, for it follows closely the mythical Greek story of Icarus, the son of Daedalus, who after escaping from the labyrinth on Crete by using wings his father built for him that were made of wax and feathers, became intoxicated with his success and the thrill of flight.

This led him to fly higher and higher until the heat of the sun melted the wax that was holding the feathers to his wings, and he plummeted to his death in the sea below.

The legend of Icarus displays amazingly close similarities to the organizational process Miller described (Miller chose his metaphor well). And when Miller's insights are combined with the material from Weick, the combination can help explain the use of the past to enact the present and future, and in so doing help explain how success leads to failure. To interpret the Icarus Paradox in these terms, assume that an organization has already experienced the initial complex of success, attribution, and confidence. In the process of organizational sense-making, the people in the organization can now take this complex as its past and use it as a metaphor to explain the present and imagine the future. In terms of the figure-and-ground analysis, the way the organization did things and its success in the initial complex constitute the figure. The ground would be a residual category of factors (e.g., general economic conditions) now deemed irrelevant to the organization's functioning and success. With each succession of success-attribution-confidence complexes, the figure becomes more and more dominant, the ground less noticeable. But the ground holds the secret of the organization's existence and success. For in the ground is to be found the organization's environment, and that environment is always changing—sometimes faster, sometimes slower—but changing nevertheless. The only point at issue is how fast it is changing.

But this change in the ground goes unnoticed, especially when those in the organization reach a state of overconfidence, of hubris. To them there is no need to change because in their exalted state those in the organization cannot even imagine a possible need to change, hence they have no reason to attend to the organization's environment and engage in a process of robust enactment, creating new environments that will require the organization to change. If the process continues long enough without the antidote of humility, the organization comes to be so out of step with environmental factors that serious problems develop precipitously. And when people in the organization attempt to interpret what's happening, they have a hard time doing so, perhaps even experiencing the ground-figure reversal that makes the situation uninterpretable for a time. Weick described it this way: "As that ground enlarges unnoticed, people who still see what they've seen before and still write the same old histories are seeing less and less of what is there and are becoming more vulnerable to a figure-ground reversal" (1979, p. 201). In past-as-metaphor terms, they see similarities, but they do not see the differences, and that tunnel vision is hazardous to the organization's health; sometimes it is even fatal.

One difference between the story of Icarus and this organizational process is that once the wax melted in Icarus's wings and the feathers fell away, Icarus was doomed. He had no hope of survival. However, unlike Icarus, an organization that is beginning its death spiral can recover before it crashes, a point illustrated by several examples in Miller (1990). Thus this difference is an example of observing a key dissimilarity among similarities.

So simply looking to the past is insufficient. The past must be used wisely and must not be interpreted as a simple recipe for success, a single recipe that can develop in organizations and be used for a long time (Butler 1995, p. 929), which means the past cannot be used simply as input for a manager seeking to make a programmed decision. It cannot be used this way because, as a metaphor for both the present and the future, the past must be interpreted sagely. This is so because there are key differences among the similarities: "Although the past may not repeat itself, it does rhyme" (attributed to Mark Twain in Least Heat Moon 1982, p. 10).[3] History does not repeat itself, which means the differences between past and both the present and future must be identified along with the rhyme, the similarities. History may be, as Michael Crichton had one of his characters say, "the most powerful intellectual tool society possesses" (1999, p. 480), but it is only powerful managerially and over the long term when it is used metaphorically, when people who look to it deliberately find its similarities to and differences from the present and the future. In this quest managers share much with poets, indeed they are poets, indeed all people are poets because all employ metaphor in the comparisons they make between the past and the present and the future. And some grammatical forms may promote this temporal poetry more readily than others.

As If the Future Had Already Happened

To consider the future, it may help to treat it like the past, that is, as if it had already happened. This is the premise Weick proposed in his discussion of future perfect thinking (1979, pp. 195–200). Future perfect thinking is a grammatical prescription instructing managers and planners and all who consider the future to do so in the future perfect tense. Thus rather than the simple future tense as used in a statement like "We shall overcome," the future perfect tense would have us say, "We shall have overcome." Alfred Schutz believed that the *"planned act bears the temporal character of pastness"* (Schutz's emphasis), because the actor projects the act as completed and in the past, a paradox that places the act in both the past and the future at the same time, something the future perfect tense makes possible (1967, p. 61). These were insights that Weick both noted (1979, p. 198) and built upon to explain why future perfect thinking may make it easier to envision possible futures.

Several studies (Bavelas 1973; Rollier and Turner 1994; Webb and Watzke as cited in Weick 1979) have revealed a consistent finding about the impact of tense on people's ability to imagine events. In all three studies participants were asked to imagine events (i.e., trips, football games, car accidents), and the experimental manipulation in all the studies had half of the participants imagining the event in the past, because they were told the event had already occurred, and the other half were told that the event would happen, so they envisioned it in the future. In all three studies, the participants who envisioned the events occurring in the past envisioned them with significantly more detail than the people who envisioned them in the future. Of course, participants in both conditions were equally ignorant of real details, so the difference in tense led to the difference in details.

Weick extrapolated these findings to the difference between the simple future tense and the future perfect tense. He argued that just as thinking about events occurring in the past makes them easier to visualize than thinking about them in the future, because the future perfect tense is more like the past than the simple future tense, thinking in the future perfect tense should make visualizing future events and scenarios easier than doing so in the simple future tense. Among the reasons this may be so is that the simple future tense is more open-ended than the future perfect tense, the latter seeming to convey a sense of closure and a focus on specific events, which is unlike the simple future tense in which anything is possible (Weick 1979, pp. 198–99). It is well to note that although Weick did not explicitly frame his argument in terms of metaphor, it is really another example of the past-as-metaphor-for-the-future idea developed in this chapter, albeit a more precise manifestation of it. The precision comes in Weick's conclusion that some futures are more like the past, are more similar to it than others. In his argument, the future described in future perfect terms is more similar to the past than the future described in simple future terms. And if he is right, the future should be easier to envision in terms of detail when it is cast in the future perfect frame. But whose future? Whose past?

The Primacy of Experience

The line of reasoning just presented may also explain why an individual's or an organization's history plays such a central role when contemplating the future. Regardless of how constructed it is, the history of the specific individual or organization will be seen as more real and less imaginary than that of any other individual or organization. And it would seem that this greater sense of reality, this verisimilitude, would come in large part from actually experiencing the history. In the case of organizations, the actual experience is reinforced by the received experiences from the organization's past, which are made more real by the imprimatur of formal and informal authority. Although vicarious learning is possible, learning from direct experience and historical continuity, if relevant, seem likely to trump experiences reported by other people and about other organizations. Personal and organizational histories occupy prominent figure positions in the figure-ground dichotomy, and that such histories are used to cope with the future is indicated by several pieces of evidence.

First, some companies and executives within them consciously use the past to deal with the future. Emerson Electric Company has used "5-back-by-5-forward" charts that contrast the past five years' financial data with five-year projections. Among other things, when combined with the current year, these data helped the company detect trends (Knight 1992, p. 62)—a point that anticipates Gregory Benford's thinking, which will be presented later in the chapter. As another example, when he was CEO at Intel, Andy Grove used the past to help him see the future: "I have a rule in my business: To see what can happen in the next ten years, look at what has happened in the last ten years" (1996, p. 68). This rule seems straight out of El Sawy's findings. Indeed, both it and Emerson Electric's five-back-by-five-forward practice, especially their temporal symmetry, are consistent with the positive correlations between past and future temporal depth presented earlier (i.e., five years back, five years forward; ten years behind, ten years ahead).[4]

And those positive correlations are a second form of evidence that the past may be used to deal with the future, especially when El Sawy's (1983) findings are added to those correlations. The correlations reveal a connection, and El Sawy's findings indicate the connection is for the past to affect the future's depth. By extension, this combination of findings would indicate that the past is used to think about the future, and not the other way around. Thus in their study of organizational visions, Laurie Larwood et al. (1995) found a significant positive correlation between how long firms had held their current visions and how long those visions extended into the future. So regardless of whether general future and past temporal depths are considered, or the past and future temporal depths of organizational visions are involved, the past is connected to the future. But it is not just connected, because El Sawy's (1983) findings and the several positive correlations point to longer past depths as a way to generate longer future depths. And longer depths may confer advantages.

EFFICACIES OF LONGER VIEWS

Advantages may accrue to older organizations in several ways. For one, the older the organization, the longer its past temporal depth (Bluedorn and Ferris 2000). This means that, ceteris paribus, older organizations will tend to have more history, more examples to draw upon from which guidance may be obtained for dealing with the future (see Butler 1995, p. 929, about long memories and variety of analogues). Having more history to choose from makes the metaphorical task even more challenging because the decision makers must choose between competing historical episodes or find creative ways to synthesize them. Having more historical material to choose from makes the metaphorical task of creating good metaphors by observing "similarities among dissimilarities" even more challenging, hence making metaphorical skill even more strategic.

This potential albeit challenging advantage conferred by greater age also suggests the importance of the organization's founding date, in particular, how that date is constructed in the organization. Its importance stems from the founding date's function as a marker defining the boundary of the organization's temporal frame. Given the salience of the organization's own past vis-à-vis the pasts of others, the date organizational members take for the organization's founding establishes a past boundary beyond which decision makers will not look for instructive episodes from the organization's history. However, unlike the organization's future temporal depth, its time line into the past is always bounded by the organization's founding date. And as with paradigms generally (Barker 1992), the frame bounded in the past by the organization's founding date limits the search for information from the organization's history to the domain of events that occurred on or after that date.

So longer organizational histories confer potential advantages, maybe even competitive advantages. And because longer past temporal depths are associated with longer future depths, do longer future depths confer advantages too?

Organizational Performance

To the extent that this question has been addressed, the received wisdom appears to hold that a long-term future temporal depth is better than a short-term depth. William Ouchi's best-seller, *Theory Z* (1981), associated long-term future depths with the success of Japanese organizations and contrasted them with the short-term depths of their then-struggling American counterparts. James Collins and Jerry Porras reached a similar conclusion in another best-seller, *Built to Last* (1997), that the long term is better than the short term. And Terrence Deal and Allan Kennedy were critical of "short-termism," especially its effects on human resource practices (1999, pp. 43–62), an issue also present in Ouchi's and Collins and Porras's analyses.

Examples of how long-term future orientations influence behavior often present striking contrasts to behaviors guided by short-term future orientations. One such example was IBM's decision, actually Thomas Watson's, to pay a large portion of employees' salaries to the employees' families while the employees were in the American Armed Services during World War II (Carroll 1993, p. 48), military pay being much lower than the remuneration at IBM. This decision made no strategic sense whatsoever from a short-term future perspective, but from a long-term perspective one would anticipate that it helped create a generation of extraordinarily dedicated and loyal employees— as evidenced by the account of this practice being "one of the stories most often told over the years as IBMers explained their loyalty to their employer" (Carroll 1993, p. 48).

A similar example occurred in 1995 when Aaron Feuerstein, the CEO of textile manufacturer Malden Mills, made a similar decision after much of his company's physical plant was destroyed in a catastrophic fire. Feuerstein decided to pay the company's idled workers their full salaries and wages while the plant was being rebuilt (Calo 1996). Just as at IBM, such a decision makes economic sense only from a long-term perspective, the behavior of employees being a major consideration in such a perspective. Indeed, one of the employees who benefited from Feuerstein's decision commented, "I owe him every-thing I have, and everything I'm gonna have. I'll pay him back" (Calo 1996). Interestingly, Malden Mills was a privately held company at the time when Aaron Feuerstein made the decision that made him famous (e.g., he was in-vited to attend the 1996 State of the Union address), so he had much more freedom to make this type of decision than would a CEO at a publicly traded company. As such, his behavior as the CEO of a privately held company sup-ports Deal and Kennedy's (1999, pp. 43–62) view that an increased emphasis on shareholder value during the 1980s and 1990s led to even more of an em-phasis on the short term.

Another example of a decision with long-term consequences also illustrates the connection between the past and the future. Lincoln Electric Company, a Cleveland, Ohio–based manufacturer of arc welders, has used a well-chronicled (e.g., Sharplin 1998) pay system for many years. An important part of this sys-tem is a major annual bonus paid to all employees based on merit ratings and company profits.

In 1993, Lincoln paid the bonus with $55.3 million in *borrowed* money (Hast-ings 1999, p. 178). The firm's top management made this decision because the bonus system was such a substantial portion of each employee's pay, but also because the bonus had been paid regularly for so many years, hence having be-come a part of the psychological contract between the employees and the company, an almost taken-for-granted part of Lincoln's culture. To have not paid the bonus because the company had a loss that year, a loss owing totally to an unsuccessful attempt to expand operations internationally, would have been to break faith with the company's workforce. This in turn would have put at risk employee loyalty and dedication developed over decades, loyalty and dedication that had developed in part owing to the company's unique incen-tive system based on the bonus. So a decades-old tradition to which manage-ment and line workers alike looked for guidance concerning the future influ-enced a major and risky decision. In terms of the metaphorical processes already discussed, the differences between past and future were identified (i.e., the company's major loss that year), but the differences were not quite enough to invalidate the metaphor, and the annual bonus, the similarity, was main-tained—as was a de facto commitment to the long-term future.

Thus IBM, Malden Mills, and Lincoln Electric, all historically successful manufacturers but in very different industries, illustrate how a long-term future temporal depth can be implemented and contribute to organizational effectiveness. But is this the case for temporal depth generally? Steve Ferris and I found the relationships between temporal depth and measures of organizations' financial performance so mixed (i.e., some were significant, some were not) and contingent (Bluedorn and Ferris 2000) that the best answer that can be given at present is *sometimes*. And a pioneering study focused on future temporal depth and managers' ethical beliefs and values suggests that sometimes a *shorter* temporal depth may even be better.

Principles of Right and Wrong

T. K. Das, no stranger to investigations of future temporal depth (e.g. 1986, 1987), conducted the study (2001). He gathered data from 585 vice presidents of American companies and measured their future temporal depths and their beliefs and values concerning fourteen ethical principles (e.g., the categorical imperative, the golden rule).

When I first encountered this research, I was still under the sway of the received wisdom that a long-term depth was generally better than a short-term depth. As indicated, I was certainly not unique in holding this view, because the Long Now Foundation, which was "established in 1996 to foster long-term responsibility," has been involved in a project that extends ten thousand years into the future (Brand 1999, p. 4). And why ten thousand years? This future depth was chosen as a result of a suggestion that because "10,000 years ago was the end of the Ice Age and beginning of agriculture and civilization; we should develop an equal perspective into the future" (pp. 4–5). If some attributes of that reason seem familiar, they should, because the reason reflects some key points from earlier in the chapter; to wit, past temporal depth affects future temporal depth (i.e., looking back over the span of human civilization to the Ice Age for guidance about the future), and past and future temporal depths are positively correlated (i.e., they looked back ten thousand years so they are going to look ahead ten thousand years). We will encounter the Long Now Foundation again in Chapter 9.

So, reinforced by my own beliefs and those of others such as the Long Now Foundation, I expected that if Das found any relationships, he would find that executives with longer future temporal depths would have stronger beliefs about ethical principles than those with short-term horizons. Actually, Robert Axelrod's (1984) famous prisoner's dilemma research similarly led me to anticipate this (i.e., Axelrod's finding that strategies which confer mutual benefits rather than mutual pain or harm are most successful in the prisoner's dilemma game *in a condition where many rounds of the game will be played*, where the game is played with a long-term future depth, also led me to anticipate what I thought Das would find).

But as sometimes happens in the social sciences, Das found exactly the opposite (of what *I* thought he would find). He found that the relationship between having a near-future rather than a distant-future orientation and seven ethical principles was statistically significant, that the executives with the near-future orientations felt more strongly about the ethical principles than did their distant-future-oriented counterparts *for* all seven of these ethical principles. If the significance level is relaxed somewhat, two more significant relationships can be added to the original seven, and just as with the original seven, these two also show that the near-future-oriented executives felt more strongly about each of the two ethical principles than did their distant-future-oriented counterparts. (The five other ethical principles tested did not reveal statistically significant differences between executives with near- and distant-future orientations.)

That the short-term is associated with stronger feelings, beliefs, and values about ethical principles than is the long-term leads to the question of whether the greater strength of these feelings leads to different behaviors involving the ethical principles. This would seem to follow from the logic of values-attitudes-behaviors complexes (e.g., Fishbein and Ajzen 1975), but Das did not collect data that would allow this relationship to be tested. Indeed, his explanation for why the distant-future-oriented executives felt less strongly about ethical principles, that having such a perspective would reveal more of the obstacles and complexities involved in trying to follow and implement the ethical principles (Das 2001, pp. 3–4), could not be tested with his data either. Nevertheless, his findings revealed a clear pattern of relationships between future temporal depth and strength of feelings about many ethical principles, findings with important albeit unexplored behavioral implications.

Das's method for measuring future temporal depth asked the executives in his sample to consider important events they expected to happen in their "own *personal* life in the future" (Das's emphasis; 2001, p. 30), in other words, while they were still alive. But there are more times still to be considered, because people recognize that there was time before their memories began and there will be time after their earthly existence ends.

And people do think about events in such times, both events that occurred before their births and ones that may occur after their deaths. But at least in the West, considerations about things before one's birth, and especially after one's death, take on a special quality. Such considerations occur in a different time than one's lifetime, which is a fundamental temporal distinction (e.g., Cottle 1976, pp. 105–6).

Still Longer Views

The Iroquois have extended planning considerations beyond the lifetime of any planner by considering the impacts of plans and decisions on their descendants *seven generations* ahead (Lyons 1980, pp. 173–74). Similarly, the CEO alluded to earlier in the chapter, Matsushita founder and CEO Konosuke Matsushita, presented a corporate plan to his employees in 1932 that would also have an impact on many future generations. The corporate plan covered the next 250 years (Lightfoot and Bartlett 1995, p. 82). A 250-year plan extends at least ten generations into the future, and given the time capsule demonstration discussed earlier, is it surprising that if someone were to develop a 250-year plan, it would be a Japanese rather than an American CEO?

Both the Iroquois and Matsushita have given consideration to matters not only beyond quarterly earnings reports, but beyond the life span of any living tribal member or company employee, indeed beyond the likely lifetime of the children of any current member or employee. Compared with the temporal depths presented in Table 5.1, the Iroquois and Matsushita were dealing with qualitatively different time frames, time frames at least one order of magnitude greater than those of the respondents in the Table 5.1 sample, different probably from those of most Americans. They are dealing with matters that on the scale of human lifetimes occur in *Deep Time*.

Deep time refers to the immense temporal vistas, scores of millions, often hundreds of millions, occasionally billions of years long that geologists use to chronicle the earth's physical history, including the history of life on it. John McPhee (1981, p. 20 and other pages) coined the term, and it has been used by other writers with an ear for a well-turned phrase (e.g., Gee 1999), including Gregory Benford (1999), who made it the title of his book dealing with the topic on a more human timescale. Among other virtues (see Chapter 7), thinking in a deep-time frame forces one to recognize the ubiquity of change. And Benford's account of the WIPP project illustrates this well (1999, pp. 33–85).

WIPP stands for Waste Isolation Pilot Plant, a U.S. government underground facility near Carlsbad, New Mexico, used for storing moderately radioactive nuclear waste. The federal government created teams of natural and social scientists to address several questions associated with the project, including the likelihood of humans gaining access to the underground facility sometime during the future, as well as how to provide a warning to future humans that the location was dangerous, a warning that would also communicate the nature of the danger. The relevant deep-time future was defined as the next ten thousand years.

But with the exception of archaeologists, social scientists seldom employ time horizons of this magnitude—if they explicitly employ time horizons at all. Some natural scientists such as astronomers and geologists do deal with such time frames, often much longer time frames, but even so, the teams had to deal with matters seldom encountered in everyday life as they addressed the issues about the WIPP project that concerned the government (i.e., dangers, warnings for future generations). For example, when will the United States no longer exist? This question was raised because the issue of political control of the WIPP site was obviously important in assessing the likelihood of future humans breaking or blundering into the facility. When the teams turned to history, they found no record of any political entity in human history that had existed continuously for ten thousand years; in fact, recorded human history does not extend back quite that far.

Another question was equally startling: At what point will no one on the planet be able to understand English (or any other contemporary language) as it was spoken and written at the end of the twentieth century? The relevance of this question is obvious to the issue of providing a warning and explanation that will be understandable for ten millennia. Languages change subtly, incrementally, yet change they do, and after enough changes accumulate, they result in different languages. How many readers today can read *Beowulf* in the original? Benford (1999, p. 76) used this example to illustrate the point about how much language can change, and in the case of *Beowulf*, only a little over one thousand years are involved. This illustrates how daunting the challenge is of even formulating a plan to plausibly communicate across deep-time intervals.

A noteworthy point, and one consistent with this chapter's analysis of temporal depth, is that the teams used deep history to help deal with the deep human future. The past again served as metaphor for the future: deep-past metaphors for deep-future concerns. The further behind they looked, the further ahead they were able to see. And looking in either deep-time direction, but especially toward the past, confers a potential advantage beyond simply being able to see further. In an interview I conducted with Gregory Benford (March 28, 2000), Greg indicated that an advantage a deep-time perspective confers is an enhanced ability to detect *patterns*. So it is not just that change becomes more evident, but patterns in the change may be detected more readily.

Jennifer George and Gareth Jones (2000) emphasized the importance of detecting such patterns in their discussion of cycles and spirals as important temporal elements of social science theory development. Similarly, Srilata Zaheer, Stuart Albert, and Akbar Zaheer (1999) demonstrated the important impact of differing temporal depths on statistical relationships among phenomena (see also Mitchell and James 2001). Perhaps one reason that the theory of long-wave economic cycles proposed by N. D. Kondratieff (1935) seems striking, even daring, is that social scientists seldom look for patterns spanning temporal depths of such lengths—fifty years being what Kondratieff proposed. Yet if one does not think and look for cycles in time spans of one hundred years or more, one cannot detect fifty-year cycles, because to be a cycle, a pattern must repeat itself. This is Benford's point: Looking at things over longer temporal depths than are usually employed increases the chances of spotting patterns with longer wavelengths, patterns whose wavelengths are too long to allow detection with either short-depth or atemporal viewpoints. All of this indicates that Ralph Waldo Emerson was right: "The years teach much which the days never know" (1983, p. 40).

Thus deep-time perspectives, temporal vistas extending beyond the length of human lifetimes, provide perspectives from which new insights can potentially be drawn. And more people may actively engage deep time than the data presented early in the chapter (e.g., Table 5.1) would suggest. For within what at first glance appears to be a large segment of deep time, John Boyd and Philip Zimbardo (1997) discovered that many people do actively attend to a deep-time-like domain, a region they dubbed the transcendental future. They defined this temporal realm as "the period of time from the imagined death of the physical body to infinity" (p. 36). Moreover, it is not just there, but people imagine themselves as active participants in the transcendental future and may see it as the time in which they attain goals such as reunions with deceased loved ones, eternal life, reincarnation, and the end of current suffering (p. 36). Boyd and Zimbardo (pp. 41–46) developed a scale to measure an individual's orientation toward this form of time and found that the young and the old tend to believe in it and be more oriented to it than those of intermediate age, that members of some religious groups are oriented to it more strongly than others, that belief in it varies by ethnicity, and that women are more oriented to the transcendental future than men. But there was no relationship between a transcendental-future time perspective and any components of the five-factor model of personality (see Chapter 3 for more about the "Big Five").

As the label indicates, Boyd and Zimbardo's concept and research concerned the transcendental *future*. They did suggest the possibility of a transcendental past as well (the expanse of time before a person's birth [Boyd and Zimbardo 1997, p. 36]), and this possibility is consistent with the connection between the past and future developed in this chapter. For example, writing of her people's relationship to the past, South African Miriam Makeba described that relationship like this: "But in my culture the past lives. My people feel this way in part because death does not separate us from our ancestors. The spirits of our ancestors are ever-present. We make sacrifices to them and ask for their advice and guidance. They answer us in dreams or through a medium like the medicine men and women we call *isangoma*" (Makeba's emphasis; 1987, p. 2). And regarding the connection between past and future she wrote, "But for us, birth plunges us into a pool in which the waters of past, present, and future swirl around together. Things happen and are done with, but they are not dead. After we splash about a bit in this life, our mortal beings leave the pool, but our spirits remain" (p. 2).

So deep time, or at least a portion of it, may extend to the considerations of many more people than just the historians, astronomers, geologists, and archaeologists among us—if one allows the transcendental future as part of the same time line associated with the same reality such scientists address. One could object, and point out that the transcendental future seems to be about more than a different time; it seems to be about a different place, a different reality altogether (i.e., the hereafter). Nevertheless, the transcendental future is a worthy addition to the catalog of human times, regardless of whether it fits neatly as a region of deep time or not.

The concept of a transcendental future grew out of research about temporal perspectives in general, specifically general orientations to the past, present, and future. However, Boyd and Zimbardo's interest was not in comparing short-, mid-, and long-term temporal depths; rather, it was in examining the degree to which people were oriented to a transcendental future, and in examining the extent to which this variation covaried with other factors such as age, gender, and ethnicity. This is a natural extension of the questions involved in research on general past, present, and future temporal orientations (e.g., Kluckhohn and Strodtbeck 1961, pp. 13–15), orientations that at first glance appear similar to issues of temporal depth. However, as I have argued elsewhere in opposing the use of the temporal orientation label, these general orientations are more an issue of the general temporal direction or domain that an individual or group may emphasize (Bluedorn 2000e) than the distance into each that the individual or group typically uses. The latter is the issue of temporal depth; the former, what I have called *temporal focus* (Bluedorn 2000e).

TEMPORAL FOCUS

Temporal focus is the degree of emphasis on the past, present, and future (Bluedorn 2000e, p. 124). Research on this topic has been conducted under a variety of labels, including time orientation (Kluckhohn and Strodtbeck 1961), Confucian Dynamism (Hofstede and Bond 1988), focus (Settle, Alreck, and Glasheen 1972 as cited in Kaufman-Scarborough and Lindquist 1999, p. 290, and 1978 as cited in El Sawy 1983, pp. 277–86h; and Settle, Belch, and Alreck 1981 as cited in both El Sawy 1983, pp. 129–40, 277–86h, and Kaufman-Scarborough and Lindquist 1999, p. 303), and time perspective (Lewin 1951, p. 75). And that societies differ in temporal focus is widely noted. For example, in comparing the West's temporal orientation to that of her own people's, Miriam Makeba wrote, "In the West the past is like a dead animal. It is a carcass picked at by the flies that call themselves historians and biographers" (1987, p. 2).[5] But the West is a big place, and temporal focus varies within it too. So Gregory Benford would write, "Englishmen were fish swimming in this sea of the past. For them it was a palpable presence, a living extension, commenting on events like a half-heard stage whisper. Americans regarded the past as a parenthesis within the running sentences of the present, an aside, something out of the flow" (Benford 1992, pp. 208–9).

Such descriptions are consistent with James March's observation that "the evidence from studies of organizations is overwhelming in substantiating a general organizational tendency to favor the present over the future" (1999, p. 73). They are consistent with March's thinking because much—not all by any means—of the published research on organizations has been based on American organizations. The observation about organizations favoring the present is important because different temporal foci indicate different organizational priorities. For example, March suggested that a focus on the present will be reflected in organizational attention to short-term efficiency, whereas a focus on the future will direct attention more to long-term adaptability (p. 73).

Individuals differ in temporal focus too, and these differences have received significant attention over the years (e.g., Cottle 1976; Doob 1971; Fraisse 1963; Usunier and Valette-Florence 1994; Zaleski 1994). Consistent with Makeba's description of the past as a picked-at carcass, Philip Zimbardo and John Boyd concluded that most research on temporal focus has concentrated on the individual's future or present orientation, "with relatively little attention to past orientation" (1999, p. 1272). Within this limitation they summarize this research as relating an individual's future orientation to several outcomes such as better academic achievement and engaging in fewer health risk behaviors, whereas a primary present orientation has been related to behaviors such as crime and addictions (p. 1272).

So both cultural and individual temporal-focus differences have been associated with other important matters, albeit the question of causality is difficult to sort out. Here the issue is constrained more tightly: Is temporal focus related to temporal depth? And if so, how? As already noted, I concluded elsewhere that temporal focus and depth are conceptually distinct (Bluedorn 2000e). In that discussion, I presented evidence that the two variables were empirically distinct as well. El Sawy investigated a similar question (1983, pp. 126–27). He used different conceptual labels and different measures, but his results, like mine, suggested that "the two dimensions are separate" (p. 126). These findings have now been joined by new results presented in the Appendix that strongly support this conclusion (see the Appendix).

The results presented in Bluedorn (2000e) and the Appendix consistently support the distinction between temporal depth and temporal focus. Conceptually the two terms refer to different phenomena, and empirical measures of the two share so little variance in common that for practical purposes they can be regarded as orthogonal. Temporal depth is the distance looked into past and future. Temporal focus is the importance attached to the past, present, and future. And how much importance do people attach to past, present, and future?

One way to address this question is to compare people's responses to Usunier and Valette-Florence's (1994) orientation toward the future and past scales. Although this method does not offer comparisons with the importance of the present, it does provide a straightforward method for gauging the importance people attach to the past and the future. Results from two large samples of college students in the United States revealed almost identical results indicating that the future is considered important in absolute terms (an average score around six on a seven-point scale in both samples) and is regarded as significantly more important than the past (averages a little over four on seven-point scales in both samples).[6] Although these results may simply confirm one's expectations for American samples, they provide a quantitative assessment that reveals the large margin by which the future is considered more important than the past. They also indicate that the past, though regarded as less important than the future, is not seen as completely irrelevant either.

A more fine-grained approach to this question is to ask people, How important are each of the three future (short-term, mid-term, and long-term) and past (recent, middling, and long-ago) depths? Doing so allows the importance of the three future regions to be compared with each other just as it allows the importance of the three past regions to be compared. Further, the importance of each component in the three sets of parallel items (e.g., short-term future and recent past) can be assessed and compared with its counterpart. And because of the results just presented from the general orientation-toward-the-past-and-future comparisons, one would expect the future item in each pair to receive a higher importance rating than its past counterpart.

The respondents in the same sample that produced the results presented at the beginning of the chapter (i.e., Tables 5.1 and 5.2, Figure 5.1) were asked about the importance of each of these six temporal depths, and their responses are presented in Figure 5.2.

Among the three future depths, only the importance of the mid-term and long-term futures differ significantly from each other. And as a visual inspection of Figure 5.2 reveals, even that difference is not large in substantive terms, at least not compared with the differences among the three past temporal depths.

A quick glance at the left side of Figure 5.2 reveals immediately that one is no longer looking at the same results depicted on the right side. Major differ-

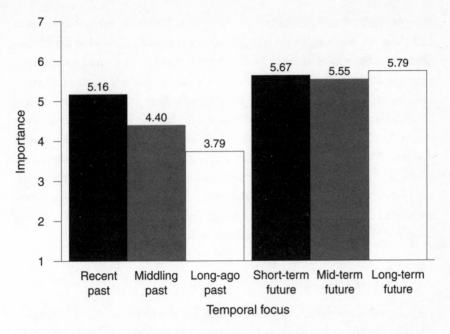

FIGURE 5.2. Average importance given to three regions of the past and three regions of the future by a sample of college students

ences exist among the three temporal zones of the past, and as would be expected from their depiction in Figure 5.2, all three differences are statistically significant. The recent past is clearly the most important of the three zones, followed by the middling and long-ago pasts in descending order of importance. Although not quite as high as the importance ratings given to the short-term future, the importance rating for the recent past (5.16) is fairly close to its future counterpart (5.67). The same cannot be said for the mid-term/middling and long-term/long-ago pairings. The middling past has an importance ranking only slightly above 4.0, the midpoint of the rating scale, and the rating of the long-ago past (3.79) falls below the midpoint, both of which are ratings well below those of their future depth counterparts.[7] Perhaps Miriam Makeba had the West's attitude about the long-ago past in mind when she described that attitude so grimly as a picked-at carcass.

Overall, the sample ratings revealed small differences between the short-term, mid-term, and long-term futures, but within the past, the sample rated the recent past as most important, much more important than either the middling or the long-ago pasts. Perhaps the most important point to emphasize is that these are the results from a single sample that was collected in the United States, and the results from the orientations-toward-the-past-and-future comparisons are from American samples too. As such, the results from all three samples are values constructed by the processes that produce American personalities and culture in general (although each sample included international students, all three samples were overwhelmingly composed of students born and reared in the United States). For this reason it seems reasonable to anticipate major differences if comparable samples were collected elsewhere in the world. Not that every sample would produce different results, but it seems reasonable to expect that many results would differ, which is all the more reason to extend this research outside the United States.

CAVEAT EMPTOR

Although the caveat just issued was directed at the temporal focus findings, it could be applied just as easily to the temporal depth findings. Temporal depths are socially constructed just as temporal foci are. But until the temporal depth findings presented in this chapter are replicated outside the United States, we cannot know for sure whether the positive correlation between the lengths of past and future temporal depths is a unique manifestation of people socialized in late-twentieth-century American culture, or whether these relationships are universal. The same concern would seem to apply to organizational age and its positive correlations with past and future temporal depths. All of these temporal depth findings require investigation in other cultures.

And what of the descriptive data about the intervals defined as the short-, mid-, and long-term regions of the past and future? Of all the findings reported, these would seem most likely to vary from person to person, from organization to organization, from country to country.

Some of the most important theoretical interpretations presented in this chapter involved the concept of metaphor and the proposition that the past is a metaphor for the future. And though more broadly based than the empirical findings, from Aristotle to Weick, the theory too is mainly from the Western tradition. Yet of all the findings presented and ideas developed in this chapter, I suspect that the one most likely to be universally true is that the past is a metaphor for the future. Even in cultures where the past is relegated to the status of a picked-at carcass, some of it is still used to explain the present and cope with the future. And when this task of metaphorical interpretation is performed carefully and wisely, distinguishing valid similarities among true differences, humanity may do more than merely cope. For at least cope it must because the eternal challenge is to construct a future humans can live in, and better yet, to construct a future in which humanity will want to live.

6

Convergence

Tempori aptari decet.
(The right thing is to fit the times.)
—Seneca, *Medea*

For decades the guards at Trenton State Prison, New Jersey, began their eight-hour shifts at 6:20 A.M., 2:20 P.M., and 10:20 P.M. They did so because the trolley would stop near the prison at about these same times (Hirsch 2000, p. 87). Both the prison and the trolley system had schedules, and the convergence of the two schedules worked to their mutual advantage. The prison benefited by having its workforce arrive reliably at the set times, and the trolley system benefited by having a loyal and regular clientele.

Schedules are rhythmic templates, and rhythms are one way times differ. Indeed, recurring schedules might well be defined as templates for rhythms because they prescribe behavior in terms of patterned repetition. Just as the sheet music provides a template for playing the tune, the schedule in the workplace for behaviors such as the arrival times of prison guards similarly provides a template for performing work. Through daily repetition, the schedule imparts a rhythm to the lives of individual workers as well as to the organization as a whole. To some extent any schedule will impart a kind of protorhythm to the day's activities, but when the schedule is repeated day after day, a genuine rhythm emerges, a repetitive pattern of behavior.

These emergent rhythms vary in important ways, such as their speed and phase patterns, and these differences must be accommodated to successfully coordinate the phenomena displaying them. Such phenomena are the subject matter of this chapter, specifically, How do rhythms differ? How can phenomena displaying different rhythms be integrated? What happens when such integration occurs? Is such integration necessarily a good thing? And "good" from whose perspective? To explore these questions the concept of entrainment will be introduced and used extensively.

ENTRAINMENT

In one of those scientific coincidences that may reflect a widespread concern with the same general questions, the concept of entrainment was first employed in social and behavioral science analyses in 1983, and by researchers from very different homes in the social science community. One investigator was Edward Hall (1983), whose work on polychronicity was discussed extensively in Chapter 3. The other investigators were Joseph McGrath and Nancy Rotchford (1983), and their work, along with subsequent investigations conducted by McGrath and Janice Kelly will inform important discussions later in this chapter.

These investigators' simultaneous introduction of entrainment into the social sciences is itself ironic, given what entrainment is. Entrainment is about rhythmic phenomena and the possibility that their rhythms may converge. So, were the rhythms of anthropology (Hall) somehow convergent with those of social psychology (McGrath and Rotchford), resulting in the simultaneous introduction of the entrainment concept in 1983? This would be tricky to determine, but the idea that rhythmic patterns come into alignment and then behave in a parallel fashion describes the essence of entrainment.

Such may be the essence of entrainment, but this does not describe its agency, for as McGrath and Rotchford explained, not only do rhythms come to oscillate together, but one rhythm is often more powerful and captures the other rhythm (1983, p. 62). The less powerful rhythm is captured and adjusts to the rhythm of the more powerful. So in the case of entrainment between the shift schedules at Trenton State Prison and the city's trolley schedules, the trolley schedules acted as the entrainer, the behavioral oscillation that captured the rhythm of the shift schedule. At least this is James Hirsch's conclusion, albeit not phrased in entrainment terminology: The trolley schedule "dictated the prison schedule" (Hirsch 2000, p. 87), and "dictated" sounds a lot like a reference to the dominant rhythm in this situation, the rhythm that is doing the entraining.

So entrainment is the process in which the rhythms displayed by two or more phenomena become synchronized, with one of the rhythms often being more powerful or dominant and capturing the rhythm of the other. This does not mean, however, that the rhythmic patterns will coincide or overlap exactly; instead, it means the patterns will maintain a *consistent relationship* with each other. The consistent relationship could be an exact overlap, but it could also be that one pattern will consistently lag or lead the other. For example, the shift schedule at Trenton State Prison likely *lagged* or followed the trolley schedule by a few minutes. That is, the shift would have been scheduled to begin a few minutes *after* the trolley arrived. If the two rhythms had overlapped exactly (synchronously), that is, if the shifts had been scheduled to begin at exactly the same time that the trolley arrived, the arriving guards would always have been a few minutes late for their shift. Both patterns—an exactly synchronous overlap and what was probably the actual pattern, one in which the entrained rhythm (the work-shift schedule) lagged the entraining rhythm—represent entrainment. The difference is only in the way the two patterns are related consistently and repetitively.

If two rhythmic patterns are related consistently, which is to say, if they are entrained, this relationship can be described precisely in mathematical terms as maintaining a distinct phase-angle difference between the two rhythmic patterns (Aschoff 1979, pp. 5–6).[1] So the mathematical definition of perfect entrainment would be when the phase-angle difference between two rhythms is constant throughout a complete cycle of each rhythm. When the rhythms overlap exactly, the phase-angle difference is zero (as would be the case if the trolley arrival times and shift starting times were identical). When the entrained rhythm lags the capturing or driving rhythm in its phase (e.g., if the shift schedule's phase were to follow the trolley's), the phase-angle difference is negative; and when the entrained rhythm leads or precedes the driving rhythm, the phase-angle difference is positive (Aschoff 1979, p. 5–6). In the case of the trolley-shift example, this would mean the shifts would have been scheduled to begin *before* the trolley was scheduled to arrive, if for some reason the prison's managers had been foolish enough to create such absurd schedules. In all three cases, if the two rhythms were perfectly entrained, the *difference* in these two times (trolley arrivals and beginning of shifts) would be

TABLE 6.1

Possible entrainment relationships.

Type of entrainment relationship	Description	Example
Lagging (negative phase-angle differences)	The phases of the entrained rhythm *follow* the corresponding phases of the entraining rhythm.	The bus arrives at 5:00 A.M., 1:00 P.M., and 9:00 P.M. (the *entraining* rhythm), and the work shifts at a nearby company begin daily at 5:30 A.M., 1:30 P.M., and 9:30 P.M. (the *entrained* rhythm).
Synchronous (no phase-angle differences)	The phases of the two rhythms occur *at the same time*.	The bus arrives at 5:00 A.M., 1:00 P.M., and 9:00 P.M. (the *entraining* rhythm), and the work shifts at a nearby company begin daily at 5:00 A.M., 1:00 P.M., and 9:00 P.M. (the *entrained* rhythm).
Leading (positive phase-angle differences)	The phases of the entrained rhythm occur *before* the corresponding phases of the entraining rhythm.	The bus arrives at 5:00 A.M., 1:00 P.M., and 9:00 P.M. (the *entraining* rhythm), and the work shifts at a nearby company begin daily at 4:30 A.M., 12:30 P.M., and 8:30 P.M. (the *entrained* rhythm).
Independent (random phase-angle differences)	*No correlation* between the two rhythms' corresponding phases.	Daily random fluctuations between the bus and work-shift schedules.

NOTE: The labels for the "lagging" and "leading" types of entrainment relationships were taken from Aschoff's definitions of positive and negative phase angles, respectively (see Aschoff 1979, pp. 5–6, and the text). Also, what I label "synchronous" entrainment, Ancona and Chong called "synchronic" entrainment (1996, p. 258). The example of a lagging entrainment relationship is a more contemporary and generic version of the example in the text, which was based on Hirsch's (2000, p. 87) description of work shifts at the Trenton State Prison. The example in the text was likely an example of a lagging entrainment relationship, because the work shifts at the prison (the *entrained* rhythm) would probably have started a few minutes *after* the trolley stopped near the prison (the *entraining* rhythm).

constant for the phases throughout the twenty-four-hour periods of both rhythms. These three possibilities and a fourth, no entrainment, are summarized in Table 6.1.

The details of rhythmic oscillating phenomena, their depiction in the form of sine wave–like figures, and how two such waves can be related through entrainment can be found in a variety of sources (e.g., Ancona and Chong 1996; Aschoff 1965, 1979; Pittendrigh 1981). But the four possibilities summarized in the preceding discussion and in Table 6.1 provide a sufficient conceptual base for examining the convergence of rhythms displayed by social phenomena—if this conceptual base is joined by a few additional considerations.

Zeitgebers

Zeitgeber is now officially an English word, pronounced "tsite-gaber," (see both *Merriam-Webster's Collegiate Dictionary*, 10th ed., and the *Oxford English Dictionary*, 2nd ed.), but it originated as a German word literally meaning *time* (*zeit*) *giver* (*geber*). In the context of entrainment, the term is often used in the sense of a pacing agent or synchronizer (Ancona and Chong 1996, p. 253; Whitrow 1980, p. 142), which makes a zeitgeber the entraining force, the rhythm that captures another rhythm. (More precisely, Whitrow and Ancona and Chong used the terms *signal* and *cues*, respectively, in their descriptions of zeitgebers, thereby identifying the signals and cues as the zeitgebers rather than the entities that produce the signals and cues. However, owing to a concern for felicitous prose, I will not explicitly make this distinction in my subsequent use of zeitgeber.) An everyday example of a zeitgeber occurs literally every day: the two-phase cycle of lightness and darkness that has been repeated so many billions of times throughout the earth's history. Throughout this four-and-one-half-billion-or-more-year history, untold physical, biological, and social rhythms have become entrained to this fundamental cycle. And human rhythms are no exception to this ubiquitous entrainment to the fundamental cycle, a point that will be examined shortly.

So a zeitgeber is a rhythm external to the system whose rhythm is being entrained to the zeitgeber's. And these rhythms have two properties that must be aligned for entrainment to occur. These properties are the speed and phase pattern of the rhythms.

Deborah Ancona and Chee-Leong Chong defined entrainment as "the adjustment of the pace or cycle of an activity to match or synchronize with that of another activity" (1996, p. 253). And as such, their analysis indicated that a zeitgeber may influence either or both of these two properties of any activity (the cycle properly being the *pattern* of phases exhibited within a single cycle such as the bus-arrival and the work-shift phase patterns in Table 6.1). This is a major departure from traditional analyses of entrainment, which more or less assumed similar speeds of the rhythms involved, an assumption revealed in Aschoff's conclusion: "A self-sustaining oscillation [rhythm] can be entrained by a zeitgeber only to those frequencies that do not deviate too much from its own natural frequency" (1979, p. 6), albeit entrainment can possibly also occur around "multiples and submultiples of the natural frequency" (1979, p. 6; see also Pittendrigh 1981, p. 106, and Wever 1965, pp. 53–54). Indeed, in traditional entrainment theory, entrainment to a zeitgeber "means that the phase of the entrained oscillation is corrected at least once during each period" (Aschoff 1979, p. 6). Thus in traditional entrainment research the issue was basically what Ancona and Chong classified as phase entrainment, aligning and adjusting the phase patterns of two rhythms operating with the same or similar frequencies. Therefore, Ancona and Chong's description of a qualitatively distinct form of entrainment, what they label tempo entrainment (the pace or speed of an activity) is a major extension of the entrainment concept. And given the assumptions of the traditional work on entrainment, tempo entrainment would appear to be a necessary condition, perhaps a boundary condition, for phase entrainment to occur. As several examples will show, zeitgebers can influence both phase patterns and the speed of rhythms.

In Search of Zeitgebers

Zeitgebers organize human life, and organizational life is no exception. Thus examples of zeitgebers entraining organizational rhythms should be plentiful in organizational contexts, and they are. For example, one of the foundation studies in organization science is Paul Lawrence and Jay Lorsch's (1967) study of organizations and their environments. Building on Burns and Stalker's (1961) classic study, Lawrence and Lorsch extended the field's understanding of how environments affect organizations. And one of the ways they discovered was through entrainment, albeit they did not describe the mechanism within this conceptual frame.

What Lawrence and Lorsch discovered and conceptualized was the time span of feedback (sometimes they use the longer phrase "time span of definitive feedback" (1967, p. 28), which is the length of time it takes for decision makers to learn about the consequences of actions (pp. 24–29, 94–95). They found that this time varied systematically by organizational activity, that the time it took for managers in the marketing department to learn about the consequences of their decisions was far shorter than the time it took their counterparts in research and development to receive similar feedback. Not that such times were invariant for managers in each functional area, nor should one expect them to be, a point suggested in work on biological entrainment. Writing in reference to biological rhythms, Colin Pittendrigh and Victor Bruce concluded, "No biological rhythms are strictly periodic in the mathematical sense, and in practice one takes some arbitrary and well-defined statistic of the rhythm as a measure of the periodically repeated events" (1959, p. 483). Thus one would expect social rhythms to vary too.

What makes all of this relevant to a discussion of entrainment is that the time span of feedback was systematically related to the future temporal depths (see Chapter 5) found in different departments (Lawrence and Lorsch varyingly referred to what I have labeled future temporal depth as either "time orientation" or "time horizons" [1967, pp. 94–95]). The longer the time span of feedback, the longer the department's future temporal depth, meaning the longer it took to learn the outcomes of decisions, the further into the future people in the department tended to look. Hence future temporal depths were entrained to the speed with which the relevant environment provided feedback about decisions, making this attribute of the environment the zeitgeber. This seems like a specific case of Michael Hay and Jean-Claude Usunier's proposition that the time horizons of key constituents generally have powerful impacts on corporations' time horizons (1993, pp. 324–27).

But temporal depths do more than entrain to their environments; they entrain to each other as well. I first suggested this interpretation as an explanation of the positive correlation between past and future temporal depths (Bluedorn 2000e). And given the consistent set of positive correlations between past and future reported for the individual level in Chapter 5 and the Appendix, and a similar positive correlation at the organizational level (Bluedorn and Ferris 2000), this entrainment pattern seems consistent. Thinking of each temporal depth as a single period, the two periods—past depth and future depth—are consistently aligned, hence entrained, and the past depth appears to be the zeitgeber. For, as described in Chapter 5, El Sawy's (1983) research strongly indicated that the length of past depth influences the length of the future depth, but not vice versa.

Another zeitgeber for organizational rhythms is the fiscal year, and as Ancona and Chong noted, fiscal years are especially powerful pacers for many organizational activities (1996, p. 253). What they did not note is that the fiscal year itself is often entrained to other rhythms, the rhythms of the annual business cycle. This is suggested in the description of fiscal years presented in Chapter 1, which indicated that fiscal years were originally scheduled to end during periods of slow business activity, thus affording companies the time to perform the required accounting tasks. In entrainment terms, this means the phase patterns of the fiscal year and the business cycle were aligned, with the pattern of annual business activity being the capturing force, the zeitgeber. And because the fiscal year had to be scheduled in advance, such attempts to schedule the fiscal year to end when companies believed business activity would be slow indicates that annual cycles existed in the business cycle and were perceivable by the people in the companies who scheduled the fiscal years.

Examples of similar attempts to deliberately entrain rhythms to the organization's advantage are easy to detect once one adopts the entrainment lens. Thus earlier analyses (Bluedorn and Denhardt 1988) described how manufacturing companies in the former Soviet Union seemed to have aligned their rhythms to each other, thereby creating a ubiquitous and curious production pattern known as "storming" (a repeating production cycle in which little is produced in the early phases of the cycle, a bit more in the middle phases, and most of what is produced in the entire cycle is produced during the cycle's final phases; see Berliner 1970).

Storming primarily involves phase entrainment, but organizational environments can serve as the zeitgebers for tempo entrainment too. For example, conditions of hyperinflation in Brazil forced Brazilian companies to adopt electronic banking in order to transfer funds hourly. Otherwise, the high inflation rate would have diminished the value of payments received earlier in the day if companies waited to make a single transfer at the end of the day (McCartney and Friedland 1995, p. A8). In this case the speed with which the value of money was changing (diminishing) in the environment acted as a zeitgeber to increase the speed with which companies deposited payments.

Although environmental rhythms and tempos are usually more powerful, hence performing the capturing zeitgeber role, sometimes organizations, singly or in concert, will capture or adjust environmental rhythms. This happened in Missouri during the 1980s when the lobbying efforts of the state's tourism industry nudged the summer tourist and school vacation seasons closer to synchronous alignment (Bluedorn and Denhardt 1988; Ganey 1983; Lindecke 1983). And it happened for all of the United States in the mid-1980s when the Daylight Savings Time coalition successfully lobbied the U.S. Congress to shift the onset of daylight saving time to the first Sunday in April. The members of this lobbying group were companies such as fast-food chains, greenhouses, and sporting goods manufacturers, all of whom believed shifting that hour of daylight from the morning to the evening for several additional weeks would enhance their sales—and Congress went along with them (Varadarajan, Clark, and Pride 1992, p. 44). Karl Weick interpreted these actions as manipulations that created an environment which made more sense to the people in these businesses (1995, p. 165). It would certainly make more economic sense to them because their hours of operation would be aligned, that is, entrained more closely, to the preferred shopping phases in their customers' activity cycles.

Vacation seasons, production cycles, and fiscal years all have end points, and end points are deadlines. Deadlines have powerful motivational effects on human behavior (Locke and Latham 1990), often becoming Swords of Damocles hanging over the heads of individuals and organizations alike as people press and strain to complete their activities as deadlines draw ominously near. As deadlines approach, activity on deadline-relevant tasks increases, speeds up (Bluedorn and Denhardt 1988), meaning that deadlines operate as zeitgebers influencing the speed or tempo aspect of entrainment. And as Connie Gersick's (1988, 1989) benchmark research showed (see Chapter 4), groups reveal distinctive phase patterns, often remarkably consistent phase patterns, *if* they are working on projects with explicit deadlines, albeit alternative mechanisms responsible for these phase patterns have been proposed (cf., Lim and Murningham 1994; Seers and Woodruff 1997). But regardless of the details of the mechanisms producing the phase patterns within these groups, the key point is that these patterns seem to be a response to the deadline, which means deadlines also affect the phase pattern component of entrainment in groups, and likely the phase patterns of individuals too.

As these examples indicate, individuals and groups both entrain to zeitgebers, and sometimes the zeitgeber's power to entrain is very widespread.

Pervasive Entrainment

The daily cycle of lightness and darkness may be the most powerful and pervasive zeitgeber for life on the surface of our planet. It certainly has been for humanity, the hominids, throughout most of our lineage's history. In a manner eerily reminiscent of an early definition of a deadline—"A line drawn around a military prison, beyond which a prisoner is liable to be shot down" (*Oxford English Dictionary*, 2nd ed.)—this daily cycle provided a temporal deadline, a portion of this period within which a hominid, especially a solitary hominid, would be likely to be, not "shot down," but simply unwillingly consumed (see Chapter 2).

Hominids, including the contemporary version, function poorly after the sun goes down, and really the only way we can function well after sunset is to use artificial suns to transform portions of night into day. (Admittedly, a full moon on a clear night would allow for some limited activities.) And despite the use of other forms of artificial lighting, primarily electricity-based forms, which permitted "the colonization of night," Murray Melbin's (1987, p. 10) wonderful metaphor, the daily cycle of light and dark is still pervasive and may still be the most pervasive zeitgeber in human life. This pervasiveness is revealed by how so many activities are still entrained to be performed during the daylight phase of this twenty-four-hour cycle. To illustrate this point, I will use an example from my professional life.

During the late 1980s, the MBA program at my university decided to move some of its classes to the evening in the hopes of attracting to the program students who worked during the day (note the entrainment to daytime activities of the prospective students). Now, as far as journeys into the night go, this effort was not a daring incursion, because it dealt with only the earliest part of the night. (When Melbin studied nighttime activities, he focused on the period from midnight until 7:30 A.M. [1987, p. 30], what might be called the deep night.) Nevertheless, even the early portion of the night proved to be a very different time than time during the day. (All times . . .)

I was involved with this experiment for several semesters before it was abandoned. During that time my classes began around 6:00 P.M. and would end around 7:30 P.M.—very early night as I said. But just as Melbin described for expeditions into the night generally, support services not only diminished or broke down, they were occasionally even worse than nonexistent. For on at least two occasions custodians entered my class *during* class and began to clean the room!

After this happened the first time, I reported the events to my chair and dean, who contacted the appropriate people in the central university administration to deal with this problem. We learned that information about my class had somehow not been added to key scheduling forms the university maintained, and that the problem would be taken care of. Unfortunately, the same thing happened at my next class. And this time the custodians called the university police. When they arrived I had to show them my university identification card and justify our presence in the room. Fortunately, after a series of additional conversations with university administrators, the custodians and constables stopped their disruptive visits to my classes.

The custodians and university police officers had been trying to do their jobs—maintain and protect the university's facilities—and I had been trying to do mine—teach the university's students. None of us were in the wrong, so the point is not to find someone to blame. Instead, the point is the power of entrainment, and in this case, long-standing entrainment, perhaps an entrainment pattern generations long. Most of the university's activities, virtually all of its regular class activities, were entrained to the daylight hours, Monday through Friday, and I suspect they had been for generations. Evening classes were nearly nonexistent, so the evening is when cleaning took place. The class phase of activities is during the day; the cleaning phase, during the night. And the MBA program's small expedition into the early night did not fit with that long-established classroom-cleaning cycle, a cycle so strong that it ignored the scheduling of a class in the early evening even though MBA program administrators had scheduled the class at this time by properly following the university's scheduling procedures.

I strongly suspect that what happened to me at my university is not unique, that similar things would have and actually have happened at most universities where the teaching phase of the activity cycle takes place mainly during the daylight hours. And if this has been so for many years, the entire organizational system, not just a few individuals in it, will have become entrained to the day-night pattern, and changes in that pattern are unlikely to be either instantaneous or seamless. This may be useful information for managers who will be managing changes involving the colonization of time, because such changes will disrupt long-established entrainment patterns.

My adventure also illustrates Melbin's observation that as human activity moves into the night, among the changes, especially in organizations, is a regular shift in power. During the night, power and authority often shift to positions lower in the hierarchy and then shift back to higher positions during the day (Melbin 1987, pp. 90–91). This is clearly at least a two-phase oscillation, and one speculates about the possibility of three or more phases if three shifts are involved. Because my course was but the vanguard of expansion into this temporal frontier, meaning the frontier had not yet been colonized by daytime activities, issues of power and authority, such as who had the right to use the room, were contested that had been resolved long ago during the day.

But the night course problems are just a recent example of the day-night cycle's pervasive impact on the rhythms of human life. Many other examples of entrainment have occurred, examples that involve other zeitgebers. Some of the most profound of these examples involve the invention and diffusion of mechanical clocks as well as the increasing accuracy of mechanical timekeeping. The first of these examples occurred about a century after the mechanical clock was invented in Europe.[2]

In 1370 a clock was completed that Charles V, king of France, had ordered constructed, and it had been built in a tower of his palace. The clock is sometimes referred to as de Vic's or De Vick's clock (Bolton 1924, plate between pp. 56–57; Crombie 1959, p. 212) in reference to Henri de Vic, the German clockmaker who constructed it. Henri de Vic—the German version of his name was Henrich von Wick, but the French versions are used more often (Usher 1929, p. 159)—had been "caused to be brought from Germany" (Bolton 1924, p. 55) by Charles V for the purpose of constructing this clock. As noted, it is sometimes referred to by its builder's name, but being an especially notable clock, its more usual designations are the Horloge du Palais (e.g., Dohrn-van Rossum 1996, p. 188) and especially, in reference to the structure housing it, the Tour de l'Horloge (Bouvet, Malécot, and Sallé 2000, p. 41): the Tower of the Clock. This clock was the first public clock in Paris (Bouvet, Malécot, and Sallé 2000, p. 41), which is why it may play a central role in a case of pervasive entrainment.

It may play a central role because of a royal decree issued by Charles V in 1370, a decree requiring all clocks in Paris to strike the hours when the clock in the Tour de l'Horloge struck its hours. In the words of Arno Borst, "From 1370 onwards, all Parisian church clocks had to keep in step with its [the clock in the King's palace] somewhat capricious chime" (1993, p. 97); in the words of

David Landes, "Charles V of France decreed in 1370 that all clocks in the city should be regulated on the one he was installing in his palace on the Ile de la Cité" (1983, p. 75). The phrases "to keep in step with" and "be regulated on" both speak to the heart of the matter. The king's decree made the Horloge du Palais the zeitgeber for the other clocks in Paris, a fourteenth-century version of contemporary phone companies' time information services. And by entraining the other clocks in the city (synchronous entrainment in this case), the sounds of the hour striking in the Tour de l'Horloge would indirectly influence the rhythms of life throughout Paris. It would influence these rhythms by setting the rhythms of the city's clocks, which in turn shaped the patterns of people's lives. (And once again we see a clear example that "In the form of time is to be found the form of living" [Jaques 1982, p. 129].)

Zeitgebers provide stimuli (Aschoff 1979, pp. 6–7), and the striking of the king's clock would certainly qualify as an audio stimulus for the clock keepers of Paris. Indeed, W. Rothwell (1959) has emphasized the importance of the striking clock in general because of its ability to communicate the hour to people "some distance from the clock itself" (pp. 242–43). So using the clock in this way certainly supports its status as a zeitgeber. It entrained other clocks; the other clocks did not entrain it.

But why would Charles V want all the clocks in 1370s Paris to "keep in step" with his? Different accounts provide different explanations. David Landes (1983, p. 75) speculated that Charles V issued the decree because he wanted to avoid conflicts about the starting and stopping times of work throughout the city, conflicts that would arise if different clocks struck different times. Carlos Cipolla (1978, p. 41) indicated the king acted because he was concerned that not everyone in Paris would be able to hear the sounds of his own clock striking, an explanation that may complement rather than contradict Landes's. And several accounts have described the king's motivation as a desire to break from the church's liturgical practice of keeping time with the canonical hours (Crombie 1959, p. 212–13; Usher 1929, p. 169; Whitrow 1988, p. 110).[3]

The canonical hours were temporal hours (see Chapter 1), so an important component of many of these explanations is the interpretation that actions such as Charles V's represented the beginning of a definitive transition from temporal to equal hours (e.g., Crombie 1959, p. 212; Usher 1929, p. 169; Whitrow 1988, p. 110). But part of such an interpretation's appeal may be a tendency to project the twenty-first century's emphasis on equal hours to an era over six hundred years earlier in which equal hours were just beginning to appear in everyday life. Rothwell provided just such a warning in the conclusion of his analysis of hour reckoning in medieval France: "In interpreting what they [writers from the period being studied] wrote, the modern reader must guard against the danger of transferring to their age the mechanical apparatus and national uniformity of his [her] own times" (1959, p. 250). And there may be even more to the appeal of the temporal-to-even-hour interpretation than this natural tendency to project one's own experiences and meanings to differing contexts and circumstances.

This additional appeal may stem from what Stephen Gould described as Whiggish history, "the idea of history as a tale of progress, permitting us to judge past figures by their role in fostering enlightenment as we now understand it" (1987, p. 4). And if equal hours qualify as "enlightenment as we now understand it," a not implausible impression, then it would be natural to search history for those enlightened pioneers who saw the light and not only began using equal hours but began changing their societies so that others would use them too. So to look for figures at a time when the technology had made the widespread use of equal hours feasible would be natural, and to look for vanguards who took action, like Charles V, equally so. But there is a danger in doing so, for to look this way not only invites transferring the "apparatus" and "uniformity" of one's own times to what one finds in history, but also may make one more likely to accept less critically accounts of "enlightened behavior" that seem to anticipate the road to the present.[4]

This may have happened in the story of Charles V. Part of the story appears correct and unchallenged. He did have Henri de Vic build a clock in a tower of the Royal Palace. What has been challenged is the story about him issuing the decree. Based on a thorough analysis of documents from Charles V's reign and the years shortly thereafter, Gerard Dohrn-van Rossum (1996, pp. 217–20) concluded that Charles V probably never issued the decree to set all of the clocks in Paris marching to the beat of the drum in his palace tower. The evidence is not definitive, it being harder to prove that someone did not do something than that someone did, but at least some serious scholars accept Dohrn-van Rossum's conclusion (e.g., Bartky 2000, p. 230). Much of the case for this conclusion rests on the absence of any records of such a decree from sources written during and after Charles V's reign, such as a biography written about twenty-four years after his death that mentions nothing about such a decree. As Dohrn-van Rossum stated, "But we find no trace of it [the decree] either in her work or the other biographers of the king" (1996, p. 218). He also questioned whether the king would even have had legal authority over "the church bells of Paris" (p. 219).[5]

So did Charles V really issue the decree in question, a decree ordering the entrainment of all the clocks in Paris to his zeitgeber in the Tour de l'Horloge? It may be impossible to ever answer this question beyond a reasonable doubt, so Alexander Waugh's use of the word "perhaps" (1999, p. 58) as a preface to the decree story may be the fairest conclusion. But perhaps there is another explanation. Perhaps the clock keepers of Paris, on their own, simply started matching their clocks to the Horloge du Palais. After all, prominent people tend to be trendsetters, and few people are more prominent than kings—especially in the fourteenth century. So clock keepers might have simply started having their clocks keep time in step with the king's as a way to be in fashion—dare I say, to keep up with the times[6]—and this practice may have developed into a tradition over the years; to wit, for the correct time, consult the Horloge du Palais, just as we call the phone company today. Or maybe the clock in the Tour de l'Horloge just had a louder chime or rang more distinctively than the rest, its salience making it easier to detect as a zeitgebing signal. All of this is speculation, of course, but if the clock constructed by Henry de Vic for Charles V did serve as a Parisian zeitgeber, it may have done so for reasons other than royal fiat.

But *papal* fiats are another matter, and one in particular provides a second example of pervasive entrainment. And this time, no one questions whether that decree was actually issued. Pope Gregory XIII issued it on February 24, 1582, a papal bull putting into effect the recommendations for calendar reform presented to him by a commission he had established for that purpose (Richards 1998, pp. 244–45), thereby culminating centuries of concern with calendar problems, for problems there were.

The Julian calendar, named for Julius Caesar, the famous calendar reformer who instituted it, was used in much of what was then called Christendom, and its problem was that "the average number of days in its year was 365.25, whereas the true length of the tropical year was about 365.24219 mean solar days" (Richards 1998, p. 239).[7] Although the Julian calendar's accuracy might have been good enough for government work, over the long term its divergence from the true tropical year meant that periodic natural events such as the vernal equinox would stop falling on the same calendar day year after year. And although many people would never have noticed this discrepancy during their lifetimes because the discrepancy amounted to "an extra day every 128 years" (Whitrow 1988, p. 116), astronomers became aware of it, and they in turn informed others who might do something about it, others such as popes. For this discrepancy was also causing problems with the date of Easter, because the date of Easter depended upon the date of the spring equinox (Duncan 1998, p. 53; Richards 1998, p. 349)—and Easter was moving *seasonally* toward summer (Steel 2000, p. 166). This made the calendar problem especially germane to popes.

With the date of Easter based on a March 21 date for the spring *ecclesiastical* equinox, by the late sixteenth century the discrepancy between the Julian calendar and the tropical year had the spring *astronomical* equinox now "oscillating by almost eighteen hours around midnight on March 10/11" (Steel 2000, p. 166). This is an apt illustration of one of the virtues of a deep-time perspective (discussed in Chapter 5), the ability to spot patterns that are invisible over short time spans. As mentioned, many people would never have noticed the shifting date of an equinox during their lifetimes because it happened only once every 128 years. But over a millennium-and-a-quarter, which by the standards of human life is definitely a deep-time perspective, the pattern of divergence became more and more apparent. The recommendations from Gregory XIII's commission proposed eliminating this divergence and adopting a specific mechanism for minimizing it in subsequent years.

The divergence would be eliminated by canceling ten consecutive days in a single year, the papal bull of February 24, 1582, specifying that October 4, 1582, would be followed by October 15, 1582 (Richards 1998, p. 251), thereby *canceling* October 5–October 14 in 1582. To minimize the problem in the future, fewer leap years would be used such that every fourth year evenly divisible by four would continue to be a leap year (a calendar year with 366 days), except that among these years, the years evenly divisible by one hundred also had to be evenly divisible by four hundred to be leap years (Richards 1998, p. 250). By this rule the year 2000 was a leap year, but 2100 will not be. This change in leap year determination reduced the discrepancy from the one day in 128 years of the Julian calendar to about one day in 3,300 for the reformed calendar (Duncan 1998, pp. 202–3), which is over a twenty-five-fold increase in accuracy.[8] If any of this sounds familiar, it should, because these reforms produced what is called the Gregorian calendar, the calendar used throughout most of the world today, ironically, for *secular* matters.

This account reveals at least two examples of entrainment. The first is the discrepancy problem and the attempt to achieve synchronous entrainment between the calendar and tropical years, the latter being the zeitgeber entraining the calendar year. Thus the interaction of the sun and the earth plays a zeitgebing role once again, just as it does in the daily cycle of light and darkness. And in a manner exactly analogous to Charles V and his new clock over two centuries before (if we accept the legend as true), the second example places Pope Gregory XIII and his new calendar in the role of zeitgeber vis-à-vis the Julian calendars around him. So as Pope Gregory shifted his own calendar to the Gregorian calendar, a zeitgeber shifting its own phase pattern, he provided a very explicit zeitgebing signal (the papal bull of February 24, 1582) for others to shift theirs as well.

What happened next followed the general pattern when a zeitgeber shifts its phase pattern (canceling the ten days in 1582). For when a zeitgeber changes its phase pattern by advancing or delaying a phase (e.g., deleting the ten days), it usually takes several periods known as "transients" before the typical phase-angle differences are reestablished between the zeitgeber and the entrained rhythms (Aschoff 1979, p. 7). In this case the typical phase-angle difference to be reestablished would be zero, with calendars in Rome and Paris, for example, experiencing March 21 on the same calendar day. But the general principle indicates that several periods are usually needed to reestablish the typical phase-angle difference (a zero degree difference in this case), and taking Europe as a whole, several transient periods were required indeed.

Although many of the Catholic countries followed the papal bull and implemented the changes described therein (canceling ten days and instituting the new way to determine whether every fourth year qualified as a leap year) during the same year it was issued (e.g., Italy, Portugal, Spain, Poland, France), many of the Protestant countries in Europe took over one hundred periods (years) to reestablish a zero phase angle, England and its colonies in 1752 being one of the last to do so. And countries within the Eastern Orthodox sphere of Christianity took even longer: Romania did not change until 1919, and Greece changed in 1924. Outside the European sphere, Japan changed to the Gregorian calendar in 1873, and China adopted it in 1912 (see Richards 1998, pp. 248–49).

As indicated, today most of the world follows the Gregorian calendar for secular affairs, and it is easy to imagine the problems of coordination that would result from trying to convert back and forth between two or more calendar systems, problems that would result in problems of poor entrainment themselves. For example, E. G. Richards noted, "In 1908 the Imperial Russian Olympic team arrived in London 12 days too late for the games" (1998, p. 247). In 1908 England had been using the *Gregorian* calendar for over a century-and-a-half, but Russia was still using the *Julian* calendar. Problems of coordination indeed.

Not only did Pope Gregory XIII initiate what became a planetwide entrainment effort, thereby illustrating pervasive entrainment, but this effort also illustrates some very important themes. Canceling ten days in a year—countries adopting the Gregorian calendar later had to omit more days because the divergence continued to grow (e.g., the Russian Olympic team was *twelve* days late in 1908)—clearly illustrates human agency in matters of time, and certainly in the case of the calendar that it is a human construction, albeit one intended to reflect important periodic events in nature. And as the Imperial Russian Olympic team learned to its chagrin in 1908, time under the Julian calendar was not quite the same as time under the Gregorian.

The final example of pervasive entrainment returns from the calendar to considerations of the hour and a timekeeper discussed in Chapter 1. That timekeeper was a watch made by Henry Ford, which kept both local time and "railroad time." Although the phrase fell out of use many decades ago, today the entire world is organized into "railroad time," perhaps even more thoroughly than it is coordinated by the Gregorian calendar.

This form of time, which would eventually encompass the globe, was implemented in the United States on November 18, 1883. More precisely, on this date Standard Railway Time began going into effect across the country (Bartky 2000, p. 142). And what was Standard Railway Time? It was a system of time zones extending across the country based on a zero hour defined as midnight at the Greenwich meridian in England. The zones progressed in segments of approximately 15 degrees, each one differing from adjacent zones by one hour (see the discussion of longitude in Chapter 4). Within each segment, though, the time was uniform throughout. Thus in the Eastern zone, when it was eight o'clock in Boston, it was also eight o'clock in New York. This uniformity replaced the babel of local times defined by the local position of the sun, a crazy quilt of times that made it difficult to know exactly when trains were supposed to arrive and depart—for passengers, engineers, and ticket agents alike.

Standard Railway Time had no legal standing when its implementation began on "the day of two noons," a phrase many commentators applied to it because noon had often come and gone as reckoned by the local times in the eastern segments of each zone before the time in the zone shifted to the new standard-time noon (Bartky 2000, p. 142). Nevertheless, lobbying efforts succeeded in convincing many municipalities to convert to Standard Railway Time, either on November 18 or during the following months. So by April 1884, seventy-eight out of one hundred "principle American cities had adopted the new time standards" (Bartky 2000, p. 146). That several months were required for the seventy-eight cities to adopt the new time reveals the existence of many transient periods required to establish a regular phase-angle difference, just as many transient periods were required to do so after the Gregorian calendar reform began. Interestingly, the system of standard time zones was not codified into law for the entire country until the Standard Time Act of 1918 was passed by the U.S. Congress, with Canada enacting similar legislation in the same year (Stephens 1994, p. 576).

Efforts to extend the principle of Standard Railway Time to a planet-circling set of twenty-four time zones received important support from the International Meridian Conference, which convened in Washington, D.C., on October 1, 1884 (Bartky 2000, pp. 150–51), and was attended by delegates from twenty-five countries (O'Malley 1990, p. 109). This conference dealt with the issue of selecting a meridian to define zero degrees of longitude universally, which would have important implications for cartography as well as for temporal matters. After some wrangling, the conference voted for the Greenwich meridian as the zero longitude meridian, twenty-two to one, with two abstentions (O'Malley 1990, p. 109).[9]

But this group of international representatives had only advisory authority, so their recommendations could be accepted or rejected or ignored by any country at its pleasure. Further, Ian Bartky has suggested that, though important, this conference's importance in the road to worldwide standard time may have been overemphasized in the historical (Whiggish?) record because he could identify only one country, Japan, that had adopted "a meridian indexed to Greenwich" as a direct result of the conference (2000, p. 152). In the words of Carlene Stephens, "the meeting resulted in the gradual worldwide adoption of the time-zone system based on Greenwich as prime meridian in use today" (1994, p. 575).[10] The key word here is the modifier *gradual*, which also supports Bartky's interpretation.

How gradually, and in this case how begrudgingly, can be illustrated in France's adoption of Greenwich as the prime meridian. The French did not adopt the Greenwich meridian until 1911, and even then they camouflaged it, adopting what Michael O'Malley called a "transparent ruse" by defining the prime meridian as "Paris Mean Time, retarded by nine minutes twenty-one seconds" (1990, p. 109). Retarding Paris Mean Time by nine minutes twenty-one seconds places one at the Greenwich meridian! Time is truly socially constructed.

Time Is Money

The examples of Charles V's clock, Pope Gregory's calendar reform, and Standard Railway Time all illustrate the importance of zeitgebers in the way humanity organizes its existence. Indeed, zeitgebers are so important that people have often been willing to pay money for zeitgeber information. Almanacs are a good example.

Michael O'Malley wrote of almanacs, "Readers went to the almanac's time-tables . . . to discover the most appropriate time for doing some task" (1990, p. 16). In this sense, almanacs contained and still contain catalogs of zeitgebers. For example, *The Almanac for Farmers & City Folk 2001* presented a table (p. 104) that specified the days of each month in 2001 that were "considered to be the most favorable for various activities" (p. 103). These activities included the "most favorable" days of each month for dental work, and the "most favorable" days to cut hair, harvest fruit, and conduct business affairs (p. 104).

The advice given by this almanac casts the moon's phases and its positions in the zodiac in the role of zeitgebers, phenomena with rhythms that signal the best times to do things such as visit the dentist. So its advice is to entrain synchronously the specified activity with the respective zeitgeber. And people did and still do buy almanacs to obtain such information.

But people have been willing to pay for more generic zeitgeber information as well. They have certainly been willing to buy calendars, either for their personal use or to give to current and prospective customers to promote their businesses. But perhaps the most systematic sale of generic zeitgeber information involves the time of day. In the nineteenth century, at least twenty-two public time services delivered time signals to the public in the United States, most of which were associated with college and university observatories and many of which charged a fee for the service (see Bartky 2000). And when the telephone companies began their operations, they found a useful source of revenue by providing a number people could call to learn the time of day, a service for which there was so much demand that on the first day (in 1928) that their telephone company offered the time of day for sale, at a price of five cents a call, "New Yorkers ponied up 10,246 nickels" for it (Gleick 1999, pp. 46–47). Phone companies have continued to provide this service to the present day; so when I dial the time number provided by my local phone company, it not only provides the time but also gives its source: "From the United States Atomic Clock." And who would argue with that authority?

But one no longer needs the phone company to keep up with the Atomic Clock. For offered in *The Voyager's Collection* catalog one finds "the award-winning Atomic Watch," which contains "a micro-receiver and antenna circuitry that gets signals right from the U.S. Atomic Clock deep in the Rockies, making the time accurate to one-millionth of a second."[11] Beside pictures of two models of the watch, a heading enthusiastically proclaims: "exactly what you need to be RIGHT ON TIME." Zeitgebed to one-millionth of a second! One has to wonder at this conspicuous temporal consumption. Has anyone ever had a problem for being a few *millionths* of a second early or late *in everyday life*? Even a few *thousandths* of a second? But even if not, such excesses do provide another example of zeitgebers.

Perhaps these examples have created the image of a zeitgeber as a kind of temporal Big Brother, a dominant rhythm controlling and correcting the behaviors of entrained rhythms. That this happens is certainly true. But zeitgebers need not always operate continuously for entrained rhythms to persist.

The Persistence of Memory

Joseph McGrath and Janice Kelly conducted a series of studies involving individuals and very small groups (two and four members) (see McGrath and Kelly 1986, pp. 96–103, for an overview and references to the specific studies). The task performed was the solution of five-letter anagrams in a sequence of three differing time intervals. Of direct relevance to the matter of zeitgebers and their continuous operation, the order of the time periods in which the anagrams were solved was varied as part of the experimental design. Some groups first performed the task in a five-minute interval, whereas others performed the task in either a ten-minute or a twenty-minute interval first. If the five-minute interval came first, it was followed by ten- and twenty-minute intervals; if the twenty-minute interval came first, it was followed by ten- and five-minute intervals. When the ten-minute interval came first, it was followed by two more ten-minute intervals. The experimental design also varied the groups' workloads (i.e., how many anagrams they were to solve). Among McGrath and Kelly's findings was support for Parkinson's Law: The more anagrams the groups were given during a work period, the more they tended to solve, meaning that the less work a group was given, the less it did during a period, suggesting that work expands and contracts to fit the time available, McGrath and Kelly's proposed expansion of Parkinson's Law (1986, p. 97). (The original statement of Parkinson's Law was, "Work expands so as to fill the time available for its completion" [Parkinson 1957, p. 2].)

The time available, the length of the work period, is also another example of a deadline (the end of the work period), which has already been identified as a general type of zeitgeber. And McGrath and Kelly found a very important entrainment effect associated with the length of the first work period each group experienced:

> People in the conditions that began with the shortest time limit and had longer time limits in successive work periods performed at faster rates, in every condition (that is for every group size, every task load), than those in conditions that began with a long time limit and had shorter and shorter work times for successive trials. The people who had three successive work periods with the same time limit had rates between the other two conditions for the comparable loads and intervals. (McGrath and Kelly 1986, p. 98)

So the length of the initial work period, which can be interpreted as how close the group is to a deadline, affected the pace, the speed at which the group worked, it "established . . . a 'temporal entrainment' of the task performance process" (McGrath and Kelly 1986, p. 99). And the speed established during the initial work period tended to persist after the initial zeitgeber was removed (or altered, depending on how one conceptualizes it). Somehow the group or its individuals collectively learned a pace of work and remembered it in the subsequent work periods. The behavior of the groups that started with the shortest work period (five minutes) is especially informative, because the enlargement of their subsequent work periods could be considered the removal of the zeitgeber. Nevertheless, the tempo entrainment that occurred during the first period tended to maintain itself even though the specific entraining zeitgeber had been removed (or relaxed in a major way). That the tempo entrainment, the pace of work, persisted could be an example of free running, the "natural" frequency exhibited without active entrainment to a zeitgeber (Aschoff 1979, see pp. 2–6, especially p. 6), which, following Ancona and Chong's (1996) distinction between tempo and pace entrainment, would extend the concept of free running to the "natural" pace of work (i.e., the pace of work without active entrainment to a zeitgeber).

Now, McGrath and Kelley's groups continued their work for only two work periods after their original entrainment, and the continuation of the same or similar pace could be interpreted as a free-running phenomenon. It could also be interpreted as behavior in one or two transient periods before a new pace would have developed in alignment with the new zeitgeber (i.e., the new dead-lines). So to see whether socially entrained speeds or phase patterns, once es-tablished, can demonstrate free-running behavior if the entraining zeitgeber is removed or modified substantially, a much longer sequence of periods without the original zeitgeber should be examined. And the trolley schedule in Tren-ton, New Jersey, provides just such a sequence.

As described at the beginning of this chapter, the guards at Trenton State Prison began their shifts at 6:20 A.M., 2:20 P.M., and 10:20 P.M., and they did so at these times because the trolley stopped near the prison at about these same times. The trolley schedule was a zeitgeber that entrained the phase pattern of the prison's guard schedule. But this illustration does more than just illustrate phase entrainment; it also illustrates free running. It illustrates free running because in 1967 the guard schedule specified three shifts begin-ning at 6:20, 2:20, and 10:20, even though the trolleys had stopped running in *1934* (Hirsch 2000, p. 87). The zeitgeber had been gone for thirty-three years, which means over twelve thousand periods (days) had passed to test whether the entrained rhythm (the guard schedule) would run freely. Clearly the en-trained rhythm passed the test, because it had become an institutionalized part of the rhythms of organizational life at the prison. In the words of James Hirsch, in 1967 Trenton State Prison was "a fortress frozen in time" (2000, p. 87), a poetic way to say that the prison had been entrained by a zeitgeber that had disappeared, but the entrained rhythm had continued as a free-running rhythm.

One suspects that many institutionalized social rhythms are free running, separated by years or decades from the original zeitgebers that entrained them.

And research conducted by Marcie Tyre and Wanda Orlikowski (1994) may explain why.

Tyre and Orlikowski studied technological adaptation in organizations, which they defined as "adjustments and changes following installation of a new technology in a given setting" (1994, p. 99). Extending their findings to organizational change in general makes them relevant to the issue of institutionalized, hence free-running rhythms. Their key finding was that once a major technological change occurred, organizations would adjust to that change during a relatively short period of time (such periods are "windows of opportunity") and thereafter devote little or no attention to further adjustments. Generalizing more broadly, this suggests that once a change occurs, organizations and individuals adjust to it, and if the results equal or exceed the satisficing threshold, they move on to other matters. Entrainment to a zeitgeber would be such a change, and once the rest of the system adjusts (schedules for other activities in the case of the prison), the change and adjusting to it pass into the background and concerns shift to other matters. The rhythm becomes institutionalized and will free-run if the zeitgeber goes away.[12]

Institutionalized rhythms are echoes of zeitgebers past, and regardless of whether the zeitgebers are still present, the institutionalized rhythms, as with all institutionalized phenomena, create expectations. Such expectations themselves can become zeitgebers for new people entering the organization as well as for other organizations dealing with it. And if the expectations are not met, the newcomers and their work tend to be regarded unfavorably, regardless of what they do; but when expectations are met, the work and its producer are evaluated more favorably. An important experiment provides strong evidence for this conclusion.

D. Lynne Persing (1992) conducted an experiment in which she varied the quality of the work produced (a computer program) and the time-use pattern the programmer used to produce the computer program. A variety of time-use patterns are possible, but Persing focused on three: the "early starter," who begins work immediately and intensively, thereby completing the project well before the deadline; the "pacer," who begins work immediately but works on the project about the same amount of time daily and completes it near the deadline; and the "late starter," who does little work on the project initially but whose effort intensifies as the deadline approaches and who just completes the project by the deadline (1992, pp. 3–4). The "late starter" is basically the opposite of the "early starter." Both quality and time-use patterns were experimental manipulations presented in descriptions of the computer programmer given to participants in the study.

When asked to evaluate both the quality of the program and the attributes of the programmer, Persing (1992) found that (1) participants rated the program produced by pacers as better than the program produced by early starters or late starters, with no statistically significant difference between the latter two time-use patterns; and (2) participants rated the programmer described as following the pacer pattern more favorably (e.g., more dependable, more careful, etc.) and as having expended more effort creating the program than either early or late starters, with the latter two patterns again showing no statistically significant differences between each other on these evaluations (Persing 1992, pp. 102–5). These findings from over 150 participants (undergraduate business students) suggest that strong expectations may have developed in the United States for a steady, evenly paced approach to work along with a belief that this pattern of work produces the best quality. For when the quality of the work was controlled experimentally, work produced under a pacing pattern was judged superior to that produced under two uneven patterns, and the producers themselves were evaluated most favorably if they were described in the experiment as having followed the pacer time-use pattern.

This cultural interpretation of Persing's results resonates with a recently developed theory of operations management, the Theory of Swift, Even Flow (Schmenner 2001; Schmenner and Swink 1998). Among other things, this theory postulates that productivity falls as variability increases in either demand on or the steps in the production process. Moreover, Schmenner expressed an expectation that "those elements of human character that support low variability (steadiness) . . . should be valued and encouraged in both the companies and the nations that are high-performing" (2001, p. 89), an expectation that certainly supports a belief in the superiority of the pacer's work pattern. This is so because, as Persing described them, there would be less variability in the pacer's pattern of work than in either the early or the late starter's. Schmenner adds even more support, based on work by Landes (1998), by emphasizing values and attitudes toward time as important cultural factors that support low variability, especially temporal values associated with greater consistency (Schmenner 2001, p. 90). So the Theory of Swift, Even Flow and what appear to be its cultural correlates, combined with the analysis of punctuality in Chapter 4, all lend support to the proposition, albeit they do not directly demonstrate it, that Persing's results were produced by an American cultural belief in the efficacies of the pacer's pattern of work. However, to what extent such work patterns are genuinely associated with individual productivity remains to be seen.

Before leaving Persing's findings, one important implication of them should be noted. Not only does her research suggest the importance of entraining to expected phase patterns, it also has important implications for performance appraisal. It suggests raters' judgment of the quality of the work being evaluated will be influenced, perhaps inappropriately, by the pattern of time use employees practice in their work.

Such expectations, and their effects on people's responses to whether they are met or not, extend well beyond the organizational domain. In the political sphere, George Washington entrained the United States to an expectation that a president would seek and serve only two terms—thereby becoming not just the father, but the zeitgeber of his country. It is noteworthy that this institutionalized pattern then free-ran for almost a century-and-a-half after the zeitgeber's passing, until Franklin Roosevelt ran for a third term in 1940—and then only because of the extraordinary confluence of economic and military emergencies.

And if the theory of writing based on reader expectations (Gopen and Swan 1990; Williams 2000) is correct, you are exercising expectations while reading this sentence. This is a structural theory of writing based on two fundamental premises: (1) readers have developed expectations about where in units of discourse (e.g., sentences, paragraphs, etc.) they expect certain types of information to appear, and (2) the wise writer will locate the expected types of information in the expected locations. Using the sentence as an example, structurally there are temporal positions, *phases*, in a sentence (e.g., beginning, middle, and end— note that these are terms that define elements, phases, in temporal sequences), and readers have developed expectations and preferences about the kind of information that should be located in each. An example of a type of information presented in a sentence is important new information. But where is the best location in a sentence for the most important new information? According to this theory based on reader expectations, readers place particular emphasis on the material at the end of a sentence. Thus the best location for the most important new information presented in a sentence is at the end of the sentence. This is what I just did in the previous sentence. The most important new information

I wanted to communicate in that sentence was that the end of a sentence is the best place to put the most important new information. So I put that information at the end of the sentence because doing so emphasizes it.[13]

This theory of writing argues that in general readers find writing clearer, easier to read, and easier to understand if writers place information in units of discourse where readers expect to find it. In entrainment terms this means writers should seek synchronous entrainment, zero phase-angle differences, between reader expectations and where writers put information in sentences and paragraphs. When writers fail to achieve this convergence with readers' expectations, readers rebel and describe the writing as "unclear," "confusing," "hard to follow," and "awkward"; they describe it as bad writing—even though they may not be sure why it is "bad." This is an example of how failing to entrain behavior with expectations produces the feeling that something is wrong, even if it is unclear exactly what is wrong.

In general, a shift away from entrainment with expectations can be seen not just as an indication that something has changed but, if unanticipated, as a warning that something may be wrong. For example, when I send an e-mail message, I click on the send button and move on to other things. My expectation is that even if the person I sent the e-mail message to happened to be on-line when I sent the message, it should still take at least thirty to sixty seconds before I would hear my computer chime to signal the arrival of a potential reply. Thus my expectation is that at least thirty to sixty seconds are required for a reply *if things are working properly*. But if I hear the chime within one to three seconds, I know one of two things has happened. Either I have just received an e-mail message that has nothing to do with the one I just sent, or I typed the address wrong on the message I sent and it has just been returned as undeliverable. So hearing the chime within a few seconds of sending an e-mail is a warning that something may be wrong. This also includes receiving an automatic out-of-office reply, which often takes a few seconds more to arrive than the message returned as undeliverable. In a sense an out-of-office reply is also an indication that something is wrong, in this case that the person to whom the e-mail was sent won't be around for several days, which is probably a longer period than desired for a reply. In either case, the too-rapid reply differs from my expectation for a good situation.

Similarly, when things take too long, when they take longer than prior experience with them indicates is typical, that too is a sign that something is wrong. For example, when a researcher submits a manuscript to a journal, hoping to have it reviewed and accepted for publication, worry increases if weeks and then months pass without receiving an acknowledgment that the journal received the manuscript. In the organization sciences, an acknowledgment is expected within two to three weeks, and once a month passes, inquiries at the journal office ensue. This example generalizes to any circumstance when something or someone is late, late in this case meaning *longer than my experience when things turned out well.* Thus parents concerned about their teenage children being out "later than usual" interpret this lack of synchronous entrainment with parental expectation as a sign of potential trouble, that something undesirable is happening. Entrainment with expectations or the lack thereof is a useful control signal, an almost universal rule of thumb about whether things are all right or whether there is reason to worry.

But entrainment with expectations can do more than signal (e.g., Bluedorn 1997); indeed, people often behave proactively, attempting to take control of entrainment, to deliberately align rhythms to achieve their ends by getting important rhythms to entrain with others in a manner they believe will produce more desirable outcomes (e.g., the summer vacation and daylight saving time examples). And such efforts may be some of the most strategic managerial actions that occur anywhere in life.

STRATEGIC CONVERGENCE

The appearance of theories of timing (Albert 1995; Smith and Grimm 1991) suggest a dawning awareness of the importance of aligning rhythms, of entrainment, even if the entrainment frame is not used in these theories. Entrainment's importance stems not just from its own existence but from its relationship with important human outcomes, a point Mary Austin (1970, p. 5) understood early in the twentieth century: "Thus we represent, each one of us, an orchestration of rhythms which, subjectively coordinated, produce the condition known as well-being."[14] Saying that rhythms are orchestrated suggests their entrainment by human agency, a point she makes more explicit in the phrase "subjectively coordinated." So timing is important not only in its own right but even more so for its impact on important outcomes such as human well-being. And from an organization science perspective, a traditionally important outcome has been group and organizational effectiveness.

Converging on Effectiveness

As discussed in Chapters 3 and 5, group and organizational effectiveness is the degree to which a group or organization achieves its goals (see Price 1972, p. 101), and the extent to which an organization is coordinated appropriately promotes its effectiveness (e.g., Lawrence and Lorsch 1967). And entrainment is a major element of coordination. So when Eviatar Zerubavel described "three patterns of temporal coordination" (1979, p. 60) in a hospital, he was also describing three versions of entrainment. One form, temporal symmetry, is synchronous entrainment, as it revealed a consistent phase-angle difference of zero between two shifts, whereas the two other forms revealed consistent nonzero phase-angle differences, but because the phase-angle differences were *consistent*, those forms of temporal coordination represent entrained rhythms as well. Indeed, one of the two forms, "temporal complementarity," involved the maximum possible phase-angle difference, because when one shift was on duty the other shift was off duty. But because this phase-angle difference was constant, this is still entrainment. The other pattern, "staggered coverage," maintained a constant phase-angle difference somewhere in between the other two as the two shift patterns partially overlapped by a constant amount (Zerubavel 1979, pp. 60–61).

It is instructive that Eviatar Zerubavel concluded that "the maintenance of continuous coverage in the hospital would be impossible without *temporal coordination* among physicians and among nurses" (Zerubavel's emphasis; 1979, p. 60). Important parts of this "temporal coordination" were the three forms of coordination Zerubavel identified, which we have seen are examples of entrainment. The reference to "continuous coverage being impossible without temporal coordination" is then clearly an indication that "temporal coordination," entrainment, is necessary for at least minimal levels of organizational functioning and effectiveness. Stephen Barley later employed Zerubavel's concepts to analyze technological change in hospital radiology departments and the importance of "temporal coordination" in the functioning of the departments. He noted, "Temporal symmetry between the work worlds of technologists and radiologists moved toward an even closer *isomorphism* in special procedures" (Barley's emphasis; 1988, p. 153).

Both Zerubavel's and Barley's work emphasized phase entrainment, but the relationship between entrainment and effectiveness is also apparent in studies of tempo entrainment. For example, Mary Waller (1999) found that flight crews' performance increased the faster they were able to prioritize tasks and distribute activities after a nonroutine event. The effectiveness of flight crews was positively correlated with faster speed for these two key activities (prioritizing tasks and distributing activities), suggesting the environment in which flight crews operate required faster prioritizing and distribution for effective results. And this interpretation is consistent with research on decision-making speed in organizations, which was discussed in Chapters 3 and 4.

To briefly review that research, Kathleen Eisenhardt (1989) found that the speed with which top management teams made decisions was positively correlated with organizational effectiveness. Her sample of organizations was drawn from organizations in dynamic ("high-velocity") environments, and William Judge and Alex Miller (1991) found that this relationship was indeed limited to organizations in rapidly changing environments. These findings too suggest that this environment required faster responses for effective outcomes.

Thus Waller's, Eisenhardt's, and Judge and Miller's studies illustrate a possible positive relationship between tempo entrainment and effectiveness. And in all three studies, the effectiveness is that of work groups (i.e., flight crews) or entire organizations. An even more explicitly temporal example of organization-level entrainment and its consequences for effectiveness occurs in Joseph Ganitsky and Gerhard Watzke's (1990) analysis of joint ventures and the time horizons of the organizations involved. But what of individuals? Does entrainment—phase, tempo, or both—affect individual outcomes? Work by a variety of scholars suggests that it does.

Some scholars, of course, recognize the mutual influence individuals, organizations, and their entrainment have on each other. T. K. Das's (1986, 1987) research revealed that managers' personal time horizons (i.e., future temporal depths) were positively correlated with the length of the time horizons they used when planning for their organizations. Obviously individuals bring personal time horizons to organizations, but as organizational experience becomes a part of an individual's general life experience, that time horizon may be modified by experiences in the organization just as an individual's propensity to use a particular time-horizon depth will influence the depth the individual uses while engaging in organizational work.

Consistent with Das's finding of a positive correlation between individuals' personal time horizons and the planning horizons they employ for their organizations is an entire theory of individual and organizational time horizons and the importance of their convergence. The theory is stratified systems theory, which was developed by Elliott Jaques (1998a). This theory identifies variance among individuals' time horizons as well as the time horizons required by different positions in organizations, the latter being Jaques's famous time span of discretion, which is "the targeted completion time of the longest task or task sequence in a role" (Jaques 1998a, glossary; see also pp. 37–40).[15] Similarly, the individual has a capability for successfully dealing with certain distances into the future, which Jaques (1998a, p. 24) labeled the individual's *time horizon* and more formally described as "the longest time-span s/he could handle at a given point in their maturation process" (Jaques 1998a, glossary).

Without getting into additional detail (see Jaques 1998a, 1998b), Jaques's stratified systems theory informs the discussion of outcomes from individual-organization entrainment because it puts the position (time span of discretion) and the individual (time horizon) together: "My proposition is that to fill successfully, say, a vacant 7-year time-span . . . role, you will require an individual of a 7-year time-horizon" (Jaques 1998a, p. 24). In entrainment terms, positions should be held by individuals such that there is a zero phase-angle difference between the position's time span of discretion and the individual's time horizon. And the word *should* is used deliberately, because Jaques said "to fill successfully," successfully indicating positive outcomes for both the individual and the organization. Thus *successfully* suggests that the synchronous entrainment prescribed by Jaques will result in better performance, ceteris paribus, than other possible combinations, and that good performance by the individual will result in continued employment, pay increases, and so forth for the individual. Further, the good performance will help the organization function better, help it achieve a higher level of goal attainment. Thus stratified systems theory indicates that both individuals and organizations benefit when, in entrainment terms, individuals and their positions are entrained synchronously. But entrainment seems likely to have even more consequences for individuals than suggested in Jaques's theory.

Fitting the Times

Some analyses focus more exclusively than Jaques's on entrainment's effects on individual outcomes, albeit not conceptualized as entrainment. For example, Edgar Schein noted, "Polychronically driven work always has the potential for frustrating the person who is working monochronically" (1992, p. 114). This statement suggests that individuals respond more positively (i.e., with less frustration) when their free-running behavior along the polychronicity continuum converges with that in their work context (e.g., coworkers, supervisors, the culture in general). And as with Jaques's work, synchronous entrainment is probably the specific form implied.

Although not focused exclusively on individual outcomes, Carol Kaufman, Paul Lane, and Jay Lindquist (1991b) described "time congruity" possibilities between individuals and organizations along several dimensions, including polychronicity. And they also described a wide range of potential outcomes resulting from congruity and incongruity. Noting that polychronic people may have problems in a monochronic environment, they suggested that heavily polychronic people "in a monochronic work group will probably not be near as effective as when they are grouped with others who also are comfortable and capable of combining several tasks at the same time" (Kaufman, Lane, and Lindquist 1991b, p. 99). They would probably not be as comfortable because they would not be entrained well, if at all.

Polychronicity describes a general rhythm for engaging life, and especially if one focuses on the extent to which one moves back and forth between different tasks and events, it is easy to see the polychronicity continuum as describing a wide range of rhythms that characterize people's behavior. Seeing polychronicity as a fundamental rhythm is also consistent with the work of Eliot Chapple (1970, 1971) and Rebecca Warner (1988), which involved a theory of human interaction based on the rhythms of individuals' periods of activity and inactivity, especially language activity, the characteristics of those patterns, and the similarities and differences between individuals' activity rhythms and those of their interaction partners. And just as Schein and Kaufman, Lane, and Lindquist suggested about polychronicity, Chapple and Warner's theoretical work suggested that some activity rhythms can be coordinated, intermeshed, more readily than others, that the ease or possibility of entrainment depends upon characteristics of the rhythms that are to be aligned. Further, there is a strong implication, sometimes made explicit (e.g., Warner 1988, p. 82), that entraining rhythms results in positive outcomes for the individual involved, but a failure or inability to entrain results in less positive outcomes. Schein's reference to "frustrating" and Kaufman, Lane, and Lindquist's to "not be near as effective" are similarly explicit.

Since polychronicity describes rhythmic behavior patterns as well as attitudes about them (see Chapter 3), from an entrainment perspective it makes perfect sense to investigate the degree of congruence between an individual's polychronicity and the level of polychronicity displayed or preferred by the people with whom one interacts, and this point comes through well in a story Edward Hall told when I interviewed him.

Hall described how he had worked with a husband and wife, one of whom was polychronic, the other monochronic, and a French manager (polychronic) whose immediate supervisor was a German (monochronic). In the case of the couple, they "just couldn't get along" (Bluedorn 1998, p. 112), and the situation between the two managers had become "nearly intolerable" (p. 113). Hall reported helping these people by explaining the way they differed in this most basic of behavior patterns and orientations. And once he had explained the differences to them, they could understand what the other was doing and stop misunderstanding it and, most important, stop taking "the other's behavior personally" (pp. 112–13).

Hall's examples illustrate the potential of studying the congruence between an individual's polychronicity and the polychronicity of the people in the individual's workplace, and under the leadership of my colleague Tom Slocombe, this is what he and I did (Slocombe and Bluedorn 1999). In a questionnaire sent to business school graduates, respondents were asked to describe their own levels of preferred polychronicity as well as the level of polychronicity collectively displayed by everyone else in their work units. We anticipated that the closer individuals' personal polychronicity preferences and behaviors were to those of the people in their work units, the more positive the outcomes would be, which is what we found. The closer respondents' polychronicity preferences were to the level of polychronicity behavior they perceived exhibited by other people within their work units, the more committed the respondents were to the organization (i.e., the more they wanted to remain a member, the more effort they were willing to expend for the organization, and the more they accepted the organization's goals), and the more favorably *and* fairly they believed their performance was evaluated. Thus our results were consistent with an entrainment interpretation, one that suggests synchronous entrainment of polychronicity patterns produces the most positive responses.[16]

BE CAREFUL WHAT YOU WISH FOR

Perhaps one reason for the intuitive appeal of congruence propositions is the general intuition that they result in positive outcomes for individuals and groups, including organizations. Yet for decades one of organization science's boogeymen has been groupthink (Janis 1972), and what is groupthink but too much congruence? The emphasis on congruence is explicit in Irving Janis's description of the groupthink concept: "a mode of thinking that people engage in when they are deeply involved in a cohesive in-group, when the members' strivings for unanimity override their motivation to realistically appraise alternative courses of action" (1972, p. 9). The phrase "striving for unanimity" is simply another way of saying the group tries to converge on a single decision with every member's public viewpoint and opinion so congruent as to be virtually identical. And temporal factors may be involved in groupthink too. Sally Blount and Gregory Janicik (in press) have suggested that members of a cohesive group will be reluctant to disrupt the group's prevailing pace of work and, further, that this reluctance during group decision making is likely to promote groupthink.

Another example of too much congruence involves a form of entrainment known as resonance. Physical systems are said to resonate when impulses of energy are applied to them at or near the one or more frequencies at which they vibrate naturally, and the amplitude of the system often increases "manyfold as exact synchronism is reached" (Considine and Considine 1989, p. 2430), the "synchronism" suggesting synchronic entrainment as described earlier. But there is danger in "synchronism," in resonance, just as there is potential danger in too much agreement between members of a group. There is danger because the energy applied can resonate too well, the amplitude of the system can increase too much, destroying the system. This is what happens when sound waves shatter a glass, and it is what happened to the Tacoma Narrows Bridge in Washington in 1940. After this bridge was completed it "began to exhibit an unusual natural resonance in which its surface twisted slowly back and forth so that one lane rose as the other fell. During a storm, the wind slowly added energy to this resonance until the bridge ripped itself apart" (Bloomfield 1997, p. 351).

So both groupthink and the resonance examples illustrate the point that neither convergence in general nor its specific temporal manifestation in entrainment are universally "good" from any particular value position. Joanne

Martin (1992) has similarly argued against an overemphasis on cultural homogeneity (i.e., congruence) in the organizational culture domain. But the opposite is also true as revealed by the point that diversity in small groups (Earley and Mosakowski 2000; Watson, Kumar, and Michaelsen 1993) and societies (Bluedorn 2001) does not automatically produce positive outcomes. Sometimes diversity produces what people want, sometimes it does not, and congruence is the same way.

Two thousand years ago Seneca recommended fitting the times (Tempori aptari decet; 1834, p. 10), but this advice would be better yet if it were tempered with a qualification to take care about which times one fits oneself to. For not only are all times not the same, they are not all equally important. The sunflower seems to have chosen wisely in picking the time to fit itself to, getting its name "because the flower follows the sun's path across the sky each day" (Perry and Perry 2000, p. 85). As the history of life on this planet shows, one could do much worse than entraining one's activities to the apparent motion of the local star. Indeed, of all the strategies of life, such entrainment appears to be almost ubiquitous—and almost ubiquitously successful. It is the right thing to fit the times—if one picks the times wisely.

The Best of Times and the Worst of Times

O, call back yesterday, bid time return.
—Shakespeare, *Richard II*

A paradox developed at the end of the last century involving some of the worst of organizational times. Research on meetings, which are often some of the worst of times, resulted in some of the best of times for the people who conducted the study. Not only was the research published in a prominent journal, but it led to a modicum of fame—if not fortune—for the research team (which we shall meet shortly). So some of the worst of times were also closely involved with some of the best of times, a paradox. Although paradoxical thinking does not come easily because it requires thinking about contradiction (Quinn and McGrath 1985, pp. 316–17), paradoxical thinking will be necessary in this encounter with the best and worst of times.

For such times, the best and the worst, and what makes them best and worst, are the topic of this chapter. Obviously not every good and bad time can be discussed in a single chapter—or even in a single book—but several prominent good and bad times can provide a basis for understanding what makes times good and bad. To develop this understanding requires us to address issues of connections and meaning; it requires us to develop a better understanding of the relationship between how rapidly time seems to pass and the quality of the experiences associated with different speeds of those passings, the received wisdom about this association requiring significant revision. Such an understanding requires us to learn about not just the need to let go, but how to move on; and in the final analysis, we must recognize the choices we often make unknowingly, lest all our days become infamous. All of this we shall do and more as we proceed through this chapter, beginning with the paradox about the best and worst of times.

THE IRON LAW OF COUNCIL

Nearly a century ago Robert Michels forged the Iron Law of Oligarchy: "Who says organization, says oligarchy" (Michels 1962, p. 365). But the tendency for those in power to maintain that power is not the only regularity in organizational life. Another phenomenon pervades organizations, perhaps even more universally than oligarchic tendencies, and it suggests the need for social science's foundry to forge another law, the Iron Law of Council: Who says organizations, says meetings.

As ubiquitous as any organizational activity, meetings could even define organizations themselves—as open-ended meetings—if one grants a little conceptual license. Without pursuing that thought further, one can still note how meetings permeate organizational activity regardless of whether that observation is based on personal experience or formal research. For example, Henry Mintzberg studied five chief executives and found they spent 59 percent of their days in meetings, 69 percent if unscheduled meetings are counted (1973, pp. 39–41). But personal experience makes abundantly clear that one does not have to be a CEO to encounter meetings. So it is surprising that so little research has been conducted on meetings (Schwartzman 1986), because as a non-contrived, naturally occurring organizational activity, the meeting would seem to provide an ideal laboratory for small-group research. Perhaps meetings have gone unstudied because they seem to be an ordinary, everyday phenomenon, one too mundane to generate much interest, let alone great passion.

So my colleagues Daniel Turban and Mary Sue Love and I were taken completely off guard when our research on meetings generated not just a Warholian fifteen minutes of fame, but several years' worth, and counting. We thought we had done something creative, something that would generate a modest interest within organization science circles, but the thought that the press, let alone the world press, would have any interest never crossed our minds—until I was asked whether I thought any of my research might be of interest to the general public, research that my university's media relations office could publicize.

As I inventoried the various projects I was involved with, I was drawn to the meeting project because its novelty seemed like something the general public might find both easily understandable and interesting. That and the fact that the manuscript reporting the research seemed close to acceptance by a major journal led me to answer the inquiry with a description of the research.

Shortly thereafter the *Journal of Applied Psychology* accepted the manuscript for publication, the university's office of media relations interviewed me and issued a press release about the research—and my phone started ringing. A story appeared in the Science Times section of the *New York Times*, and I was interviewed on the BBC—twice.[1] What had we done that generated such attention?

We asked people to stand during their meetings. And we compared those meetings with more traditional meetings in which people sat around a table. More specifically, we had 555 students from an undergraduate management course form 111 five-member groups. Randomly assigned to the stand-up or sit-down conditions, each of these groups held a meeting to solve the same problem, a problem requiring at least moderate amounts of judgment and creativity, but a problem for which the quality of solutions could be evaluated quantitatively and objectively.[2] The same meeting rooms were used for both conditions (each group met by itself), but all of the furniture was removed from the rooms beforehand when groups were assigned to the stand-up condition; for the sit-down meetings, the furniture consisted of a table and five chairs (see Bluedorn, Turban, and Love 1999, for details).

Why did we do this? We conducted this experiment because advice proffered in the time management literature directed managers to increase meeting speed by having participants stand throughout the meeting (e.g., LeBoeuf 1979, p. 159; Mackenzie 1972, pp. 102–3; Reynolds and Tramel 1979, p. 117). But this advice made no allowance for whether decisions would be made during the meeting or whether the meetings would just be used to give instructions and pass on information. We were *sure* that if the meetings were used to make decisions, the decisions would be better if they were made in the sit-down condition because the participants would take more time and—we thought—more carefully consider information relevant to the decision.

Thus when we examined our results we were flabbergasted, because there was no statistically significant difference between the average quality of the decisions produced in the two conditions. On average, the fifty-six groups in the stand-up meetings produced decisions that were just as good as those produced by the fifty-five groups that conducted their meetings in the traditional sit-down posture. This finding was so contrary to our expectations that we re-checked and re-rechecked our data and records and analyses, but the original results were correct. Even though the sit-down meetings took significantly longer on average—34 percent longer—than their stand-up counterparts, there was no average difference in quality between the two conditions. So our results supported the stand-up imperative given in the time management literature, which is a likely reason the press took such an interest in the study. Had we found what we expected to find, I suspect the press would have ignored the study.

But we had produced results supporting a way to reduce the length of meetings without harming an important and widespread instrumental meeting function, decision making. The reaction our study received as well as the manifest discussions in the time management literature both suggest that people want fewer meetings, and of the meetings they have, they want them to be shorter. A naive scientific management interpretation of this motivation would be that people desire more efficient operations, but one knows that would just be a rationalization. The real reason is more basic, that within organizational life for much of the last century, people have developed an aversion to meetings. They hate them. And the question is, why?

It seems unlikely that hominids innately dislike meetings. After all, much of the several million years of hominid history has involved meetinglike gatherings, from the daytime foraging expeditions to hunt for and gather food, to meals themselves, to the several hours spent together at dusk and into the early evening before sleep would come. So it is doubtful that hominids, including the contemporary model, are genetically hardwired to abhor meetings; if anything, just the opposite may be true. And this suggests there may be something wrong with meetings in organizations, rather than with meetings generally, that makes them so repugnant to most participants, that makes the Iron Law of Council a description of some of the worst of times.

What makes the meetings in organizations so infamous?[3] There are likely several reasons, and I suspect the smaller the group having the meeting, the less salient these reasons become, but reflecting on my own experience and my perceptions of others' experiences, I believe there are at least three factors involved in our distaste for the organizational meeting.

The first factor is the agenda. The purpose of the meeting is usually someone else's, which means the meeting may not be relevant to the goals of most participants—at least as they perceive them. This combines with the second factor, that employees of the organization are expected to do certain things and are often rewarded symbolically and tangibly according to how well they do them. Put these two factors together and one has a prescription for frustration when it comes to meetings. People believe they are rewarded for performing certain tasks, and they often are, but along comes an activity, one for which they will not be rewarded for taking part in—or for at least being physically present—but for which they may be punished if they do not attend. The meeting and its specific agenda items may have little if anything to do with participants' agendas, personal or professional, yet attendance is required. From the participants' perspective this makes the meeting, or at least major portions of it, increasingly frustrating because it seems like wasted time, time people would rather be devoting to activities that promise either greater personal fulfillment or progress toward greater extrinsic rewards. The combination of working on someone else's agenda at the expense of one's own is not a prospect designed to generate enthusiasm.

What may be involved in all of this is a third factor, locus of control. Locus of control refers to an individual's general beliefs about the factors responsible for events, and the most general distinction is between factors under the individual's control, internal, and beyond the individual's control, external (Rotter 1966, p. 1). As just described, the organizational meeting tends to be beyond the individual's control (i.e., someone else's agenda), thus making it largely a set of forces to which the participants must succumb, but a set of forces over which they have little or no influence. The meeting puts most participants in an external locus of control field, which contrasts with most of their regular organizational activities, activities that are likely to be seen as more under the individual's control, hence as involving more internal locus of control than the typical organizational meeting. This contrast in locus-of-control balance is likely to make the meeting seem even more frustrating because the typical attendee often feels powerless in such circumstances.

Consistent with the perception and reality of an external locus of control in meetings is Karl Weick's description of at least some meetings as proceeding with "autocratic leadership, norms that encourage obedience, unwillingness to risk embarrassment by disagreeing with superiors" (1995, p. 186). Ironically,

Weick argued that "people need to meet more often" (p. 185), at least about certain types of issues, but such meetings will not work well if they are conducted the way he described them as typically being conducted (i.e., with "autocratic leadership" and so forth). So Weick, as so many others, recognized a major problem with meetings in organizations, and he identified some meeting processes as candidates for change.

Meetings clearly are not the best of times, far from it, but it would seem that they could be made better times, or at least more palatable times. Interestingly, the type of issues about which Weick felt "people need to meet more often" are the issues of ambiguity, the ones about which clarity is lacking and sense-making is required (see Weick 1995, pp. 185–87). Making sense of something is, of course, either to give it meaning or to alter its meaning, and the meanings of things and events, as we shall see again, come from their connections with other things and events. These connections, hence meanings, make times good or bad.

CONNECTIONS WITH MEANING

The discussions of meaning in Chapters 2 and 5 established the basic premise that significance, hence meaning, originates in the connections among things, requiring, in Whitehead's phrase, "a knowledge of their [things'] relations" (Whitehead 1925a, p. 12). But for the human experience of time, what kind of connections are involved? The answer is likely narrative connections.

Narrative

A narrative consists of three essential elements: past events, story elements, and a temporal ordering (Maines 1993, p. 21). According to Jeffrey Bridger, constructing the plot, which helps create story elements (Maines 1993, p. 21), may be the most important of the narrative tasks because it transforms the events from at most "a chronicle," a list of events arranged in sequence, "into a temporal whole," for he concluded, "A singular occurrence is not particularly meaningful; events take on meaning to the extent that they contribute to the development of the plot" (Bridger 1994, p. 605). So when your companions ask you, "What is your point?" they are asking you to connect your thought or idea to other thoughts and ideas, ones they hope to find relevant (i.e., other thoughts and ideas with which they are connected). Your companions want to know other elements of the story, especially as those elements relate to their stories.

These conclusions reinforce Whitehead's views about significance by emphasizing the point that an event has meaning only when it is linked to a plot or story, that is, to a greater whole. This is very similar to the concept of temporal context, which Joel Bennett used in his analysis of intimacy in human relationships. Bennett referred to this context as "the dynamic weaving of events, interactions, situations, and phases that comprise those relationships" (2000, p. 27), the dynamic weaving of events, interactions, and situations being very similar to narrative.

So taken together, Whitehead's, Bennett's, and Bridger's conclusions indicate that events and things must be related to other events and things to give them meaning, and this meaning increases if the relationships form a coherent whole that provides an ongoing interpretation, a story with a plot. Without relationships there can be no plot, and without a plot there can be no meaning. And without meaning what can be hoped for in the way of experience? Friedrich Nietzsche stated this point differently but so very well: "If we have our own *why* of life, we shall get along with almost any *how*" (Nietzsche's emphases; 1968, p. 468).

And as the literature on alienation suggests, when why is connected to how, all of experience becomes more meaningful.

Meaning and Alienation

Melvin Seeman distinguished several forms of alienation, one of which—meaninglessness—refers to the lack of meaning, the lack of understanding of the events in which the individual is engaged (1959, p. 786). Although Seeman explained how the different forms of alienation were conceptually distinct, he also suggested their empirical connections and described how they might be related to each other. Hence the lack of meaning, the inability to understand events—especially those in which one directly participates—is likely to make it harder to control them (powerlessness), a point at the heart of a famous statement made by Kurt Lewin, only the final clause of which is usually given:

> Many psychologists working today in an applied field are keenly aware of the need for close cooperation between theoretical and applied psychology. This can be accomplished in psychology, as it has been accomplished in physics, if

the theorist does not look toward applied problems with highbrow aversion or with a fear of social problems, and if the applied psychologist realizes that there is nothing so practical as a good theory. (Lewin 1951, p. 169)[4]

Lewin's famous final clause, "there is nothing so practical as a good theory," means that understanding (theory) can guide useful action (the practical), so theory (understanding) can be empowering. A lack of meaning and understanding of events also makes it very hard to know what one *should* do (normlessness or anomie), because it is nearly impossible to know what to do in a situation if one cannot comprehend it. Lack of meaning is also involved when an individual does not value goals or beliefs held by the larger group and experiences isolation from that group. The lack of meaning in this case results literally from the lack of connection between the individual and the group. In the final form of alienation Seeman identified—self-estrangement—individuals engage in behavior because of rewards they will receive for performing it, not because the individual achieves intrinsic satisfaction from the behavior. Here, too, the agency of connections is in play, for when self-estrangement occurs, the connections between individual and behavior become indirect because the reward is distinct and separate from the behavior itself. The behavior comes first, the reward later.

Although one could plausibly argue that each form of alienation can and probably does influence the others, connection-engendered meaning is given the central part in this analysis; and it is given the central part because of the fundamental social science and linguistic principles presented in Chapter 1. Put succinctly, these principles indicate that the definition of the situation guides human behavior, and according to the Sapir-Whorf Hypothesis, language is a necessary prerequisite for defining any situation (see Chapter 1). Language provides the elements from which meaning is constructed—the names of things and their qualities and the manner in which they may be related—and the definition of the situation combines these elements to construct sense-making explanations of events, definitions of the situation. Both the definition of the situation and the Sapir-Whorf Hypothesis are about meaning first, not power and control, not what should be done, not about contingent rewards. Sometimes such matters are directly linked, even simultaneously so, but until the basic "is" of the situation is defined, the questions "What should I do?" "How can I influence things?" and "What's in it for me?" are impossible to answer. So connections and the meaning they generate are fundamental, which is why the loss of meaning is so troubling—the systematic loss of meaning even more so. And to illustrate the fundamental temporality of connections, hence of meaning, two topics discussed in Chapters 4 and 5, respectively, will be examined in this light: speed and temporal depth.

Speed

Americans value speed, and at least at the level of surface values regard it as a general good. So if fast is good, faster is better. If not, how else could "Do everything faster!" (Cottrell and Layton 2000, p. 34) have been unabashedly offered as time management advice? The prescription is for everything, not just the right things, or judiciously selected candidates for acceleration, but *everything*. This belief in the unlimited virtues of acceleration does have a connection, and that connection is to efficiency, a connection and matter discussed in Chapter 4, but for now, finding other connections, hence meaning, for the belief that speed is a general virtue is difficult, but not impossible.

For example, the idea of entrainment presented in Chapter 6—the adjustment of the pace or cycle of an activity to match or synchronize with that of another activity (Ancona and Chong 1996, p. 253)—does suggest a reason why faster could be better, but it also suggests that faster could be worse. To match the pace of another activity might mean increasing speed if the other activity is moving at a faster pace. But what if the activity one wishes to match is moving *slower*? To speed up in such a situation would decrease the match, making things worse from the perspective of matching the two activities. Deborah Ancona and Chee-Leong Chong (1996, pp. 262–63) provided an example of just such a problem in the case of several Japanese computer companies whose rate of innovation was faster than that desired by the market—so the companies needed to slow down. The companies certainly did not need to accelerate. So what at first might have seemed like a confirmation of the acceleration imperative illustrates instead the contingent nature of speed and its connections to desired and undesired outcomes.

And the *contingent* nature of speed may be more widely realized than pronouncements like "Do everything faster!" indicate, even in the United States. James Gleick certainly seemed ambivalent about speed in his book *Faster*, at one point noting that, at least in some physical activities such as races, "Statistical trends over time suggest that we are, as a species, approaching asymptotically a true maximum speed" (1999, p. 109). If so, at some point doing everything faster becomes impossible in any meaningful way. It is noteworthy that Gleick made this statement in the same chapter in which he described and questioned the premise that intelligent brains are faster (1999, p. 113). He presented his overall conclusion about this premise by quoting Robert Sternberg: "If anything, the essence of intelligence would seem to be in knowing when to think and act quickly, and knowing when to think and act slowly" (quoted in Gleick 1999, p. 114). This, as the entrainment phenomenon and examples illustrate, supports the contingency view of speed, that the appropriate speed varies by activity and context. There is no universally best speed, and faster is not always better. In fact, it is often worse.

Even in the speed-oriented United States, people seem to recognize this at deeper levels. For example, D. Lynne Persing's (1992) results indicate that the participants in her experiment recognized that faster was not always better (see Chapter 6 for a detailed description of her experiment). Her results indicate this because the participants rated the quality of identical work as worse when they were given information that the work was done in shorter amounts of time than when it was done over a longer time.[5] A reasonable interpretation is that taking more time, hence working at a slower pace, was believed to produce better-quality results. The message would seem to be that speed is more positively meaningful when the right speed is selected, "knowing when to think and act quickly, and knowing when to think and act slowly." Otherwise speed tends toward the meaningless, or at least loses its potential for positive meaning. Like the driver who weaves back and forth between lanes, passing cars right and left but gaining only thirty or forty feet on the traffic flow by the next stoplight, greater speed for its own sake is a fast track to nowhere. For meaning, especially positive meaning, speed needs to be connected to things, some of which may be in either the past or the future.

Temporal Depth

Connections to things in the past or the future involve questions of temporal depth, and as the discussion of temporal depth in Chapter 5 indicates, the number of things to be connected to may be shrinking because, at least in the United States, people's temporal-depth intervals may be getting smaller. Temporal depth has two general components, past and future (see Chapter 5), and connections can be made with elements in both components. But as mentioned, temporal depth may have shrunk and become quite shallow. If true, this is important because it limits the distance fore and aft that people can search when they try to make connections, when they search for meaning. A shallower temporal depth provides fewer possibilities.

But how do connections with elements in the past or future provide meaning, regardless of how far away they may be temporally? In Chapter 2 I quoted Quy Nguyen Huy about the relationship between the past and the present, and that quotation bears repeating here: "Since one cannot distinguish a figure without a background, the present does not *meaningfully* exist without a past" (emphasis added; 2001, p. 608). As the background, the past provides a benchmark for the present against which comparisons can be made. And such comparisons indicate whether the present is the same as the past or different from it. Edmund Husserl has described the nature of this relationship between the past and present by analyzing what makes a sequence of musical notes form a melody, an important part of which is "a direct apprehension of identity, similarity, and difference" (1964, p. 41). If the past appears to be the same as the present, then the interpretation and understanding of the past can simply be employed to interpret the present. But if the past and present differ, the question of how the differences developed helps interpret the present. Thus the past provides a context, a frame, for the present, and the linkages with the past provide an explanation for the present by suggesting how the present came to be, which makes the present more understandable, more meaningful. (See the discussion of the past as metaphor for the present and the future presented in Chapter 5.)

So connections to the past help make the present more meaningful, but connections to which past? As the data regarding temporal depth presented in Chapter 5 indicate (see Tables 5.1 and 5.2), people gravitate toward different temporal depths when they think about the past, and since none of them appeared to go much further into the past than twenty-five or thirty years, at least in that sample, the data also suggest that the totality of the human past is seldom used. People appear to establish a referent past, possibly referent pasts, to create plots that help explain the present. Exactly how this is done in organizational contexts and how such plots are used is just now beginning to be studied by researchers such as Ellen O'Connor (1998, 2000). Nevertheless, even such pioneering work indicates the past is used to interpret and understand the present, and at times to help anticipate the future and cope with it. But which past, which future?

Deep Connections

Gregory Benford's recent work indicates that modern organizations, and a significant portion of contemporary humanity in general, may be shortchanging themselves by looking into the past too shallowly. As discussed in Chapter 5, Benford identified one of the virtues of a deep-time perspective as an increased ability to detect trends, trends invisible when only shorter intervals are considered. Identifying such trends, almost by definition, establishes linkages between the past and the present, linkages that may confer even more meaning than the average linkage with the past because long-term trends somehow seem more powerful. They seem more powerful because they have withstood more rigorous tests of time; that is, by extending over longer intervals, they have been subject to more potentially disruptive influences but have been able to maintain themselves, which makes them a more credible force, a more credible explanation. And connections with deeper times seem involved in generating and maintaining profound meaning, the absence of which Benford was insightfully aware: "Whipsawed by incessant, accelerating change, the modern mind lives in a fundamental anxiety about the passing of all referents, the loss of meaning" (Benford 1999, p. 3). And, "When hatred and technology can slaughter millions in months or even minutes, such terrors deprive life of that quality made scarce and most precious to the modern mind: meaning" (p. 204).

The loss of meaning is a contemporary dilemma—Benford and I are certainly not the only analysts to discuss it—and the two Benford statements support the idea that meaning is lost as connections are severed. At a minimum, severing connections alters meaning, and when enough are severed—the possibility of slaughtering millions in a few minutes—meaning may disappear entirely.

Hence Benford concluded, "A yearning for connection also explains why ancestor worship appears in so many cultures; one enters into a sense of progression, expecting to be included eventually in the company" (1999, p. 3). The idea of progression in the phrase "a sense of progression" is a form of trend, and being "included eventually in the company" is a form of profound connection with a larger segment of humanity, be that segment a literal company, a profession, a tribe, a country, or even humanity's complete family tree back to the earliest hominids—or before. (Miriam Makeba revealed a similar perspective in her descriptions of her people's relationships with their ancestors and the relationships among past, present, and future; see Chapter 5.) To sever or ignore the possibility of such connections forfeits much of the possibility for meaning in modern life. And when combined with an equal disconnect from the future, little meaning is left, because, as Benford suggests, "I suspect that deep within us lies a need for continuity of the human enterprise, perhaps to offset our own mortality" (1999, p. 3).

No one enjoys thinking about the prospect of one's own death. Yet in the age of nuclear and biological warfare, contemplating the end of humanity is even worse. But why? We inevitably die as individuals in any case, so why does the thought of us all dying together seem worse? Benford's analysis explains why this prospect, the end of humanity, is so depressing. It is so depressing because it would remove so much meaning from one's own life by eliminating any possibility of a deep-time connection with the future, and as described earlier, these connections across deep time seem so very profound.

This is why Mary Leakey's interpretation of the trail of hominid footprints from nearly 4 million years ago is so moving to us today (see Chapter 2). Without it, the trail of footprints is reduced to merely an important archaeological relic; but with it, a much deeper connection is created between us and them: These were beings with whom we shared common feelings, with whom we shared a common humanity. And that such a connection can bridge a gap of 4 million years makes the connection that much more profound, which is what deep-time connections do: "Deep time in its panoramas redeems this lack [of meaning], rendering the human prospect again large and portentous. We gain stature alongside such enormities" (Benford 1999, p. 204). We gain stature, but only if we connect with such deep-time enormities.

Because Benford's work on deep time proved so insightful to me, I wanted to learn more. So I secured an introduction and asked if I could visit him to discuss deep time.[6] He graciously accepted my request, and I flew out to visit him in his physics department office at the University of California-Irvine. We spent a marvelous four hours together—I think he was slightly amazed that someone would travel across half a continent to discuss time for a few hours, but then, all times are not the same!—and I gained several insights that had not been covered in his book.[7] An especially germane point about connections with the deep future arose in the discussion, which I present now:

Benford: Fundamentally, I mean deep time to be the scale upon which there is no ensured continuity of human culture—of a particular culture. And therefore the very context or meaning in your life is lost. That seems to be the working definition of the time scale that produces, shall we say, the most anxiety. Because it's beyond the loss of life for a person; it is beyond the scale of a century. It's the loss of life of any society that anyone could know or fathom. It's the time scale in which you have to worry about issues of meaning in a very general sense.

Bluedorn: Not only will I be gone, but I won't be remembered and my work will have gone for nothing.

Benford: Exactly.

Meaning is lost when there is, in the case of the future, an anticipated loss of continuity—going too far ahead to where too many things have changed, where too few things, perhaps nothing, will be familiar. (This could happen with the past too.) And the loss of meaning owing to missing connections with either the past or the future, and especially the deep past and the deep future, creates a present that is one of the worst of times. Henry Wadsworth Longfellow captured the importance of connecting with this great continuity in this stanza from his poem *A Psalm of Life*:

Lives of great men all remind us
    We can make our lives sublime,
And, departing, leave behind us
    Footprints on the sands of time.

(Longfellow 1883, p. 13)

Following the arguments about connections and meaning, our lives become sublime when we make more connections with the sands of time, both the strata laid down before us and the sands yet to come. By doing so, our present becomes at least a better time. But we also have beliefs about what leads to the best and worst of times, beliefs that have been accepted uncritically, and these beliefs require closer examination, lest they lead us to worse times when we think we are heading for better.

THE TEMPORAL QUALITY OF EXPERIENCE

A resident of Missouri once asked me whether I would like to hear a story, which he then told me. This is that story:

A patient was visiting a doctor to receive the results of several important tests, and the doctor began the conversation by saying, "I'm afraid that I have some very bad news. The tests indicate that you have only six months left to live."

The patient exclaimed, "That's terrible! Isn't there anything you can do?"

The doctor replied, "I'm sorry, but your condition is something medical science has no cure for right now. However, I have one suggestion that might help."

The patient looked up hopefully and asked, "What's that?"

The doctor answered, "Move to Kansas."

Puzzled, the patient inquired, "Why? How would that help?"

The doctor explained, "You won't live any longer; but it will seem like forever."[8]

We laugh at this joke because of something we believe about time and how we perceive its passage: that time seems to pass quickly while we are experiencing something pleasant and that it seems to pass slowly while we are experiencing something unpleasant. And we have held this belief for a long time. In the year 105, Pliny the Younger wrote, *"nam tanto brevius omne quanto felicius tempus* (the happier the time the shorter it seems)" (1969, pp. 36–37). Nineteen centuries later people say it differently, "Time flies when you're having a good time," but it is exactly the same idea as Pliny's, and though unstated it also implies that the sadder the time the longer it seems, that time drags when you're having a bad time. We believe this uncritically for at least two reasons. First, this has been accepted wisdom for a long time, sometimes even buttressed with evidence from experiments (e.g., Gupta and Cummings 1986). Second, we can all think of times when we have experienced exactly what the accepted wisdom tells us, which seems to confirm these axioms once again.

But these beliefs are not axioms, at least not all of the time, because they only partially describe the relationship between the pleasantness of the experience and the perceived passage of time. For at other times unpleasant experiences seem to pass quickly, whereas pleasant ones may seem to linger. These are the findings reported by Michael Flaherty based on data from 705 descriptions of situations in which time seemed to pass so slowly that the difference was noticeable to the individual involved in the situation (1999, p. 41). Some of these situations were pleasant, others unpleasant. Flaherty also reported accounts of situations that appeared to participants to pass more quickly than normal, but that also consisted of both pleasant and unpleasant experiences.

A New Theory of the Perceived Passage of Time

Flaherty (1999) wanted to discover the determinants of the perceived passage of time, which made the relationship between pleasantness of the experience and the perceived speed of time an important secondary issue for him rather than his primary concern. And his work on the perceived passage of time appears to have produced a theoretical breakthrough, one that moves well beyond the thinking on this issue that goes back at least as far as William James's conclusions: *"In general, a time filled with varied and interesting experiences seems short in passing, but long as we look back. On the other hand, a tract of time empty of experiences seems long in passing, but in retrospect short"* (James's emphases; 1918, p. 624). And not only does it move beyond James, it appears to advance beyond more contemporary theories as well (e.g., Hogan 1975; Ornstein 1997).

Flaherty's data and theory indicate that rather than the amount and nature of the objective experiences in a situation, what makes time seem to pass extra slowly or quickly is *the extent to which the individual engages in conscious information processing during the time.* When the amount of conscious information processing is about average for the individual, the individual experiences time as passing at what that individual has come to perceive as the usual rate, but when the amount of such processing is high, time appears to slow down (protracted duration); when such information processing is low, time appears to speed up (temporal compression) (Flaherty 1999, pp. 84–114). Using this model, Flaherty accounted for several paradoxes about the perceived passage of time, and he also interpreted, perhaps explained, Mihaly Csikszentmihalyi's (1975, 1990) findings about the optimal state of personal experience (one of the best of times), the state described as *flow.*

One wonders whether the flow experience, which is very pleasing because it is, after all, "optimal experience" (Csikszentmihalyi 1990, p. 39), might be responsible for why the time-flies-while-you-are-having-a-good-time maxim has been accepted so readily and been believed for so long. Because when one is experiencing flow, one is experiencing, perhaps, the best of times; and while experiencing flow, "in general, most people report that time seems to pass much faster" (Csikszentmihalyi 1990, p. 66). During flow people are having a good time—this phrase trivializes what is actually a profoundly positive experience—and for them time flies. This type of experience, being quite powerful and memorable, may be what people are thinking of when they hear or say that time flies while one is having a good time.

Csikszentmihalyi does indicate that "occasionally" during flow the reverse occurs, what Flaherty calls protracted duration, and time seems to pass more slowly. But the example Csikszentmihalyi used, of ballet dancers describing decelerated time while performing "a difficult turn" (1990, p. 66), might actually be a brief break in the flow. This is because the "difficult" in the description of the turn suggests they would significantly increase the amount of their conscious information processing, which according to Flaherty's theory would produce a perception of protracted duration. And this, of course, is the perception of time's passage that Csikszentmihalyi was describing as occurring *occasionally* during flow.

The explanatory power of Flaherty's theory stems from the concept of cognitive information processing. As used in his theory it includes a wide array of cognitive activities, including the level of cognitive involvement with the self and the situation, and the individual's emotional activation and involvement, especially about one's ability to deal with the situation (Flaherty 1999, 84–114).[9]

For example, as I will recount in greater detail later in the chapter, I was once so overcome with emotion—a combination of profound sorrow and grief —in front of a large lecture class that I could not speak for what seemed like a long time, an uncomfortably long time. And given Flaherty's theory, I am sure that the length of time seemed longer to me than the amount of time that had passed on my watch. I believe the reasons for this are (1) the intensity of my emotions took me by surprise; (2) to some extent things seemed to be out of control, which generated even more emotional engagement, in this case a combination of anxiety and embarrassment; and (3) my mind was racing to find a "solution" (i.e., What should I do now?), something other than simply not talking. Without forcing it, this example fits very well—high cognitive involvement with myself (i.e., What's going on?), strong emotional activation (i.e., the original emotions of grief and sorrow combined with anxiety and embarrassment), and of course concern about my ability to find a solution. All of this produced an experience of protracted duration, the perception that more time had passed than really had, because the amount of conscious cognitive processing abruptly shifted to a much higher level.[10]

I could not manage this situation completely at the level of automatic processing (Ashcraft 1989, pp. 67–70)—because, among other reasons, I care about the classroom experience for both myself and my students. But if I had not cared about such experiences, the protracted duration I experienced might have been less severe or might not have occurred at all. Why? Because if I had not cared, my inability to speak would not have generated as much concern, and though unexpected, the surprise would have been less salient, and my indifference would have resulted in little or no search for a solution. According to Flaherty's model all of this should result in a perception of time passing normally (or closer to it) rather than at a slower pace. So it is the individual's involvement with the situation, not the objective situation alone, that determines one's perception of time passing.

What happened to me in front of that class bears a striking similarity to other situations in which people often experience protracted duration. The prisoner whose sentence spans decades, the recently widowed woman, the soldier isolated in a foxhole, the patient with a long-term illness, all of these frequently experience protracted duration, and of course, all of these are generally unpleasant experiences too, bad times (see, respectively, Brown 1998; Lopata 1986, p. 705; Ambrose 1997, p. 262; and Charmaz 1991, pp. 87–93).[11] Notably, severed connections are a common element in all of these situations, and in these cases the connections severed are those with other people. Even my example is a case of severed connections, for by not talking I had severed what were at least my usual connections with my class. But sometimes connections need to be severed to end the worst of times. Sometimes closure is required.

Closure

Dante placed an ominous sign above the Gate of Hell, the concluding statement on which is the most famous: "ABANDON EVERY HOPE, WHO ENTER HERE" (Dante 1980, p. 22). The souls in Dante's Inferno received any manner of torture and abuse, but what made the time in hell so terrible was not whatever torture was being inflicted upon any of the souls there. What made hell so terrible was that the experience of it would never end. And the sign at the gate gave fair warning: "THROUGH ME THE WAY TO THE ETERNAL PAIN" (1980, p. 22).

But pain, be it in this life or after, makes the experience worse the longer it lasts. And in our lives, too, some pain seems unending, or beyond the hope of ending. Yet for some of these experiences, closure can reduce the pain, if not end it. An example, a collective example, of this occurred immediately after World War I.

Hundred of thousands of soldiers had been killed during the war but had never been buried in marked graves. As a result, a War Graves Commission was established in England whose task was to locate and rebury such of these bodies as they could, providing proper identification when possible (Gilbert 1994, p. 528). Early on, two members of this commission, Henry Williams and commission head Fabian Ware, had an idea. The idea was to bury one of the unidentifiable soldiers in England to represent all of the vast numbers who had perished and lay in unknown graves. Thus was conceived the concept of the Unknown Solider, one of whom was buried in Westminster Abbey on November 11, 1920; another, in the Arc de Triomphe in Paris on the same day, at the same hour (Gilbert 1994, pp. 528–30).

Thus the unknown multitude would be honored, but honoring them was not the main reason for establishing such a memorial. Rather, the primary purpose for creating this institution was to provide solace for the living, for whom an Unknown Soldier's grave "could become a focal point of prayer and contemplation for the hundreds of thousands of parents, widows and children whose loved ones had no known grave" (Gilbert 1994, p. 528). For this was no lone soldier's grave; instead, "for every one of us who had his own dead could not fail to see that they too went with him; that, after two years of waiting, we could at last lay a wreath to the memory of that great company" (Lascelles as quoted in Gilbert 1994, p. 529). That "every one of us" included the parents, the widows, the children of all the unknown soldiers.

The world's Unknown Soldiers embody the worst of times, the memories of the terrible wars and the loved ones who perished in them. But they also connect to better times, to the release from unceasing grief that comes with laying "a wreath to the memory," which allows mourning to pass its course. And more, for connections to the deceased are maintained because the tangible memorial allows not only grieving but also the renewal of memories of better times when the deceased were alive.

But such combinations of the best and worst of times are not uniquely associated with Unknown Soldiers. They can be generated by other monuments too.

In 1989 my wife and I were in Washington, D.C., for a meeting, and during our stay we visited the Vietnam Memorial. As we walked into the memorial, I was struck by the quiet. The cacophony of traffic noise and people talking ended as if we had entered an oasis of stillness—intimating reverence. Only by listening carefully could one even hear the whispers.

Two of my wife's high school classmates had been killed in Vietnam, and she wanted to see their names on the monument. All of the names on the monument were listed in directories that resembled large telephone books. The directories identified which section of wall in the monument contained each individual's name. We found the names of her two classmates in the directories and walked to the appropriate sections of monument wall to see them. We read the names in the sections and came upon each classmate's name in turn. As we found one of the names, I happened to look at my wife. She was standing close to the wall, and her classmate's name was carved in it fairly high up, a foot or so above the level of her eyes. So she raised her hand above her head to the level of the name, tremulously pointed her index finger at it, and said simply, "Here."

It was a solemn moment, and we did not speak. Indeed, the only sounds were distant whispers accompanied by susurrant shoes on the summer sidewalks. I found the experience profoundly moving, but I did not realize then just how deeply it had affected me.

Several years later I tried to talk about the Vietnam Memorial in class. My reason for doing so was to use it as an example of a cultural artifact, one that genuinely connected to matters believed and valued by much of American society. I described the monument and then started to describe my wife pointing to her classmate's name on the wall—and I had to stop. That scene was just too poignant in my mind's eye. It produced too much emotion, and to regain control I stood silent before a class of three hundred students. The room fell silent too, for they understood what was happening, and all was silent for I am not sure how long. Eventually I regained my equilibrium, briefly explained that the story had been too moving for me to continue telling it, and moved on to other material.

The irony of this story is that to this day I know of no one whom I knew personally whose name is carved into that ebon stone. My connection is with people I never knew. It is indirect, through my wife. And because I shared in her moment of reconnection that summer afternoon, the monument became more meaningful to me, much more meaningful. And perhaps it helped me as well as others transcend a sense of grief and guilt over this tragedy that took place a generation ago. It may have brought some closure in that regard, but I also realize that I have never tried discussing that example in class again.

So perhaps the closure comes, not in the complete severing of connections with events and people, but in changing the nature of the connections. Rather than a complete break in continuity, what closure means in cases involving grief and mourning is that one *phase* of the relationship or continuity has ended, and one has moved on to the next. So what has closed, come to an end, is one part of a sequence, moving from A to B or from B to C, not the end of the entire sequence. We cannot continue the same relationships with the dead that we experienced with them while they were living, but we can still maintain a different kind of relationship, albeit a one-way relationship, through memory, ceremony, and ritual. And the signal that one phase has ended its dominance is a crucial part in dealing with the grief and pain accompanying profound loss that leads the mourner into the next phase of the relationship. An example of such movement involved an entire country's relationship with a single individual, and it was described eloquently in the concluding three lines of Carl Sandburg's magnum opus about Abraham Lincoln:

And the night came with great quiet.
And there was rest.
The prairie years, the war years, were over.

(Sandburg 1954, p. 895)

The country's association with Lincoln had not ended, but it had changed and entered a new phase. And just as this change in phase involved pain, as we shall see, changes in phase generally affect the experience of time—for better and for worse.

THE ESSENTIAL IMPORTANCE OF TIMING

Chapter 6 presented the concept of entrainment, describing it in some detail. Entrainment is about what people generally refer to as timing, the relationship between two or more streams of activity. Although Chapter 6 intimated a few possibilities about the consequences of entrainment, that chapter mainly considered the different forms that entrainment might take rather than the effects the different forms might produce. As will be seen here, the different forms produce different experiences.

To understand different forms of entrainment, the concept of phase-angle difference will be reviewed. As described in Chapter 6, phase-angle differences simply refer to whether the phases, the parts of one rhythmic pattern, lag, precede, or coincide with those of another. So within Ancona and Chong's (1996) larger distinction between phase and tempo entrainment, lagging, preceding, and coincident phases identify three general categories of phase entrainment. Sometimes even just a simple dichotomy—coincident or noncoincident, in phase or out of phase—will be sufficient.

A Few Nearly Intuitive Examples

Every driver who has driven onto a modern interstate-style highway (a.k.a. freeway, tollway) has used an entrainment strategy—because the driver's life depended on it. And in this example, there is definitely a right and a wrong entrainment strategy. The right strategy is an out-of-phase one; the wrong strategy, one that is normally fatal, is an in-phase strategy. The two rhythms involved are (1) the flow of traffic on the highway and (2) the rhythm of the car being driven onto the highway, and each rhythm consists of two phases: (1) a vehicle-is-present phase and (2) a vehicle-is-absent phase (i.e., space). These two strategies are diagrammed in Figure 7.1. As Figure 7.1b indicates, the in-phase strategy produces a collision between vehicles, which at interstate highway speeds could easily result in the death of the people in either or both vehicles. The out-of-phase strategy illustrated in Figure 7.1a puts the merging car onto the highway safely because this strategy matches the merging vehicle with a vehicle-not-present phase (space) of the highway's traffic flow. Competent drivers recognize which strategy works and employ it daily. The point is that whichever entrainment strategy is used makes a huge difference for both the merging driver and other drivers on the highway.

Other entrainment strategies must be crafted for unique situations—millions of drivers use the out-of-phase merging strategy daily—but idiosyncratic strategies can be just as effective for the people involved as are those used by millions. An example is the entrainment strategy Kay Napier, vice president of Proctor and Gamble's North American pharmaceutical business, developed to deal with the chemotherapy and radiation treatments she was receiving for her breast cancer. To cope with the treatments in a way that would allow her to continue working, she scheduled the treatments for Wednesday evenings. The side effects from such treatments usually took over a day to develop, so taking the treatments on Wednesday evenings allowed her to work on Thursdays, and if the side effects did begin on Friday, Napier came to work a little late and dealt with the worst of the side effects on the weekends (details from Nel-

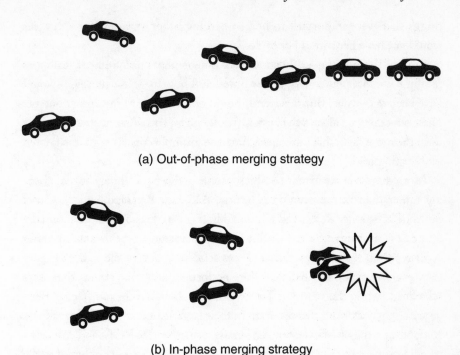

(a) Out-of-phase merging strategy

(b) In-phase merging strategy

FIGURE 7.1. Illustrations of out-of-phase and in-phase strategies for merging traffic flows

son 2001, p. A1). This is a sobering, courageous example, but any of the other choices facing Kay Napier would have been sobering too.

In terms of entrainment strategies, Kay Napier's was an out-of-phase strategy. An in-phase strategy would have had her taking the treatments on the weekend or earlier in the workweek, so she would have experienced the worst of the side effects during the workweek (side effects and work time in phase), likely making it impossible for her to continue working during that portion of the week. Her strategy was to time her treatments so that most of the side effects would occur after the workweek rather than during it (side effects and work time out of phase). The strategy was to have one rhythm, her pattern of side-effects-free and side-effects-present times, lag the pattern of workweek and nonworkweek cycles. Obviously this entrainment pattern did not result in an ideal life—life would have been unpleasant regardless of the strategy over the treatment period—but it did allow Kay Napier to continue doing many things that were important to her, something other entrainment strategies would not have permitted her to do.

Both of these examples illustrate how out-of-phase entrainment strategies produce favorable outcomes, better times, and how in-phase strategies would have been disastrous. But this trend should not be taken to mean that out-of-phase strategies are always the best. Often in-phase entrainment strategies produce the more favorable outcomes. And one example concerns what is known as "morningness."

Morningness is the extent to which people prefer to do things in the morning rather than in the evening (Guthrie, Ash, and Bendapudi 1995). Carlla Smith, Christopher Reilly, and Karen Midkiff (1989) developed a questionnaire to measure morningness and used it to learn that students who were morning types reported that they preferred classes scheduled during the morning more than evening types and believed they performed better in classes that were scheduled during the morning. James Guthrie, Ronald Ash, and Venkat Bendapudi (1995) extended this research. In their large sample of 454 undergraduate students at a major Midwestern university, using Smith, Reilly, and Midkiff's morningness scale, they found that the students who were more oriented to the morning were indeed more likely to schedule courses in the morning, were more likely to study from 6 A.M. until noon, and were more likely to sleep during the intervals from 6 P.M. to midnight and from midnight to 6 A.M. In terms of performance, "students with a morning orientation fared significantly better in early morning classes than those with an evening orientation" (Guthrie, Ash, and Bendapudi 1995, p. 189).

The data from both studies suggest that an in-phase strategy produced the best outcomes for morning-type students. These students preferred activities during the morning, and when they had the freedom to do so, scheduled their activities during the morning. Further, their performance during the early morning (i.e., in classes that began at 8:00 or 8:30 A.M.) tended to be better than that of their evening-oriented counterparts. In fact, evening-oriented students tended to do things later in the day and did better than their morning-oriented counterparts later in the day. Thus an in-phase strategy—scheduling activities when students preferred them—seemed to produce better results (for morning-oriented students, schedule activities during the morning; for evening-oriented students, schedule activities later in the day). Although these studies do not involve random samples of the general population, the quantitative measures of performance allowed the researchers, especially Guthrie, Ash, and Bendapudi (1995), to rigorously examine the potential relationship between morningness and performance.

Studies of such relationships with performance and other outcomes have been conducted on shift work, and a considerable literature has developed regarding this phenomenon. Jon Pierce, John Newstrom, Randall Dunham, and Alison Barber (1989) conducted a major review of this research and, as will surprise no one, found many more negative effects of shift work than positive. Consistent with the argument about entrainment strategies presented in this discussion, Pierce et al. concluded about the negative effects of shift work, "These problems appear to be caused by the incompatibility of the nontraditional work hours with individual and community rhythms" (1989, p. 93). Indeed, Pierce et al. concluded, "the majority of worker problems occur when the worker is out of phase with either established physiological or social rhythms. But when there is harmony between the hours of work and the employee's physical and social rhythms, the level of adjustment predictably increases and the negative consequences associated with shift work [lessen]" (1989, pp. 101–2).

Although Pierce et al. did not use the entrainment concept explicitly, their explanations clearly fall into the entrainment frame (e.g., the phrases "out of phase" and "harmony"). Further, they used as an explanatory mechanism Muhammad Jamal's (1981, p. 536) suggestions about the importance of routine formation in employees' lives, a conclusion they then linked to rotating shifts, which Pierce et al. saw as making it "difficult for people to establish routines" and leading people to "experience a more disrupted life" (1989, p. 101).

Working rotating shifts puts one out of phase with more general social rhythms—such as the 4:00 P.M. to midnight period in which the bulk of traditional community activities normally occur (Dunham 1977, p. 628). And even more than the interface between the rotating-shift worker and the cycle of general community activities, a special challenge is interacting with one's family. Several studies have revealed the increasing number and intensity of problems organizations and work are creating for family life in general (e.g., Bailyn 1993; Hochschild 1989; Perlow 1997), and when one or more family members work rotating shifts the problems and stresses seem to increase (see Hochschild 1997, pp. 145–48, for an example). Families are trying to cope through mechanisms such as day care facilities that provide their services twenty-four hours a day

(Carton 2001), but even with such support facilities, shift work, especially rotating shift work, is stressful.

So shift work in general and rotating shifts in particular stress not only the shift workers but also the communities and families of which they are a part. And as Pierce et al. found, and neatly summarized in a table (1989, p. 99), nontraditional schedules, and especially rotating shifts, are associated with a large variety of physiological and social problems. Rotating shifts would be especially disruptive because they will more frequently upset the phase relationships that are being reestablished, hence leading to more of the problems. One should note that subsequent research continues to associate shift work with these types of problems (e.g., Martens et al. 1999; Totterdell et al. 1995).

Among the problems associated with shift work are sleep problems. These problems include reduced amounts of sleep, difficulty getting to sleep, awakening during sleep more often, and not feeling as refreshed upon awakening (Pierce et al. 1989, p. 94). But as will be discussed next, shift work is not the only phenomenon that causes sleep problems, and sleep problems lead to other difficulties.

Days That Will Live in Infamy

The problems just discussed may directly affect the roughly 25 percent of the U.S. labor force that works under some form of shift work (Pierce et al. 1989, p. 92), and they undoubtedly indirectly affect many shift workers' families and friends through their effect on the shift worker. So the number of people affected is large indeed.

Larger still is another phenomenon that affects every person in the United States and many other countries. This phenomenon is the change into and out of daylight saving time.[12] Overlooked by many as but a minor inconvenience, the human costs of this shift have begun to be cataloged, and they are much more serious and pervasive than perhaps anyone even thought possible when experiments with daylight saving time began—if anyone thought there might be serious problems at all.

The modern form of daylight saving time was proposed in 1907 by an English builder named William Willett, but Germany was the first nation to actually use it, adopting it in 1915, one year before England followed suit in 1916. The United States adopted daylight saving time after it entered World War I, but in 1919 Congress repealed the law that created it. Today the United States moves its clocks ahead one hour on the first Sunday in April (in part because of the efforts of the Daylight Savings Time Coalition [see Chapter 6]) and moves them back one hour on the last Sunday in October (historical details from Stephens 1994, p. 576).[13] Although no longer the espoused motivation, at times countries adopted daylight saving time to conserve energy. Ian Bartky (2000, p. x) indicated, however, that one study found no evidence of net energy savings in the United States resulting from daylight saving time, but it did suggest a link to increased fatalities of schoolchildren during weekday mornings of January and February in 1974, the United States having shifted to daylight saving time during the winter as one response to a major energy crisis.

Even though that experiment was soon abandoned in favor of the cycle in use today—seven months of daylight saving time followed by five months of standard time—questions still remain: Does daylight saving time have effects other than just shifting daylight from the morning to the evening, thereby increasing the possibilities for daylight recreation and leisure activities for much of the population? That seems reasonably benign. But are there negative effects? Is there a cost to be paid for the extra leisure possibilities?

With each study the answer becomes a more certain yes, and the reason is the same reason that shift work, especially rotating shift work, causes problems: Both changes disrupt established entrainment patterns. In fact, the shifts into and out of daylight saving time can be thought of as similar to very slowly rotating shifts, albeit the magnitude of the shift change is only about one-eighth as great as that between rotating shifts (as measured in fungible hours).[14] Perhaps the long interval between the shift changes and the smaller magnitudes of the changes have masked the effects to everyday observers, but systematic observations are beginning to change this.

Timothy Monk and Lynne Aplin (1980) studied the effects of both the spring and the fall daylight saving time shifts on a sample of about one hundred adults in Great Britain. Two findings indicated there might be problems generated by these shifts. First, the disruption in waking time lasted for about one week after *both* the fall and the spring changes. Second, moods seemed to be affected, with the fall change seeming to improve moods, whereas moods deteriorated after the spring change. Note that sleep problems are associated not only with shift work but also with the daylight saving time shifts.

Shifts in sleep patterns and mood sound more inconvenient than threatening, but findings from several additional studies are much more ominous because they are directly related to life and death. Timothy Monk (1980) performed the first of these studies, and he investigated the impact of the spring shift into daylight saving time on traffic accidents. Using weekly traffic-accident data from all of Britain for 1972 and 1973, he compared the change in the number of accidents that occurred in the week before the shift with the number that occurred in the week following the shift. To provide a second base of comparison, he compared the accident statistics for the comparable two weeks in both 1970 and 1971, when there was no spring shift in Britain. His findings? In 1970 and 1971, when *no* shift occurred, the difference between the two weeks was a slight decrease of about 0.6 percent, whereas in 1972 and 1973, the week following the shift into daylight saving time revealed a 10.76 percent increase in the number of accidents in all of Britain.

But Monk did not examine the fall shift because both his earlier work (e.g., Monk and Aplin 1980) and that of others seemed to indicate that the fall shift was beneficial. So Robert Hicks, Kristin Lindseth, and James Hawkins (1983) conducted a study similar to Monk's on traffic accidents in California from 1976 to 1978. Using data on all accidents in the state for one week before and one week after each change in each year, they found that traffic accidents increased after *both* the spring and the fall changes. Over the three years, the number of accidents increased an average of 3.6 percent for the week immediately following the change.[15]

One more researcher took up this question and expanded these findings to yet a third country. Stanley Coren (1996a, b) used data for all traffic accidents in Canada (except Saskatchewan, which did not observe daylight saving time) in 1991 and 1992 to compare the number of accidents on the Monday preceding shifts into and out of daylight saving time with the Monday immediately following each shift. Based on 21,603 accidents for these eight days combined (four from each year), Coren found a statistically significant *increase* of about 8 percent in accidents after the spring change, and a statistically significant *decrease* in accidents of "approximately the same magnitude" (1996a, p. 924) after the shift out of daylight saving time in the fall.

All three studies found statistically significant increases in the percentage of traffic accidents after the shift *into* daylight saving time, which is consistent with the sleep problems experienced by those who work on rotating shifts, but the results for the fall shift are mixed. Hicks, Lindseth, and Hawkins (1983) found an increase in accidents in the fall, but Coren (1996a, b) found a significant decrease. Monk (1980) did not study the effect of the fall shift on traffic accidents.

Two other studies may inform this issue, although they concern daylight saving times' effect on phenomena other than traffic accidents. In the first of these studies, Mark Kamstra, Lisa Kramer, and Maurice Levi (2000) examined the association between the two annual daylight-saving-time shifts on stock market indexes in the United States, Canada, the United Kingdom, and Germany. Several indexes were examined for the United States for periods up to seventy years, with those for the other three countries ranging from twenty-six to thirty years. The investigators examined the indexes on "the first trading day following a daylight saving time change using several different indices" (Kamstra, Kramer, and Levi 2000, p. 1007). The investigators noted previous findings of general weekend effects, so comparisons with the "mean regular weekend" were especially important, and they compared returns from the daylight saving weekends with those from the "average regular (non-daylight saving) weekend," which is the average of all nondaylight saving weekends (2000, p. 1007).

There was a strong *negative* association between both daylight saving time shifts in Canada, the United Kingdom, and the United States. (Negative means the stock markets went down.) Indeed, the negative returns for the spring daylight saving weekend were much larger than those for the means of the nondaylight saving weekends—200 percent to 500 percent greater in the spring, and even larger for the fall-change weekend (Kamstra, Kramer, and Levi 2000, p. 1008). The magnitude and direction of the association for the German exchange was similar to that for the exchanges in the other three countries, but large variance added to the German data by data from 1970 kept the German results from being statistically significant (Kamstra, Kramer, and Levi 2000, pp. 1008–9).

These findings indicate the two daylight saving changes have important statistical associations with stock market behavior, but in substantive terms, how big of a difference do these effects seem to have? According to Kamstra, Kramer, and Levi, "In the United States alone, the daylight saving effect implies a one-day loss of $31 billion on the NYSE, AMEX, and NASDAQ exchanges" (2000, p. 1010).

The other study is one I conducted to see whether people themselves perceive one daylight saving change or the other as more difficult. To do this I

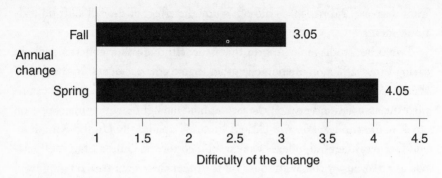

FIGURE 7.2. Average difficulty experienced by student samples for the spring change into daylight saving time and the fall change out of daylight saving time. *Note*: The spring ratings were collected on the Tuesday following the change *into* daylight saving time; the fall ratings, on the Tuesday after the change *out of* daylight saving time.

developed questionnaire scales to measure the perceived difficulty of changing into daylight saving time in the spring and changing out of it in the fall. I administered these scales to two large samples of college students, 406 students in the fall of 1997 and 313 in the spring of 1998. Consistent with designs developed in the other studies of these shifts, the questionnaires were administered to *both samples* on the first Tuesday afternoon following the shift into or out of daylight saving time. Thus all respondents in both samples completed the questionnaires about two-and-one-half days after the time change, which is well within the period in which previous research had found effects related to these changes (e.g., Hicks, Lindseth, and Hawkins 1983; Monk 1980; Monk and Aplin 1980; Monk and Folkard 1976). The results are presented in Figure 7.2.

As Figure 7.2 shows, the spring shift into daylight saving time was perceived as substantially more difficult for respondents than the fall change out of daylight saving time.[16] That the fall change would be seen as the easier of the two—relatively speaking—is not surprising given Monk and Aplin's (1980) finding about a positive mood shift after the fall change. But one would expect a more positive rating about the ease of the fall change if all that were occurring was the addition of an hour of sleep, which Coren (1996a, b) indicated should produce positive effects. The fall change was rated on the easy side of the midpoint, the average of 3.05 falling almost exactly at the slightly disagree (that the change was difficult) anchor on the seven-point scale respondents used to indicate agreement or disagreement with the statements about the difficulty of the change (see note 16). This suggests that although receiving that extra hour of sleep may help—the fall change was perceived as significantly easier to cope with than the spring change—other disruptions are likely to still be involved with that change.

Having now inventoried several serious problems associated with daylight saving time, what is to be done? Notably, the authors of two of these studies explicitly suggested that the changes into and out of daylight saving time may not be worth the costs (Hicks et al. 1983; Kamstra, Kramer, and Levi 2000). Indeed, Kamstra, Kramer, and Levi (2000), based on their own stock exchange research and other findings about daylight saving time, came right out and said it: "An obvious policy implication is to do away with the time change altogether" (p. 1010). One is tempted to agree, because the *unintended* consequences of daylight saving time correspond unsettlingly with the effects obtained or anticipated by those who would deliberately disrupt sleep patterns.

For instance, Aleksandr Solzhenitsyn described a deliberate arrest strategy employed by the KGB: "The kind of night arrest described is, in fact, a favorite, because it has important advantages. Everyone living in the apartment is thrown into a state of terror by the first knock at the door. The arrested person is torn from the warmth of his bed. He is in a daze, half-asleep, helpless, and his judgment is befogged" (1974, p. 6).

Sleep is disrupted, the prisoner is dazed, and "his judgment is befogged." This could be describing a shift worker rotating to a new shift, ergo the maladies inventoried in Pierce et al. (1989). Or it could describe all of us on the day after the shift into daylight saving time.

William Shirer recognized these effects too, and he even saw a way to turn them to military advantage. Then an American foreign correspondent working in Berlin, Shirer wrote this entry in his diary on September 18, 1940:

> Churchill is making a mistake in not sending more planes over Berlin. A mere half-dozen bombers per night would do the job—that is, would force the people to their cellars in the middle of the night and rob them of their sleep. Morale tumbled noticeably in Berlin when the British visited us almost every evening. I heard many complaints about the drop in efficiency of the armament workers and even government employees because of the loss of sleep and increased nervousness. (Shirer 1941, p. 507)

Again, the idea was to disrupt people's sleep patterns so their ability to function would decline. And in both cases, the disruption was intended, not to help them, but to produce harm to those whose sleep would be disrupted. So why do this to ourselves *voluntarily* in the case of daylight saving time?

Humanity constructs its times, and daylight saving time is no exception. Perhaps it is time to rethink the practice and sever our connections with it. Deconstruction may be in order. Should policy makers require further study before ordering its demolition, let the funding agencies support such research, but with connections to deadlines on the order of three or four years, not thirty or forty. And whatever the time frame for conducting the research and making the policy decisions, let it all proceed with a greater sense of urgency than that which moved the Board of Longitude (see Chapter 4). Otherwise, latter-day Dantes will have far too much material to draw upon as they make each new day in their Infernos the first day after the switch into daylight saving time. Mistaking the worst of times for the best of times is a nasty error.

8

Carpe Diem

> The utility of living consists not in the length of days,
> but in the use of time; a man may have lived long,
> and yet lived but a little.
>
> —Michel Eyquem de Montaigne, *Essays*

What is the most impressive project in human history? What project is most impressive, that is, if impressiveness is gauged by the project's complexity and scope, its audacity and importance, and ultimately, of course, by its success. There is no right or wrong answer to this question, value-laden as it is, but one could certainly agree on legitimate contenders: the pyramids of ancient Egypt and Mesoamerica, the Great Wall of China, the Apollo program (moon landings), and the mapping of the human genome would generate few objections. All of these projects were efforts to seize the day (carpe diem), to seize it grandly, and all of them involved *preparations* to seize the day as well—seizing the day and preparing to do so being the subjects of this chapter. But there is another project, little known to the general public, often unknown even to major portions of the professorate, that deserves to be included in this list.

The project was proposed in 1857, and what was proposed was nothing less than a complete inventory of every word in the English language, past and present. But not just a list of these words, though the outcome envisioned would certainly include such a list, but a list that would describe the origin of every word and include the first written sentence in which the word was published. Further, the list would also include additional published sentences illustrating every major meaning the word had taken on as well as the important subtleties in its use. Finally, the project was to be undertaken almost entirely by volunteers (Winchester 1998, pp. 103–7). The task of finding the *first* published use of even a single English word would provide sufficient challenge to most bibliophiles, but for *every* English word that had existed up to the time the project would begin seems almost unimaginable. Yet not only was the unimaginable conceived; it was undertaken. And almost unbelievably, it was accomplished, producing what came to be known as the *Oxford English Dictionary*. The second edition of this work is a de rigueur portion of any major library's reference collection, a blue-clad twenty-volume portion of that collection measuring—according to my tape measure—44.5 inches from A to Z (not counting two supplements). It is a dictionary nearly four feet wide. And the print is small.

The project began formally in 1858, and it did so with some fundamental management errors. For the dictionary's first editor had a pigeonholed storage device constructed to hold records of the sixty thousand to one hundred thousand words he expected to receive from the volunteer workers, the people who were to find the dictionary's words and their initial published uses. He also estimated that the first volume of the dictionary would be produced in two years.

His estimate was a little on the short side, for the first volume was not produced in his lifetime, nor was the entire dictionary produced during his successor's. (It was estimated the task would take ten years when his successor took over.) The first edition of the entire dictionary was produced, 6 million word records and seventy years later, being declared finished on December 31, 1927. It contained 414,825 listed words, 1,827,306 previously published usages, 227,779,589 typeset letters and numbers in all (project summary details from Winchester 1998, pp. 106–12, 219–20). Those who organized and managed this project knew it was a big one, but they clearly had no idea early on that they had undertaken a project of epic proportions.

Nor were they the only ones to underestimate the scope of a big dictionary project. In 1963 Frederic Cassidy proposed a plan to develop a dictionary of American folk speech, a plan that the American Dialect Society then placed him in charge of as the dictionary's editor. And Cassidy explicitly estimated how long the project would take: "We must expect that collecting, to be adequate, must continue for at least five years, and editing, though it may begin before the collecting is completed, will take another three or four years" (Cassidy as quoted in Penn 1999, p. 25). Cassidy originally made that estimate in 1963, and as of his death in 2000 the staff of the *Dictionary of American Re-*

gional English, as the dictionary is titled, was working on volume 4, which was to contain the words beginning with P, Q , R, and some of S (Cushman 2000, p. 15). That is thirty-seven years later and some of S and all of T, U, V, W, X, Y, and Z were still to come.

Do these two examples illustrate a special problem in estimating how long it will take to complete a new dictionary, that lexicographers especially are overly optimistic about their projects? No. It turns out that dictionaries and their lexicographers epitomize the rest of us and our projects. We are *all* overly optimistic about completion times, and we are so most of the time. This characteristic is called the planning fallacy.

Daniel Kahneman and Amos Tversky (1979) first applied the planning fallacy label to this phenomenon, which is defined formally as "the tendency to hold a confident belief that one's own project will proceed as planned, even while knowing that the vast majority of similar projects have run late" (Buehler, Griffin, and Ross 1994, p. 366). Intriguingly, Roger Buehler, Dale Griffin, and Michael Ross's (1994) research indicates that people estimate completion times for other people's projects with little *systematic* error, whereas they tend to substantially *underestimate* the completion times of their own projects. So the planning fallacy does not seem to be a universal planning error; instead, it seems to be focused on estimates for people's own projects. These differential estimates apparently result from using different types of information to make the estimates. For other people's projects, people tend to rely on factors such as deadlines and how long similar previous projects took to complete. But for their own, people seem to downplay, even disregard such information, and instead they focus on the scenarios they envision for their projects, scenarios that tend to be unrealistically optimistic (Buehler, Griffin, and Ross 1994).

So as already indicated, the editors of the *Oxford English Dictionary* and the *Dictionary of American Regional English* were not unusual at all when they underestimated—by an order of magnitude—how long their projects would take. Nor were the builders of the Sydney Opera House, who were off by ten years about its completion date (Buehler, Griffin, and Ross 1994, p. 366).

Such errors in estimating completion dates can be more than just troublesome and embarrassing, because they can lead to timing problems, especially if explicit in-phase or out-of-phase entrainment strategies (see Chapter 7) have been built around the estimates. For example, because key American commanders thought the war with Germany would be over before the end of 1944, winter clothing was stored rather than transported to the front, where it would be needed during the winter of 1944–45—when the war was very much not over (Ambrose 1997, p. 110). Planning forms the basis for decisions, and if the planning is wrong, so are the decisions. And the planning fallacy tends to make at least part of many plans wrong.

Buehler, Griffin, and Ross found that the planning fallacy could be counteracted if people were able to both "consider their past experiences and to relate the experiences to the task at hand" (1994, p. 376). Remembering past experience was not enough; such memories had to be related to "the task at hand" to reduce the planning fallacy.[1] This finding connects well with the general premise about the importance of linking the past with the present and the future (see Chapters 5 and 7; Neustadt and May 1986; Weick 1995). Nevertheless, if people are left to their own devices, at least for their own projects, they seem to focus on an idealized future, one they intend to create, hence days they intend to seize.

But given the planning fallacy, when people plan for themselves, they may unknowingly make plans to seize many more days, many more years, many more decades than they realize their plans commit them to seize—which may not be all bad. Because if a project's length is directly proportional to the amount of effort and resources it will require, many great undertakings might never have been accomplished because they would have been too intimidating to begin.

Carpe diem means *seize the day*, and as Chapter 7 revealed, there are some days people want to seize, to experience, and others they never want to encounter again. Chapter 7 was about *what* makes some times good and others bad, which implies that most people seek out the former and try to avoid the latter. This chapter is about *how* people try to do this seeking and avoiding, *how* they try to increase the proportion of the days that qualify as at least good times if not the best of times, as well as avoiding the opposite as best they can. As will be seen, *how* to seize the good hours and avoid the bad ones often involves complex rather than simple choices. Given finite resources, choices of which days to live imply choices of days to forgo, and such choices of paths not taken can be exceptionally important, because some of them will haunt us for the rest of our lives. But before examining the consequences of forgoing days and hours, we will consider ways people prepare to seize some days and avoid others, dealing first with planning in general, and then with time management.

PLANNING OR ORGANIZING THE FUTURE

Planning is a way we attempt to organize our days before we arrive at them, and as the findings about the planning fallacy indicate, such attempts are seldom error-free. Thus Karl Weick observed about planning, "The dominance of retrospect in sensemaking is a major reason why students of sensemaking find forecasting, contingency planning, strategic planning, and other magical probes into the future wasteful and misleading if they are decoupled from reflective action and history" (1995, p. 30). The overly optimistic completion estimates produced by the planning fallacy are certainly an example of how such "magical probes into the future" can be "wasteful and misleading," especially misleading. Yet Weick did not dismiss planning out of hand because his statement included an important contingency: if the plans are "decoupled from reflective action and history," a point supported by Buehler, Griffin, and Ross's (1994) planning fallacy research.

Buehler, Griffin, and Ross (1994) found that people exhibited no significant planning fallacy tendencies if they were instructed to recall and use their past experiences to construct a plausible scenario for completing an upcoming task. Conversely, people who either were given no instructions about using past experiences or were simply directed to think about past experiences with projects similar to the one they would be estimating revealed significant planning fallacy effects. The plausible-scenario-condition results eliminated a systematic error from the planning process, but not all error.[2] Nevertheless, by demonstrating that some planning errors can be reduced or eliminated by properly integrating the past into the planning effort, Buehler, Griffin, and Ross's results are consistent with Weick's more general point that plans need to be linked to "reflective action and history."

But planning fallacy errors aside, does planning help people and organizations move on to better days? Does planning help them seize the days they intend to seize? One way to answer these questions is to examine the relationships between planning and various measures of performance. For individuals the evidence is mixed, and because just about all of that evidence comes from research on time management practices, it will be discussed later in this chapter in the section on time management. For groups, Gregory Janicik and Caroline Bartel (2001) found that both strategic planning (evaluating the performance environment and developing strategies based on the evaluation) and temporal planning (discussion of time allocation, deadlines, etc.) were positively correlated with group performance, the latter by fostering the development of group temporal norms about matters such as deadlines and being on time. And at the organizational level, until recently reviews of planning research have revealed mixed relationships between planning and organizational performance (e.g., Rhyne 1986; Pearce, Freeman, and Robinson 1987).

To resolve such ambiguous results, C. Chet Miller and Laura Cardinal (1994) performed a meta-analysis on forty-three samples reported in twenty-six studies. This approach allowed them to calculate the average correlations between planning and organizational growth and between planning and organizational profitability, both of which were statistically significant and positive, though relatively small. The authors concluded, "These mean correlations support two conclusions: planning positively influences growth, and planning positively influences profitability" (1994, p. 1656). Additional analysis revealed that variance in the planning-performance correlations was significantly related to a variety of methodological factors.[3]

These are important findings because they help explain three decades of accumulated contradictory findings about planning's impact on organizational performance. And they indicate planning is positively correlated with performance, that planning is associated with higher levels of organizational performance. And given these findings, it is not surprising that J. Robert Baum, Edwin Locke, and Shelly Kirkpatrick (1998) found a similar phenomenon was positively related to organizational growth too, a phenomenon that overlaps a great deal with planning, especially strategic planning. That phenomenon was vision, and the attributes and contents of entrepreneurial CEOs' visions were positively related to their firms' growth.

So the entire planning-visioning complex seems to be positively associated with organizational performance, meaning it helps organizations seize the days more successfully. But it may also be related to something else, something even more important: the evolution of the entire human species.

Visiting the Futures

Richard Alexander examined the evolution of the human line, the hominids, and asked the following question: "What selective challenge could drive the hominid line so far away from that of other primates, and, even more puzzling, what sort of challenge could have caused this divergence to accelerate in its later stages?" (1990, p. 3). In his judgment, traditional forces of natural selection such as climate and predators were inadequate answers to this question, so he considered nontraditional forces. The answer he proposed was that the hominids provided the selective forces for their own evolution, that they had become "so ecologically dominant that they in effect became their own principal hostile force of nature" (p. 4). They provided their own hostile force through intergroup competition, a process that arose at some point in hominid development when hominids "began to cooperate to compete" (p. 4). That is, hominids learned how to cooperate within their groups, in part, in order to compete with other hominid groups. Such a process would provide the selective forces that would produce the evolution of the human brain and important aspects of the human psyche. And in this process cultural and technological differences could confer significant advantages to one group over the other.[4]

From the standpoint of organizing the future, a crucial part of Alexander's analysis is the central role he saw for foresight, planning, fantasizing, and dreaming in the development of human mentality and consciousness (1990, p. 7). As Alexander put it,

> Among other things consciousness implies the ability to think about times
> and places and events separated from our immediate personal circumstances.
> It implies the ability to use information from the social past to anticipate and
> alter the social future, to build scenarios—to plan, to think ahead, and to
> anticipate different possible outcomes and retain the potential to act in several
> alternative ways, depending on circumstances that can only be imperfectly
> represented at the time the plans or scenarios are being made. (p. 7)

Planning in general and scenario building in particular are thus given center stage in Alexander's model of human evolution. But why? What advantages could such abilities confer? From an evolutionary perspective, scenario building, which to Alexander includes "dreaming and daydreaming as well as serious or purposeful planning" (p. 7), is a form of practicing for the future that is much less expensive than actual experience. It allows one to test hypotheses about what will happen, especially in interactions with other people. In Alexander's view the development of human problem-solving abilities had largely involved developing the ability to deal with *social* problems (i.e., issues involved with living among other people in general, not just matters such as crime and poverty, though these would be included too). This makes the primary significance of scenarios their ability to help both individuals and groups deal with other individuals and groups, especially from the standpoint of competition. Competition often involves deception, so detecting and dealing with deception may be a particularly important matter that scenarios can help people address. This is suggested by the importance Alexander assigned to the role of reputation in human interaction, especially for cooperation in groups (pp. 8 and 10), and research on benevolent behaviors supports Alexander's views about reputation (Wedekind 1998; Wedekind and Milinski 2000).

But another advantage to scenario building may accrue as well, and it is social in many ways too. For these advantages come from observing others and the scenarios they construct. In Alexander's view of human evolution, scenario building is so important because it provides an increased ability to "anticipate social situations and the reactions of others to them," which through mental parodies assists the person "in developing responses most self-beneficial when the necessity for social interaction arises" (1990, p. 9). And as with so many other human abilities and characteristics, Alexander described the ability to create useful scenarios as itself varying, the variation being sufficient for some people to specialize in this task and make a living at it, people such as strategic planners.

So the task of planning, either in one's individual life or in organizations, is actually a task that may have played a central role, perhaps one of *the* central roles, in the evolution of humanity. This does not reduce the importance of this task in its organizational context. If anything, its major role in human evolution reinforces and extends its importance. But this is not to say that all planning efforts are equally successful or that all approaches to planning are equally valid. Thus what has become the dominant approach to planning in so many twenty-first-century organizations may be flawed, and a change, not to a brand new approach, but to a tremendously ancient approach may be mandated.

Alexander offered a clue when he referred to people's ability to "generate and use alternative scenarios" (1990, p. 8). The concept of generating and using *alternative* scenarios is the key, and the problem with so much traditional strategic planning is that it focuses on developing a single plan (see Mintzberg 1990 for a review). But a single plan presupposes a single dominant scenario. Hence much traditional planning works to produce a single plan based on a single view of the future, either a view that is the only view, one that is basically an extrapolation of the present, or a view regarded as the most likely future. The problem with this approach is that it leaves so little room for change and adjustment when the future that arrives differs from the future that was anticipated, and the future that arrives is always different. Perhaps because a goal provides guidance for the planning effort, a future state that is to be achieved, it may imply a single plan is needed. But the equifinality principle teaches that the same end can be reached from different starting points and with a variety of means (von Bertalanffy 1952, p. 142), meaning there are "many ways to skin the cat" (Perrow 1984, p. 94).

And there is always more than one possible future. Charles Dickens framed the issue well in the question he had Scrooge ask the Ghost of Christmas Yet to Come: "Are these the shadows of the things that Will be or are they shadows of the things that May be, only?" (Dickens 1984, p. 127). Following complexity theory (Marion 1999), more than one future is always possible, and even if a single future were somehow preordained, it is unlikely that anyone would ever know it. Yet so much traditional planning acts as if there were a single future, hence a single best plan. And by taking such a view, perhaps guided by Frederick Taylor's template that there is always one best way (Kanigel 1997), it abandons an age-old human ability, the ability to "generate and use alternative scenarios."

It is this ability that some companies and planners use, albeit a small number of them, when they engage in what is explicitly known as scenario planning. This approach to planning was first used by military organizations (van der Heijden 1996, p. 15), which fits well with Alexander's emphasis on intergroup competition and the importance of scenarios. After World War II, scenario planning came to be used by business executives as well as generals. This form of planning combines highly probable events with uncertain events to develop several scenarios that portray alternative futures. When this approach is used in conjunction with decision making about projects, those involved try to develop projects that will succeed in several of the scenarios, all of them if possible. Doing so may require some balancing of success probabilities among the different scenarios, hence some tradeoffs, but it is easy to see the difference between this approach to planning and the approach that produces a single, often inflexible plan. A crucial characteristic of this approach is that the scenarios are not ranked in terms of their perceived likelihood, with a single plan being directed toward a most likely scenario. Instead, the multiple scenarios force managers to consider a variety of possible futures (see van der Heijden

1996, pp. 16–17). And as demonstrated by Royal Dutch/Shell, the ability to consider several possible futures can lead to competitive advantages.

Royal Dutch/Shell pioneered the use of scenario planning in the private sector, and according to van der Heijden, it allowed the company to deal with the radical changes facing the energy industry in the 1970s much more rapidly and successfully than many of its competitors. This was because the information about the changes was readily interpreted in terms of the crisis scenario, one of the scenarios the company had developed during its planning to describe a possible future (van der Heijden 1996, p. 18). Not that the crisis scenario was considered the most likely future; rather, the scenario planning effort at Shell had required managers to think through that scenario, along with several others, and to take them all seriously as real possibilities. Doing so created several frames for interpreting information (frames are basically different definitions of the situation, Goffman 1974, pp. 10–11),[5] so when the information about the growing energy crisis started coming in, Shell executives were better prepared to recognize it and interpret it correctly because they had already developed a frame for doing so. They recognized what was happening because, in a sense, they had been there before. By and large, executives at Shell's competitors did not have such frames available for ready interpretation, so they were visiting what to them was unexplored territory.

One could argue that Shell just got lucky because a crisis scenario happened to be included in the set of scenarios developed during its planning efforts. Actually, this is true, but in a different sense than the way such an attribution is usually made. For by skillfully developing several plausible scenarios about the future, Shell increased the odds that it would have thought through the *general* conditions that would actually develop. There is no guarantee in scenario planning that any of the scenarios will actually describe the future, but by creating several scenarios the chances increase that one of them will be close, or at least close enough to help understand it. But as in the case of the Shell example, the point is not really to predict the future exactly; the more important and achievable goal is to provide managers with frames that will allow them to more accurately perceive and understand what is happening. In this way, the scenarios about the future function much like the past was described in Chapters 5 and 7, as a source of meaning for events.

But the key for using them this way is the ability to have several scenarios, hence several frames, available to help people recognize and interpret signals, the information being received. F. Scott Fitzgerald believed that "the test of a first-rate intelligence is the ability to hold two opposed ideas in the mind at the same time, and still retain the ability to function" (1945, p. 69). Seen from the perspective of scenario planning, Fitzgerald was conservative, because multiple frames, at least three of them (the optimal number is in dispute) (Schwartz 1991, p. 233; Wack 1985, p. 146), should be active at any time, and performance above the base "ability to function" is not only desired but usually required. Implicit in scenario planning is the understanding that all times are not the same, for to believe otherwise would mean that scenario planning would be impossible.

Yet scenario planning presents a paradox: If a critically important human attribute is the ability to develop alternative scenarios, as Alexander's work indicates, why is scenario planning, formal or informal, so rare in contemporary organizations? The answer may lie in other forms of time humanity has constructed.

Polychronicity, Clocks, and Planning

Chapter 1 introduced several important forms of time, most notably polychronicity and fungible time, the latter being the form of time that would develop from the mechanical clock and the clock metaphor. Both forms were then discussed at length in subsequent chapters, the clock-based form as fungible time in Chapter 2, polychronicity in Chapter 3, and both forms provide an explanation for the scenario-planning paradox.

Consider the example at the beginning of Chapter 1. The Kaiser wanted to reconsider the German battle plan that would begin World War I a few hours later, but his chief of staff, General Helmuth von Moltke, told him of the plan, "once settled, it cannot be altered." So the German plan remained unaltered, a crucial turning point in world history. Converging in this decision are the two temporal forms, one of which is revealed in the sequence of events. The plan was developed first, and then it would be implemented; task A was followed by task B, and the two tasks would not be mixed.

This is an extreme form of monochronic behavior, the low end of the polychronicity continuum, and as Richard Gesteland (1999, p. 55) and Edward Hall and Mildred Hall (1990) have noted, Germany has traditionally been a very monochronic society, so much so that Hall and Hall would say, "Other Western cultures—Switzerland, Germany, and Scandinavia in particular—are dominated by the iron hand of monochronic time as well. German and Swiss cultures represent classic examples of monochronic time" (1990, p. 14). So in a monochronic culture, one does one task at a time; ergo, one plans first, then one implements the plan. Frederick Taylor would promote this template in his scientific management writings, saying that in his system specialists would plan and employees would carry out the plan: "As far as possible the workmen, as well as the gang bosses and foremen, should be entirely relieved of the work of planning, and of all work which is more or less clerical in its nature. All possible brain work should be removed from the shop and centered in the planning or laying-out department, leaving for the foremen and gang bosses work strictly executive in its nature" (Taylor 1947b, pp. 98–99).

Thus in Taylor's system, a temporal template mandating a prescribed sequence of activities would be buttressed by the organization's structure, its division of labor: specialists who would plan and workers who would carry out the plan.

Notice the singular form: plan. The Kaiser was dealing with a plan, and workers in Taylor's system would each be carrying out a plan. They were not dealing with many plans based on multiple scenarios. Their approach to planning, even if in some phase it involved considerations of alternative futures, leads to a single plan, whereas scenario planning is much more open-ended as it keeps open the possibility of multiple futures rather than closing them. And a single plan, first to develop, then to implement, is more compatible with a monochronic orientation than is an approach that would have people simultaneously aware of several scenarios and moving back and forth among them as they try to make sense out of what is happening. As this description indicates, the scenario-planning approach is compatible with a more polychronic orientation. But with much of northwestern Europe as well as the United States being traditionally monochronic, one would expect planning processes in these countries to reflect a monochronic orientation; and in planning they have, which is, perhaps, a major reason why scenario planning in them has been relatively rare.

Moreover, the clock metaphor reinforced the monochronic approach to planning by providing the ultimate exemplar of good managerial performance: God created the universe, gave it a push to get it going, and because he designed and built it so well, it will operate well forever—like clockwork—without intervention (see Chapter 1). With this image framing the manager's general worldview, it is easy to see how a manager would believe that if the plan were good enough it could be passed on to others who would implement it well without further effort or contact with the manager. But if problems arose, it would reflect poorly on the manager because the plan was deficient. This is an impossible standard for any manager, who after all is not God. Yet managers would apotheosize to just such a status in Frederick Taylor's vision of scientific management. God would plan, and if he planned well in the "planning department," the "workmen, gang bosses, and foremen" would execute well. So it is not without foundation that Robert Kanigel could conclude, "Taylor bequeathed a clockwork world of tasks timed to the hundredth of a minute, of standardized factories, machines, women, and men" (1997, p. 7). And this bequest was monochronic in the extreme.

Both of these temporal forms, a strongly monochronic orientation and the clock metaphor, produce a relatively inflexible plan and inflexible attitudes about the plan, Von Moltke's "once settled, it cannot be altered" being an example of the most extreme inflexible attitude. Not that the absence of any direction is desirable either, as I experienced in Romania a few years ago.

I was in Romania to teach a version of my MBA organization theory class to the faculty who were establishing a business department in their university—the overthrow of the Communist regime having made such an enterprise possible. After I had been in the city of Sibui for four days I commented to one of my American colleagues, "In Romania, the word *tentative* in the phrase 'tentative plan' is redundant." My colleague, who had been in Sibui for several months, laughed and agreed. To check that my perception was not just the impression of an American interpreting things through American eyes, I made the same observation to a Romanian I was working with at the university there. He smiled and exclaimed, "Plans! We have no plans!" This, of course, supported my conclusion as well, the point being that plans and planning were more flexible in Romania than they were in the United States, not that they did not exist at all. Perhaps they were too flexible, but then again, this degree of flexibility may have been appropriate given conditions in Romania at the time.

Overall, the point is that plans and attitudes about them can be too flexible or too strict, and contextual factors such as the amount of change in an organization's environment (Burns and Stalker 1961) may influence what amount of flexibility will yield the best results. For example, the traditional methods of strategic planning may be effective in slow-moving industries (Brown and Eisenhardt 1998, p. 158). But given the volatility present in many industries, an approach such as scenario planning that incorporates a greater degree of flexibility may lead to better results in such environments. Shona Brown and Kathleen Eisenhardt stated this general flexibility dilemma so very well: "Whether it is airlines, space exploration, or pharmaceuticals, the dilemma of strategy in an uncertain, changing future involves balancing between the need to commit to a future while retaining the strategic flexibility to adjust to the future. The most effective managers achieve this balance by straddling between the rigidity of planning for tomorrow and the chaos of reacting to today" (Brown and Eisenhardt 1998, p. 147).

Direction is necessary, not rigidity. And another way to flexibly approach the future while still maintaining direction is to engage the future with a series of low-cost probes. Brown and Eisenhardt's (1998, pp. 147–59) intriguing image of probing the future refers to experimenting—with new markets, new products, and new business partners—but not in anything like a bet-the-company manner, for these are *low-cost* probes. Nor are these the "magical probes" Weick critiqued (quoted earlier in the chapter), for they most certainly involve "reflective action." This is so because many of them fail *usefully* and provide important information to the company, often about what *not* to do; conversely, others succeed and encourage the company to proceed with certain products, partners, and so forth. Most effective when guided by an overall vision of the company, such probing of the future provides strategic advantages to the firm by enhancing its decision making and its timing.

This approach can complement scenario planning. Conceptually there is no reason a continuous program of low-cost future probes should be incompatible with scenario planning, and at a major computer firm both approaches were used extensively. Side by side with a continuous series of future probes, four managers devoted most of their attention to envisioning "alternative future scenarios" and assessing how the company would perform in each of them. A regular monthly meeting served the vital function of integrating the work of the scenario planners with the results of the probes, and involved both planners and operating managers, which further integrated the efforts (Brown and Eisenhardt 1998, p. 153).

So a variety of methods can provide planning flexibility within the framework of an overall direction, and these can be used together effectively if integrating mechanisms are provided to unify them. And planning itself seems to be associated with greater degrees of organizational success. But these conclusions involve planning at the organizational level. What of individuals? What works or is claimed to work for individuals and their attempts to seize the day?

RUMORS OF TIME MANAGEMENT

Many people develop deliberate time management strategies, which they often pass on to others as examples of how they too can successfully manage their time. For example, "When I get up in the morning, before anything else I ask myself what I must do that day. These many things, I list them, I think about them, and assign to them the proper time: this one, this morning; that one, this afternoon; the other, tonight. In this way I do every task in order and almost without effort."

Although this advice reads like a manager's testimonial for to-do lists in the latest time management treatise, it was actually written by Leon Battista Alberti shortly before 1434 (Alberti 1971, p. 180; see Watkins 1969, pp. 3–4). Among other things, Alberti's advice demonstrates that what would today be called basic time management techniques had appeared far before the latter stages of the industrial revolution, and their appearance serves to illustrate the changing attitudes and beliefs about time that appeared during the Renaissance (Quinones 1972).[6]

But what is time management? As authors who write about time management admit, time cannot be managed in the same sense that other resources can (e.g., Mackenzie 1997, p. 13); instead, time management is "the management of the activities we engage in during our time" (Ferner 1980, p. 12). Another way to say this is that time management is self-management (Ferner 1980, p. 12; Mackenzie 1997, p. 13). It may be self-management, but a lot of people seem to be involved, for an entire industry has evolved to help people perform this task and encourage them to make the effort, an industry that includes a large literature of articles and books, a major selection of training programs, and an entire set of products, both paper and electronic, to help organize one's time.

Underlying this entire complex of beliefs, techniques, and prescriptions is the same fundamental assumption that has underlain much of organization-level planning for so long, the belief that it is best to do things monochronically.

The Monochronic Assumption

The most famous story in the time management literature is about a top executive and a consultant and the advice the consultant gave the executive to help him manage his time better. Both Alec Mackenzie (1997, pp. 41–42) and Michael LeBoeuf (1979, pp. 52–53) presented the story in their books on time management, and both accounts included the same basic information. The executive was the president of Bethlehem Steel, Charles Schwab; the consultant was Ivy Lee; and the consultant's task was to respond to Schwab's command to "Show me a way to get more things done with my time" (LeBoeuf 1979, p. 53).

As the story goes, Ivy Lee handed Schwab a piece of paper and gave him the following instructions:

> Write down the six most important tasks that you have to do tomorrow and number them in order of their importance. Now put this paper in your pocket and the first thing tomorrow morning look at item one and start working on it until you finish it. Then do item two, and so on. Do this until quitting time and don't be concerned if you have finished only one or two. (LeBoeuf 1979, p. 53).

Added to this advice for a single day was the further prescription to do this daily from then on. The story then advances to its denouement, which is that after trying it out, Schwab found the advice so wonderful that he paid Lee twenty-five thousand dollars for giving it to him.[7]

The story teaches the importance of making to-do lists and setting priorities, and to teach their importance is the way the story is normally used. But it also teaches that it is best to do things monochronically, one thing at a time. This is the message in the portion of the advice that says "look at item one and start working on it until you finish it. Then do item two, and so on." This is as purely monochronic as it is possible to get if one has more than one thing to do during a day. Interestingly, even Alberti intimated the importance of a monochronic pattern in some of his advice, or at least the dangers of being too polychronic: "And do you know, my children, what I do to prevent one task from interfering with another and finding afterwards that I have started many things but finished none, or perhaps that I have done the worst and neglected the best?" (Alberti 1971, p. 180). What he did, of course, was what was presented earlier: He created a to-do list, with different tasks assigned to different parts of the day. And he did this "to prevent one task from interfering with another and finding afterwards that I have started many things but finished none." His solution was to create a relatively monochronic to-do list, albeit it is hard to tell whether he did so mentally or on paper.

Some contemporary time management writers have even been explicit about this point, although they do not use the term *monochronic*. For example, "*Do you remember* hearing the adage 'To save time, do two or three things at once?' Times change. The wisdom gleaned from so much frenetic activity—and the resulting burnout and slipshod quality—is now: 'Do one thing at a time and do it well'" (emphasis in original; Hedrick 1992, p. 36).

I do not remember the "old adage" Hedrick refers to. Instead, the adage I have heard repeated to me and others many times is to take things one at a time. Not that doing some things monochronically is necessarily wrong—Peter Drucker lauded as "one of the most accomplished time managers I have ever met" a bank president who set aside blocks of time for single activities so he could deal with them uninterrupted (1967, p. 48). And as seen already, the idea of doing one thing and staying focused on that one thing—working mono-chronically—has been an important prescription, explicit as well as implied, in the time management domain for centuries.

But as a *universal* imperative, the monochronic prescription is simply wrong. As presented in Chapter 3, people vary widely in their preferences along the polychronicity continuum, and while advising a very monochronic individual to do things monochronically may be the equivalent of bringing coals to New-castle, telling a very polychronic person to do things monochronically will soon have that person reading the inscription over the entrance to Dante's Inferno, if not passing through it. Telling monochronic people to do everything poly-chronically would be equally wrongheaded and equally debilitating. So just as the greater wisdom about speed is knowing when to go fast and when to go slow (see Chapter 7), rather than trying to do everything monochronically, true wisdom comes from learning which things to do monochronically and which things to do polychronically; wisdom is missing from prescriptions to do every-thing one way or the other.

Prescriptions to one extreme or the other do not just lead to misery, they can also lead to excessive flexibility or inertia. Just as at the organizational level, the monochronic approach often leads to inflexibility regarding one's daily schedule, which may be the reason Carol Kaufman-Scarborough and Jay Lindquist (1999) found monochronic people more likely to defer dealing with tasks that arise after their schedules are planned: because they do not want to change their schedules. Interestingly, Alberti presented a relatively inflexible face when he described his day and plans for it: "In the morning I plan my whole day, during the day I follow my plan, and in the evening, before I retire, I think over again what I have done during the day" (Alberti 1969, p. 172). That he would state that he then followed his plan during the day not only emphasizes the plan's importance, it emphasizes the point that the plan was *followed*. If he is to be taken at his word, Alberti stuck to his daily plans, and as we have already seen, he also claimed to plan things monochronically.

How inflexible can individual plans become? The executive whose time management Peter Drucker admired would allow phone calls only from his wife or the president of the United States to be put through to him if they occurred during one of the times he had planned for focused attention on a single task (Drucker 1967, p. 48). This may have been appropriate for such times, just not for all times. But a too polychronic, too flexible approach is likely to lead to the other problem, the one Alberti described as "finding afterwards that I have started many things but finished none," which is also known as dithering (see Bluedorn, Kaufman, and Lane 1992, p. 23).

So the monochronic-is-best assumption has pervaded individual time management just as it has organizational planning. But even so, time management has been developing for at least six centuries, as the Alberti material indicates, making it reasonable to assess its results.

Does It Work?

To assess time management, two sets of information are needed. First, one needs to know what criteria to use. Second, one needs to know exactly what behaviors and practices to assess. As far as the criteria, these vary, but they seem to be a combination of individual effectiveness and efficiency, the extent to which individuals achieve their goals and how well they use their resources, especially time, to achieve them. Several sources in the time management literature have given primacy to individual effectiveness (Lakein 1973, p. 11; LeBoeuf 1979, p. 17; Reynolds and Tramel 1979, p. 13; Seiwert 1989, p. 2) while acknowledging that being efficient is important too. Others such as Jack Ferner (1980, pp. 12–13) considered both important, whereas Mackenzie (1997, pp. 14–28) included effectiveness and efficiency—identified as progress toward goals and productivity, respectively—among a set of four time management purposes. Overall, Jack

Ferner summed up the criteria by which time management should be assessed: "This [his] book is designed to help you become an effective and efficient user of your time" (1980, p. 13). So the proper measures are (1) Does following time management practices make a person more efficient? and (2) Does following them make a person more effective?

But before beginning the assessment, we need to know what to assess. Not counting the monochronic imperative, which has already been invalidated by data presented and discussed in Chapter 3, what behaviors are prescribed by the time management field? The behaviors can be divided into two categories: fundamentals and tactics. Given my reading of the traditional time management literature and my experience teaching the subject, I believe there are just two fundamentals: (1) know your goals (set them if you do not have any), and (2) use your goals to set priorities about what you will do and how many resources you will expend in the doing. (Stephen Covey [1989, pp. 149–50] reached a similar conclusion, albeit stated differently.) Tactics are easy too because they are everything else, from the universal to-do list to advice to hold stand-up meetings (see Chapter 7). Everything else is a tactic because, by definition, goals define what is important to a person, and in the time management paradigm they should serve as the arbiter for why anything is done. Thus during his tenure as Bethlehem Steel's CEO early in the twentieth century, Charles Schwab should have been using his goals (we will never know for sure—he could have been using someone else's) to decide what the "six most important tasks" were that he had to do tomorrow, for if he was using them, all six tasks would, at least in Schwab's judgment, have helped him attain those goals. Similarly, his ability to "number them in order of their importance," to prioritize them, would have been guided by goals as well, some being more likely to lead to goal achievement than others. This description is, of course, simpler than reality, but it does capture the foundation of the time management paradigm. But does it work?

If we count goal setting as part of the paradigm, the extensive body of research on goal setting presents a compelling case that when done in accordance with several principles, goal setting promotes effectiveness at both the individual and organizational levels (see Locke and Latham 1990). But goal setting did not really develop from the time management literature; this technique to enhance motivation and performance developed largely outside the time management domain, and in terms of scientific research on goal setting, entirely outside the time management domain. So goal setting works, but it is not uniquely a time management technique.

Nevertheless, a look at one of the tactics recommended in the time management literature suggests more than goal setting may work as advertised. The tactic is to identify prime time and match that time with important activities (e.g., Lakein 1973, pp. 48–50).[8] The idea of prime time is that there are certain times of the day when a person is particularly aware and competent, and if a person can learn when these times are, it makes good sense to schedule important activities for them. This time management tactic is supported by the research on morningness that shows such times do exist, although they vary from person to person, and that scheduling important activities for them does enhance performance on those activities (see Chapter 7). And Alberti seems to have anticipated this time management idea too (did he invent all of this single-handedly?):

> With strenuous effort he accomplishes the same thing that earlier and at the proper time would have been easy. Remember, my children, that there is never such an abundance of anything, or such ease in obtaining it, but that it becomes difficult to find out of season. For seeds and plants and grafting, for flowers, fruits, and everything else, there is a season: out of season the same thing can be arranged only with a great deal of trouble. One must, therefore, keep an eye on time, and plan to suit the season; one must labor steadily, and not lose a single hour. (Alberti 1969, p. 172)

The references to things being out of season and that there is a season for everything reflect a more epochal view of time (see Chapter 2), after which Alberti skillfully uses the metaphor to deal with scheduling issues generally.

The whole idea of prime time and the principle of matching the task to the time is a specific example of in-phase entrainment strategies described in Chapter 7. This is the strategy of matching one rhythm with another so that they are in phase. (Doing some things "out of season," an out-of-phase entrainment strategy can be a good one too, such as making phone calls during low-rate periods.) But sometimes prime time is not just there; sometimes it must be created, as was revealed in a study of software engineers.

Leslie Perlow (1999) studied a team of software engineers who worked at a Fortune 500 company, and it would be fair to say that the people she studied had time problems. Their work required collaboration, hence frequent interactions, but their work also required individual efforts, and these efforts were made difficult by the incessant interactions, many of which were spontaneous, that would disrupt the individual efforts. Having observed the problem associated with the team's work patterns, Perlow designed an intervention that would create a new kind of time for the team, a time Perlow and the team called *quiet time* (1999, p. 72).

Quiet time was a time of the day, sometimes two times during the day, during which spontaneous interactions and interruptions from team members were forbidden (making quiet time similar to the periods of uninterrupted time Drucker lauded the bank president for creating). A key attribute of quiet time was that it was scheduled; people knew ahead of time when it would occur. This is so important because Perlow had discovered that having a period free of interruptions was not that useful if the people who experienced the interruption-free period did not know it was coming. If such a period occurred by chance owing to the vagaries of everyone's work activities, it could not be exploited well because the engineers always expected to be interrupted by one or more of their colleagues. Uninterrupted time allowed only the kind of individual work the engineers needed to perform if they knew when to expect it.

So quiet time was not only an uninterrupted interval, but also a *scheduled* uninterrupted interval. And as it turned out, it developed into a series of intervals scheduled on several weekdays that came to be highly valued by the engineers and which seemed to substantially enhance their productivity. So quiet time did not just happen; like so many times it was constructed, socially constructed, and in this case it was socially contracted too. With Perlow's help, the engineers created their own form of prime time, a time in which they knew they did certain kinds of work best. And following the time management prescription for prime times, the engineers learned how to prepare for and plan to do certain types of work that required sustained periods of uninterrupted time.

Perlow's work with the software engineers also illustrates the point made before about the problems that can be associated with either an *unvarying* monochronic or an *unvarying* polychronic basis for work. Before the quiet time intervention the engineer's day could be described as relatively polychronic because the engineers were constantly interrupting each other, forcing each other to weave back and forth among multiple tasks. Instituting quiet time changed that by creating monochronic eyes in the hurricane of polychronic interaction, but it did not transform the workday into a monochronic desert either. Instead, some times were monochronic (quiet time), whereas others not only permitted but encouraged the interaction between team members necessary to do parts of their job.[9] Perlow called this latter temporal form *interaction time* (1999, p. 72). That overall productivity improved when both types of time alternated during a week supports the assertion that neither an unvarying monochronic nor an unvarying polychronic time is optimal, that what needs to be learned is when to do things monochronically and when to do them polychronically. This was a lesson the team of software engineers apparently learned and learned well during the quiet-time intervention. Unfortunately, this learning was less successful at the organization level because quiet time stopped being used after the study ended, even though some of the engineers wanted it to continue. Apparently key elements of the organization's culture, such as the criteria for success, had not changed and these aspects of the culture motivated behaviors that led the practice of quiet time to "disintegrate" (Perlow 1997, p. 124, also pp. 125–28). So even though the engineers had started to seize their days more effectively by using quiet time, after the study ended, quiet time disappeared and the days seized the engineers once again.

Perlow's findings support the time management principle of using and scheduling prime time wisely, and they also support the idea that time management should not be based on an invariant polychronicity. But these are but a single tactic and one major assumption, important as they may be. What about time management as a whole? Does the complex of assumptions, fundamental principles, and tactics—if followed and implemented well—lead people to be more efficient and effective? Given the number of time management books sold and the number of people who have participated in time management training, surprisingly little research has been conducted on this question, remarkably little. And of the research that has been conducted, the results are mixed.

Four studies have dealt with the possible association between the time management complex and individual effectiveness. Two of them examined the effects of the time management complex on college students, and two examined time management's effects on organizational employees.

In the first of the two studies of college students, Therese Macan et al. (1990) developed a questionnaire measure of the time management complex. The measure consisted of four dimensions: (1) setting goals and prioritizing activities; (2) mechanics, such as making lists; (3) perceived control of time, which gets at the extent to which people believe they can affect how time is spent; and (4) preference for organization/disorganization, which is how people prefer to keep their work space and approach projects (Macan et al. 1990, p. 761). Macan et al. correlated each dimension with a series of performance and outcome measures, several of which directly addressed the students' individual effectiveness. Students' self-ratings of performance were significantly and positively correlated with all of the time management dimensions except preference for organization; conversely, students' grade point averages were positively correlated with preference for organization as well as perceived control of time. Job satisfaction was positively correlated only with perceived control of time, and life satisfaction was also positively correlated with perceived control of time as well as time management mechanics. Overall, Macan et al. found that some aspects of the time management complex were positively related to two measures of student performance as well as two types of satisfaction. Interestingly, the fundamentals (goal setting and prioritizing tasks) were correlated only with students' self-rating of performance.

Using a different set of scales, Bruce Britton and Abraham Tesser (1991) measured attributes of undergraduate college students' time management behaviors while they were early in their freshman year. Then they waited four years and correlated the freshman time management scores with the students' grade point averages (GPAs). They found both short-range planning (similar to Macan et al.'s goals/priorities and mechanics factors) and time attitudes (similar to Macan et al.'s perceived control of time factor) significantly and positively correlated with the GPAs. The higher the scores were on both variables, the higher the GPAs.

Both sets of results suggest that at least some parts of the time management complex may have positive impacts on college students' lives. But what about people in the world of full-time work?

Lyman Porter and John Van Maanen (1970) studied the time management behaviors of forty city government administrators and thirty-eight industrial managers. Each manager's superior rated the manager's overall effectiveness, and Porter and Van Maanen found that among the city administrators the *less* effective managers did more planning of their time allocations; but among the industrial managers, the *more* effective managers did more planning of their time allocations and perceived more control over them. These results suggest the effectiveness of time management behaviors may be contingent on the organizational context in which they are used.

The final study of organizational employees is in many ways the benchmark study of the time management complex in the workplace. It was conducted by Therese Macan (1994) on a sample of 353 employees drawn from a public social service agency and a department of corrections system, and it used versions of the same time management scales Macan et al. (1990) used in their study of college students. Supervisors' ratings of each participant's performance served as the measure of individual effectiveness.

There were no statistically significant correlations between *any* of the time management dimensions and employee performance. Neither setting goals and priorities, nor using the mechanics of time management, nor having a preference for organization was related to individual performance. However, perceived control of time was negatively correlated with both job-induced tensions and with somatic tensions (the greater the perceived control, the lower the tensions) and positively correlated with job satisfaction (the greater the perceived control, the higher the job satisfaction). And Macan found the same results in this research when she estimated the structural equation parameters of a theoretical model linking the time management complex to job performance and other outcomes.

Although both the individual correlations and the structural equation estimates revealed no associations between the dimensions of time management and individual job performance, the structural equation model did reveal several important associations. First, the more employees reported having a preference for organization and the more they indicated they set goals and priorities, the more they perceived that they were in control of their time. This is an important link, because perceived control of time was the only variable from the time management complex directly related to any of the outcome variables in the model. It was not related to job performance, but it was related to three other individual outcomes. The relationships were negative with both job-induced and somatic tensions (the greater the perceived control of time, the lower the tensions) and positive with job satisfaction (the greater the perceived control of time, the greater the job satisfaction). Thus the data supported Macan's (1994) model that three dimensions—setting goals and priorities, using mechanics, and preferring organization—influence the fourth, perceived control of time, and the fourth has an impact on two types of tensions and job satisfaction. But nothing from the time management complex was related to individual job performance.

Even if future research should replicate Macan's finding that none of the time management dimensions are related to individual job performance, her findings still indicate that the time management complex is related to several important human outcomes. If, as her model indicates, the time management complex can lead to reduced tension and greater job satisfaction, those results themselves justify support for time management practices. Although such practices may be related to the individual effectiveness of college students, Porter and Van Maanen's findings indicate a highly context-bound association with effectiveness on the job, and Macan's findings suggest there may be no association with individual performance at all. Nevertheless, these few tests of time management's effectiveness do provide sufficiently positive results to indicate that the time management complex may indeed help people manage their activities, to help them seize and experience better days. But which days?

CHOOSING THE DAYS

Eugène Minkowski described two ways people connect with the future. One way is activity: "Through its activity the living being carries itself forward, tends toward the future, creates it in front of itself," while the other way is expectation: "In activity we tend toward the future. In expectation, on the contrary, we live time in an inverse sense; we see the future come toward us and wait for that (expected) future to become present" (1970, pp. 83 and 87). Paul Fraisse (1963, p. 173) described the same two modes. These modes of engaging the future can differ by culture as well as by personality, and Neil Altman's account of his stay in India illustrates the cultural variation well. Altman said of his reaction when he first arrived, "It took a year for me to shed my American, culturally based feeling that I had to make something happen" (Neil Altman as quoted in Levine 1997, p. 204). This is an obvious description of the activity mode, one with an obvious bias toward proactively seizing the day. But after living in India for a year, Altman had shifted to an orientation closer to the one that existed in the part of India where he was living: "Whatever work was going to get done would come to me. By the second year Indian time had gotten inside me" (Neil Altman as quoted in Levine 1997, p. 205). Rather than making things happen, the work would now come to him, which reflects more of an expectation mode, perhaps a mode where, however calmly, the day seizes the individual.

So do we seize the day or does the day seize us? As would seem reasonable, both do some seizing. For example, even Peter Drucker acknowledged that the banker he lauded as one of "the most accomplished time managers" he had ever met "had to resign himself to having at least half his time taken up by things of minor importance and dubious value" (Drucker 1967, pp. 48–49). And Drucker's overall conclusion was "the higher up an executive, the larger will be the proportion of time that is not under his control and yet not spent on contribution" (p. 49). Time "not under his control" indicates a proportion more in the realm of expectation, and speaking of proportions indicates a mixture, which is also important.

Stephen Kern concluded, "Every individual is a mixture of both modes [activity and expectation], which makes it possible for him [her] to act in the world and maintain an identity amidst a barrage of threatening external forces" (1983, p. 90). Perhaps both modes are necessary because both modes involve choices. The choices are easier to see in the activity mode because they are often reflected in overt behavior. But the expectation mode does not necessarily imply passivity (Minkowski 1970, p. 87), so it involves choices too, both about what to do and what not to do. Indeed, the ability to refuse a request, to say no, is often a political matter, especially in organizations (Izraeli and Jick 1986), and saying no or not clarifies one's relationships with others (e.g., one's power status, how powerful one is, at least vis-à-vis other people in the organization, especially compared with a specific individual who asks you to do something, a subordinate being easier to decline than one's boss, ceteris paribus).

The software engineers Leslie Perlow studied illustrate how the ability to say no as well as the lack of this ability influenced the engineers' work. Perlow found that engineers framed work as either "real engineering" or "everything else," with "everything else" meaning things that were mainly activities that involved the engineers with each other interactively, hence in interactive time (Perlow 1999, p. 64). The problem was that all time was interactive time, and because of norms of reciprocity and the knowledge that they had to cooperate with each other to do their jobs, the engineers were apparently reluctant to say no when other engineers wanted to interact. So by keeping them from doing "real engineering," the interactions presumably kept the engineers from doing what they considered their proper work. Interestingly, what quiet time did was say no for the engineers. The engineers did not have to say no individually, because quiet time was socially constructed as a time during which it was illegitimate to approach or interrupt another engineer. This made it possible for the engineers to do "real engineering."

The case of the software engineers illustrates both the activity mode of engaging the future, the creation of quiet time, and the expectation mode, interaction time, where the time dominated the individuals—they could not say no. So creating quiet time helped the engineers seize the day, and interaction time represented the day seizing the engineers. But sometimes the person and the day pass each other by, with neither seizing the other, first just in intervals of minutes and hours, then in months and years; when this happens, the people involved do wish for time to return.

Times That Might Have Been

Michel de Montaigne admonished, "what may be done tomorrow, may be done to-day" (1892, p. 69), because he was very conscious of human mortality and the uncertainty of any person's duration. So if one's life might end tomorrow, then it was best to experience more of life today because tomorrow might never come.

The opposite of Montaigne's admonition is, of course, what we call procrastination: What may be done today, may be done tomorrow. Although it can involve the deferral of tasks for years, procrastination tends to be thought of as a short-term time management strategy, albeit a faulty one, that deals with things on a scale of days and weeks. The procrastination pattern has inspired a body of research (see Van Eerde 1998, 2000), but results are mixed about this behavior's impact on individual effectiveness (Lay 1986; Puffer 1989; Steel, Brothen, and Wambach 2001). Of interest here is the point that procrastination represents a choice, regardless of whether that choice is un-, semi-, or fully conscious, and that choice represents one type of future engaged and at least one other forgone. Further, when procrastination results in a task or project never attempted, in a path never taken, the consequences can last a lifetime. And from the standpoint of the individual psyche, a major consequence is often regret.

Thomas Gilovich and Victoria Medvec have developed a research stream on the topic of regret, and they have focused on how regret is related to both action (errors of commission) and inaction (errors of omission) (see Gilovich and Medvec 1995). But the errors of omission cover a wider domain of decisions than just procrastination decisions that result in things never done. The domain includes deliberate decisions not to do something, but with no pretensions about that something being done later, as is normally part of procrastination behavior. Gilovich and Medvec have found that the things people regret the most in their lives tend to be the things they decided not to do, the errors of omission. Moreover, actions that go astray tend to generate the greater regret in the short term, but the regret associated with inaction tends to be greater in the long term (Gilovich and Medvec 1994). And a collaboration with Daniel Kahneman revealed that major regretted inactions could produce two kinds of emotional responses, a relatively innocuous wistfulness and a more powerful despair (Gilovich, Medvec, and Kahneman 1998).

But why should actions forgone produce either emotional response over the long term? Gilovich and Medvec have developed a series of reasons, and as they note, "there is unlikely to be a single answer" (1995, p. 385). In addition to the cognitive processes they suggest as answers, another explanation is implied by the cognitive processes associated with the planning fallacy. The planning fallacy seems to be produced when people estimate completion times for projects they will undertake *themselves*. In making these estimates people seem to envision idealized futures in which the project proceeds flawlessly as planned; they do not rely on base rates generated either from their own experiences with similar projects or from others' experiences with similar projects.

A comparable type of idealized envisioning may take place as people consider the consequences of actions they failed to take. As Gilovich and Medvec noted, "People tend to idealize many aspects of their distant pasts, and lost opportunities are no exception" (1995, p. 391). As people think about their past and actions they decided not to take, they envision an alternate future from the point at which the action was not taken. This type of visioning can be seen as the same type of visioning that goes on when people estimate completion times for their projects, which is based on idealized visions of the project unfolding without problems or errors. Doing the same thing for an action not taken would also develop a type of planlike scenario for what would have happened if the action had been taken, and the more idealized the plan, the more likely it would be to generate stronger regret feelings. For as Gilovich and Medvec noted, unlike actions, inactions tend to be open-ended, which means the chains of events that can be imagined following from forgone actions are "potentially infinite" (1995, p. 390). Because such event chains are "potentially infinite," the imagined consequences of forgone actions can compound and grow constantly larger in one's mind.[10]

This explanation does share some elements with Gilovich and Medvec's explanation based on differences in imagined consequences, so it may be more of an extension of that explanation than a completely distinct one, yet it is hoped that it provides some new understanding. It has not, of course, been tested empirically, but the potential similarity to the cognitive processes that seem to underlie the planning fallacy enhance its plausibility, not as "the" explanation, but as one factor in an overall set of explanations.

The phenomenon of regret associated with actions not taken, a regret that grows more intense as time passes, illustrates the importance of seizing the right day or of being seized by the right day. But if neither form of temporal engagement takes place, the connections with the day form weakly, if they form at all, and one has a sense of falling behind the time and of being profoundly late—and in the contemporary era such feelings and experiences may be increasing.

Behind the Times

In an era before most of the United States would have recognized the three-letter acronym HMO, Julius Roth described the "career timetables" (1963, e.g., pp. xviii–xix) of patients hospitalized with serous illnesses. One of the salient features of these career timetables was the undesirable experience of falling behind, or at least perceiving that one was falling behind the typical treatment schedule. As Roth observed, "Patients as a group, moreover, maintain a constant pressure to be moved along faster on the timetable, the pressure increasing sharply as patients perceive themselves falling behind schedule" (Roth 1963, p. 61). The patients wanted to catch up, because a sense of being behind the timetable is unpleasant. For example, Barbara Lawrence studied career timetables among a large sample of managers at an electric utility company. She found that those managers who were significantly behind the typical career timetable at the company (those who had not been progressing upward through the hierarchy of managerial positions as quickly as the average manager) were less satisfied with their careers than the managers who were "on time" or "ahead of time," were more likely to wish they were in a different occupation, and were more likely to see their jobs as just a source of income (Lawrence 1984, p. 28).

Roth's insight about perceptions of falling behind leading to pressure to speed up resonates with James Gleick's (1999) observations about a general tendency toward acceleration in contemporary life. It might be that a vicious cycle is created in which people begin to perceive themselves as falling behind a benchmark timetable, so they exert pressure to speed up in order to catch up, which may also accelerate the larger system of which they are a part. This then makes it harder to stay apace and leads to more feelings of falling behind, which leads to more efforts to speed up, which accelerates the system, and so on until either the individual or the system—but probably the individual—collapses.

Several recent analyses of the work and family domains and their intersection and intermingling suggest the vicious cycle may be accelerating and that many people feel they are having to work longer and harder trying to keep up (Bailyn 1993; Hochschild 1997; Perlow 1997, 1998). Whether people in general are actually working longer hours could be debated given recent research findings regarding time use (Robinson and Godbey 1997), but the issue of the number of hours worked reflects thinking in a fungible time frame: "By themselves, minutes-per-day data in the time diary almost defy interpretation. There must be an understanding of the respondent's perception of time" (Robinson and Godbey 1997, p. 229). So number of hours worked may not be the main issue anyway, for just because people may work the same number of hours does not mean they work the same hours—a point manifesting yet again the fundamental principle that all times are not the same. In this case, not all hours are the same, which is indicated by the way people feel about their hours. Many people seem to feel and believe they are falling behind—"are running out of time" (Robinson and Godbey 1997, p. 48)—and many also believe they are having trouble combining the domains of work and family (e.g., Hochschild 1997), which strongly suggests that the hours are indeed not the same. This is critical because as the definition of the situation principle instructs (see Chapter 1), these beliefs and feelings are real in their consequences.

Analyses of the work and family domains also suggest changes in times that could break this vicious cycle, or at least mitigate it if not drive a dagger into its heart. One such change has been discussed, the institution of quiet time in the team of software engineers Perlow studied (1997, 1999). This change not only helped the engineers do their own work better and helped the entire team work together better, but also helped some of them improve the balance between their work and family lives.

Balancing work and family is a challenge, one that seems to call for the creation of new times. The creation of quiet time is one example of a new time that helped people address this challenge; flextime and compressed workweeks may be others (Baltes et al. 1999). So to address problems of family and work time balance, to address the more general matter of a sense of falling behind, indeed, to address the experience of life itself, humanity can exercise the option of creating new times. And these creation efforts can be facilitated by a set of guiding principles about time and the human experience, the presentation of which make up the final chapter.

9

New Times

> Look back on time with kindly eyes,
> He doubtless did his best.
> —Emily Dickinson, *Poems*

"What is time then?" asked Saint Augustine sixteen hundred years ago. Many have offered answers to this formidable question since Saint Augustine posed it, several of which were presented in Chapter 2. But Saint Augustine's concern suggests another line of inquiry guided by two related questions: (1) How did humanity organize time during Saint Augustine's day? (2) How has the human organization of time changed over the last sixteen hundred years? Several temporal concepts and findings presented over the preceding eight chapters allow these questions to be answered with reasonable certainty.

First, epochal rather than fungible time would have dominated Saint Augustine's era (see Chapter 2). The hours of his era were temporal hours, not equal hours, and dates past were reckoned in terms of a particular sovereign's reign (see Chapter 1), all of which tend toward epochal time rather than fungible. So excepting the Julian calendar, most time reckoning occurred in an epochal frame.

Second, sixteen hundred years ago the cultures throughout most of the world likely emphasized polychronic rather than monochronic life strategies (see Chapter 3). Such was the judgment of Richard Gesteland: "Centuries ago when all societies on Earth were polychronic" (1999, p. 58). Edward Hall reached a similar conclusion (Bluedorn 1998, p. 114). So it is plausible to suggest that monochronic time may have been a product of trends that developed during the Renaissance and the industrial revolution. Saint Augustine probably lived in a society whose members interacted polychronically.

Third, one suspects the pace of life was slower given the apparent acceleration of this pace in the twentieth century (see Gleick 1999; Robinson and Godbey 1997)—and there was less concern with being on time. After all, as I quoted him in Chapter 4, Daniel Boorstin has noted, "Since no one in Rome could know the exact hour, promptness was an uncertain, and uncelebrated, virtue" (1983, p. 31).

Fourth, a greater connection with the past would have been likely, both with one's family and ancestors and with the society's past generally. For example, in -46 Cicero proclaimed, "To be ignorant of what occurred before you were born is to remain always a child. For what is the worth of human life, unless it is woven into the life of our ancestors by the records of history?" (1962, p. 395). Cicero's statement supports the idea of a greater connection to the past, hence a greater temporal depth (Chapter 5), and it also concurs with the arguments in Chapters 5 and 7 about the past's importance in creating a meaningful present (i.e., "the worth of human life"). Connections with the future are more difficult to gauge, although general findings reported in Chapter 5 that longer past depths are associated with longer future depths suggest a significant connection with the future too. Certainly Saint Augustine lived 450 years after Cicero, but by then Christianity had spread and was continuing to spread throughout much of the European and Mediterranean worlds. And Christianity included, of course, a prominent concern with the afterlife (i.e., the transcendental future; see Chapter 5). This concern also suggests a connection with a long-term—on a human scale—future.

Finally, Seneca's advice to fit the times certainly indicates the importance of entrainment (Chapter 6) and suggests the importance of phase strategies for obtaining the best of times (Chapter 7) and managing one's affairs (Chapter 8).

Of all of these, the importance of entrainment may be most similar to the times of the twenty-first century, albeit our century probably provides more times to fit oneself to. As for the rest, in many parts of our world, particularly the industrialized, bureaucratized world, temporal depths are much shallower than those in Augustine's time, the pace is faster, punctuality is a greater concern, life strategies are more monochronic, and time is believed to be not just more fungible, but absolutely fungible (even though that is not true). The times differ substantially, which is not surprising because the idea that times differ, that all times are not the same, is a principle that has provided an axiomatic foundation for this entire book.

But do these differences and the axiom they reflect lead to other principles about time and life? Do the findings and ideas presented throughout the eight chapters suggest additional principles, additional conclusions? I believe they do, and far too many to identify and discuss. But even though every potential extension can neither be identified nor be discussed, several important principles and conclusions can be inferred from the basic ideas and findings presented so far, some of which became evident to me as I thought, wrote, and thought about what I wrote.

As I reflected about how the findings and ideas might be extended, the principles and conclusions seemed to group themselves into three domains: principles about maintaining a diversified temporal portfolio, principles about temporal balance, and perhaps the most important of all, principles about creation. So I will discuss these principles by grouping them into sections corresponding to these three domains. And I present them, not as the end of time's story, but as a punctuation mark in that continuing, developing saga.

MAINTAINING A DIVERSIFIED TEMPORAL PORTFOLIO

William Judge and Mark Spitzfaden's research demonstrated that rather than a single time horizon, managers dealt with a portfolio of time horizons. Moreover, their research indicated that just as diversified financial portfolios seem to produce the best results for investors over the long term, so too did more diversified portfolios of time horizons seem to be associated with better organizational performance (Judge and Spitzfaden 1995). All time horizons were not the same in these organizations, and from the standpoint of organizational performance, they should not have been the same. Not only do Judge and Spitzfaden's findings provide an example of the general principle that all times are not the same, but they actually provide an empirical basis for extending it to a normative statement: All times *should not* be the same. And this principle applies to many temporal forms, perhaps all temporal forms, not just time horizons or temporal depth.

During eras when time was reckoned more epochally, this principle was well known, as the following eight verses reveal:

To every *thing there is* a season, and a time to every purpose under
the heaven:
A time to be born, and a time to die; a time to plant, and a time to
pluck up *that which is* planted;
A time to kill, and a time to heal; a time to break down, and a time
to build up;
A time to weep, and a time to laugh; a time to mourn, and a time to
dance;
A time to cast away stones, and a time to gather stones together;
a time to embrace, and a time to refrain from embracing;
A time to get, and a time to lose; a time to keep, and a time to
cast away;
A time to rend, and a time to sew; a time to keep silence, and a
time to speak;
A time to love, and a time to hate; a time of war, and a time of peace.
[italics in original]

(Ecclesiastes 3:1–8).[1]

Drawing on similar theological authority, Abraham Joshua Heschel (1951)
argued that the days of the week were not all alike, that one day should take
precedence over all the others, that day being, of course, the Sabbath. Thus
Heschel concluded, "The Sabbath is not for the sake of the weekdays; the
weekdays are for the sake of Sabbath" (p. 14). Not only is the Sabbath different; it should be preeminent.

But this ancient principle—that not only do times differ, but they should
differ—though not extinct, has been slowly vanishing for centuries, perhaps
since the thirteenth-century invention of the mechanical clock (see Chapter
1). Yet why does this or any other form of cultural homogenization, though
sometimes noted and lamented, receive in the end just a mental shrug of the
shoulders accompanied by the attitude that *this is just the way things are going*?

The attitude about biological diversity is far different, as the virtue of such
diversity is accepted as a truism. Yet scholars working on the challenge of
managing the earth's resources, especially its shared resources, concluded, "Pro-
tecting institutional diversity related to how diverse peoples cope with CPRs
[common-pool resources] may be as important for our long-run survival as the
protection of biological diversity" (Ostrom et al. 1999, p. 282). The "institutional
diversity" refers to methods for managing common-pool resources (e.g., the at-

mosphere, irrigation systems, etc.; see Ostrom et al. 1999, p. 279), which various peoples have developed and then institutionalized. They are important because the successful practices among them were developed to deal with resource management of a particular sort and might be transplanted or modified to deal with similar problems in similar contexts. To the extent that such practices are cultural, and anything that is institutionalized has to be, culture reinforces the group's adaptive solutions and helps transmit them across the generations, a function of culture described by Schein (1992, pp. 11–12). In a similar vein, the times and temporal practices people construct are ways of dealing with problems and circumstances, they are ways of dealing with and adapting to the world. Thus the preservation of some temporal role models may provide examples of how to deal with particular circumstances too. Robert Levine stated this point well: "All cultures, then, have something to learn from others' conceptions of time" (1997, p. 187). All cultures.

Yet ever so slowly times have become more similar, more homogenized. For example, in the United States the celebration of many holidays has been legally mandated to occur on Monday in order to create three-day weekends. But in so doing, these holidays have lost some of their distinctness, because one three-day weekend seems about the same as another. Thus the holidays that have become fused with other days into three-day weekends are not as distinct as they once were.

Another example concerns the Spanish tradition of siesta. On a recent visit to the Costa del Sol region of Spain, I learned that this tradition is slowly disappearing, although I observed it still occurring because retail businesses did seem to close from about 2:00 to 4:00 P.M. But if the siesta is disappearing, all that will replace it will be work hours during this period, making a Spanish workday much like workdays in so many other places.

So it was with dismay that I received the news that the siesta tradition was slowly disappearing, because encountering it was one of the reasons that I had come to Spain, and the image of a future in which this famous temporal pattern was no more was depressing. To have encountered a siestaless workday would have been to encounter a day much like those back in the United States, just as encountering the ubiquitous American fast-food restaurant chains abroad does not accentuate the sense that one is experiencing something different and new.

Perhaps these experiences—seeing so many holidays become indistinguishable three-day temporal archipelagos and learning that if I had waited a few more years I might have missed the siestas—led me to the concept of *temporal conservation*. Conservation is so often thought of in terms of space and material things, such as national parks to preserve the wilderness and practices to conserve energy and natural resources, but it is seldom thought of in terms of retaining temporal practices. The idea is not to save time in the time management sense; rather, it is to save *times*. Or at least to preserve some of them. All of them cannot be retained because doing so would lock life into a temporal cul-de-sac, something as undesirable (see the creation section) as it is impossible.

This is basically what Heschel advocated in the specific case of the Sabbath, and it is worth considering in the case of holidays, hours of the day (i.e., the siesta), and other times too. In this regard, one American holiday may have been spared many of the homogenizing forces that other holidays have faced. That holiday is Thanksgiving.

Thanksgiving is linked explicitly to a Thursday, so it cannot become part of a three-day weekend. Even for those whose activities permit a four-day-long holiday period, such a period is distinctively different from the three-day variety. Thanksgiving Thursday itself has long been associated with a traditional complex of behaviors that includes family gatherings and a meal, the courses of which have traditionally consisted of well-defined foods. And Thanksgiving has so far avoided the commercialization that has become so dominant for Christmas in the United States. Perhaps this is because the day after Thanksgiving has become such a prominent shopping day—for Christmas, the temporal proximity of which may have helped shield Thanksgiving from such an emphasis.[2] Perhaps Thanksgiving remains distinctive because it is the only major American holiday that is linked to a particular day of the week other than a Monday, and that day, Thursday, makes it impossible to create a three-day weekend in conjunction with Thanksgiving. To create a longer holiday period, Thanksgiving must be linked to a different temporal archipelago, thus making Thanksgiving more different from any of the three-day weekend holidays than any of those holidays are from each other.

Somehow, homogenizing the holiday experience defeats the purpose of having a holiday, for a holiday—as its origin as "holy day" reveals—is supposed to be different from the other, the ordinary days. By maintaining its distinctness, Thanksgiving provides a useful model of how a time can be made unique and remain unique. Indeed, holidays should not be the same as ordinary days, and they should not be the same as each other.

Temporal differentiation has several virtues, hence the lesson that all times should not be the same. But disadvantages are possible too, the most obvious being problems of coordination and integration, as described for differentiation in general by Lawrence and Lorsch (1967). Another readily recognizable problem is chronocentrism, the belief that one's times are the true and superior times, vis-à-vis other times (see note 4 to Chapter 6). Lewis Mumford noted this tendency when he observed, "Each culture believes that every other kind of space and time is an approximation to or a perversion of the real space and time in which *it* lives" (Mumford's emphasis; 1963, p. 18). This means chronocentrism often manifests itself in the belief that what is new or more recent is better than what is old or ancient. But despite these and possibly other potential problems, temporal diversification still has several virtues.

One virtue, or at least function, was described in Chapter 1, and that was time's ability to signal in-group and out-group differences (i.e., the different Sabbaths of Judaism, Christianity, and Islam), such differences being signaled to both in-group and out-group members alike.

Another virtue is preparation for temporal change. By definition, change is something different, so if one becomes used to encountering and experiencing different times, ceteris paribus, when one's own times change, these changes should be accommodated more readily.[3] A change in times many of us undergo involves the variable of morningness, our relative preferences for doing things earlier or later in the day (see Chapter 7). As people age, they tend to shift toward the morning, toward earlier parts of the day, and those whose cycles and preferences had once been for doing things later in the day change a greater amount than those who had tended toward the morning originally—although both shift toward the morning (Coren 1996b, p. 93).[4] So we will encounter and be forced to deal with temporal changes, changing times such as changes in the morningness complex, and having learned to deal with different times should help one accommodate such changes—but to learn to deal with different times requires different times for one to encounter.

One can encounter different times in temporal estuaries. Robert Levine used this delightful phrase ("temporal estuary" [1997, p. 203]) to describe the regular, indeed continuous intermingling of cultures that have created distinctive times and temporal practices. His example is the region of southern California and northern Mexico spanning San Diego and Tijuana. This metaphor is based on the idea of two different entities coming together and intermingling or merging. Geographically, an estuary is the region where a river meets the sea, where salt water and freshwater mix. And the cultures of Mexico and the United States have different times indeed, as the discussions of polychronicity (Chapter 3) and pace and punctuality (Chapter 4) certainly made clear.

But temporal estuaries are more than places to encounter different times and practice dealing with them; they are places where new times are born. Vicente Lopez lived in the San Diego–Tijuana region, and he described the culture the Chicano commuters had developed as an "estuary culture," one with salient temporal components: "In an estuary, nature creates a set of organisms which are not from one side or the other, but completely different. In the same way, people who live on the Tijuana border have this kind of estuarian time. It's not a Mexican time. It's not an American time. It's a different time" (Vicente Lopez as quoted in Levine 1997, p. 206).

This wonderful use of metaphor suggests that cultural estuaries can be incubators for many new cultural forms, including new times. The San Diego–Tijuana area has produced just that, and other areas of the world show similar promise. For example, the Costa del Sol region on the Mediterranean Sea in Spain was mentioned earlier in this discussion, and it is a region that is undergoing fundamental change. The creation of the European Union has made it possible for many northern Europeans to retire in this area of Spain, or at least spend their winters there. And the close proximity to Africa—one stands on the Rock of Gibraltar and sees the mountains of northern Morocco—has been a source of not-always-friendly contact for millennia, contact that continues and grows today. The intersection of these cultures in the Costa del Sol makes it a potential incubator too, as other areas likely do elsewhere around the world.

So a virtue of different times, of having times that are not the same, is that contact between them can produce yet other times, new times, times whose possibilities would not otherwise have been anticipated or explored. And because such new times would be the result of contact between parent cultures, visitors from the parent cultures who encounter the new times might be less overwhelmed by them because they would likely contain temporal elements similar to those in the parent cultures' times. If so, these kinds of encounters with new estuary times might lead to the greatest learning, or at least greater learning than if someone from one parent culture visited the other parent culture. This conclusion follows Alexi Panshin's logic in his analysis of science fiction, an analysis in which he argued that the greatest understanding comes from combining something familiar with something unfamiliar rather than combining either the familiar with the familiar or the unfamiliar with the unfamiliar (1968, p. 2). Since estuary time would combine familiar elements from one parent culture with unfamiliar elements from the other, there may be enough similarities so that visitors from either parent culture would be able to understand what was going on and learn from it rather than either seeing nothing new at all or being completely confused. The key factor determining which of these three outcomes results from contact with the estuary time may be the balance of novelty and similarity in the new time.

BALANCE

Time can take many forms, but which of its forms are best? The answer already suggested in several chapters is, it depends. So just as the universal approach of the early management theorists (e.g., Fayol 1949; Gulick and Urwick 1937) was abandoned generations ago, one would expect that the optimal mix of temporal structures and practices will vary by individual, by culture, and by context. An important example is the relative balance given to the past, the present, and the future; a description of the balance among these three was written early in the nineteenth century.

Alexis de Tocqueville (1945a, b) wrote one of the most influential descriptions of American life ever published, *Democracy in America*. Based on his travels in the United States during 1831 and 1832, de Tocqueville developed a detailed sense of Americans in the early 1830s, and one of these details was their attitude about the past, present, and future. He wrote of the typical American at this time: "He is acquainted with the past, curious about the future, and ready for argument about the present" (1945a, p. 328). The positive tone in de Tocqueville's description indicates the author's favorable appraisal of these connections, just as it indicates a relatively even balance among the three as well.

De Tocqueville's description raises two issues. First, is it necessary to assign importance and attention to the past, present, and future simultaneously? Second, is an even balance among all three always the right balance? These questions will be examined in turn.

As data presented in Chapter 5 indicated, the past component of temporal depth in American society was very shallow, at least by the end of the twentieth century. Yet as was argued in Chapters 2, 5, and 7, the past is crucial for making sense of the present, for giving it meaning. The past also provides a critical foundation for making decisions about the future. As Richard Alexander noted, "We know intuitively that to understand how we came to be may tell us things of value about modern human activities, especially those that perplex and frighten us" (1990, p. 1). So the past not only should not, but cannot be ignored. And at least some approaches to planning and large-scale change such as search conferences and future search follow this rule by systematically including attention to the past and history in their methods (Bunker and Alban 1997, pp. 35–36, 47–48).

As for the present, in twenty-first-century America little concern need be given to worries that the present will not be emphasized. For as Lewis Mumford noted about mid-twentieth-century America, "And in fact no generation before our own has ever been so fatuous as to imagine it possible to live exclusively within its own narrow time-band, guided only by information recently discovered; nor has it ever before this accepted as final and absolute the demands of the present generation alone, without relating these demands to past experience or future projects and ideal possibilities" (1970, p. 282). It would seem that the generation Mumford wrote about is no longer alone in holding this chronocentric worldview.

Ironically, even though the present receives attention, perhaps too much attention as a general temporal zone, the kind of attention it receives may be the wrong kind, a point that will be discussed later in the section on creation. But would it be wrong to completely ignore the present and completely focus on some mixture of the past and future? A number of reasons can be given for saying that it would be a mistake to do so, one of which has been given by James March.

March focused on organizations and argued that organizations must do two things: explore new possibilities and exploit old certainties (1991, p. 71). To March, exploration involved long-term future perspectives, whereas exploitation was much more short-term, short-term enough that it is reasonable to regard it as a concern falling in the present. And March used the different temporal depths associated with these two activities as part of his explanation for why organizations may focus too much on exploitation: "The certainty, speed, proximity, and clarity of feedback ties exploitation to its consequences more quickly and more precisely than is the case with exploration" (p. 73). These reasons help explain why managers frequently complain of having trouble finding time to look at the big picture (e.g., Hymowitz 2001). So because exploitation, which includes "refinement, choice, production, efficiency, selection, implementation, execution" (March 1991, p. 71), is necessary and because exploitation takes place in the present, the present cannot and should not go unattended. Although March's focus was on organizations, this conclusion should apply to other groups, and to individuals as well.

But what of the future? What is the case that it is important and deserving of attention in its own right? This case can begin with the received wisdom presented in a proverb: A society grows great when the old plant trees in whose shade they know they shall never sit.[5] And a remarkable, and literal example of this proverb's claim can be found in the society of scholars known as Oxford University. As one of the world's most famous and prestigious universities, Oxford has clearly grown great. And trees planted by those who knew they would never sit in their shade may have played a role: "The oak beams in the College Hall of New College, Oxford, needed replacing in the nineteenth century, so the college cut down some oaks planted in 1386 for that express purpose" (Benford 1999, p. 26). The trees were planted in anticipation of a need centuries in the future—what turned out to be five centuries in the future. Thus Matsushita's 250-year plan (see Chapter 5) would not have daunted the fourteenth-century dons of Oxford. But one can contrast this concern about the future, the deep future, with the temporal depths described in Chapter 5 (see Table 5.1), most of which seldom extended beyond ten years into the future.

So in the age of the mayfly and the nanosecond, the Long Now Foundation's attempt to establish an organization focused on a clock and a library that will endure for the next ten thousand years seems even more remarkable (see Brand 1999). Excepting religious organizations whose concerns can be linked to the eternal, and government projects to store radioactive waste (see Chapter 5), the Long Now Foundation's attempt to span ten thousand years likely exhibits the concern with the deepest future depth of any secular endeavor. I do not include the deep space probes attached to which are messages that may be encountered by alien species or humanity's descendants hundreds of thousands, even millions of years from now. I do not include them because messages were not the probes' *primary* purpose, whereas the Long Now Foundation's clock, library, and the organizational apparatus to support them are the primary purpose of this endeavor.

As these examples indicate, a concern for the future, especially the deep future, seems associated with profound decisions and profound consequences. Thus at the individual level, Alan Straithman et al. found that what they called "consideration of future consequences" (1994, p. 742) had significant effects on actions people favored or opposed, depending upon whether the effects of those actions would occur in the near or long-term future. As would seem to follow, people whose focus is on the short-term future were influenced more by information about what would happen in the short term, and people more oriented to a longer-term future were influenced more by information about what would happen in the long term.

Since the short-term future often is nothing more than a focus on the present, Straithman et al.'s (1994) results indicate that a short-term future focus is not enough; such people need to believe that they will sit in the shade of the trees they plant. At least some people need to be concerned about things further ahead, about trees in whose shade they will not sit for decades, perhaps never. For, as March noted in his distinction between exploitation and exploration, exploitation is about the short term, which is often really the present concern, and involves matters other than exploration, the long-term concern. Both exploitation and exploration are important, which means that neither can be ignored; hence, like the present, neither the short-term nor the long-term future should be ignored.

For this reason I propose a new way to conceive the future, and that is to think of the future as a *temporal commons*. Garrett Hardin published an influential article, "The Tragedy of the Commons" (1968), about destructive tendencies that may occur over the long term when individuals and groups pursue self-interested behaviors when using a commons (a jointly owned resource). A recent examination of large commons, sometimes planetary commons, revealed that the outlook was neither as bleak nor as simple as Hardin's conclusions indicated (Ostrom et al. 1999).

Hardin's original article painted a grim, Malthusian portrait, but Ostrom et al. (1999) noted that since Hardin's original article was published research has revealed that people and groups have developed many approaches for managing commons successfully, sometimes for thousands of years, and much can be learned from these approaches for managing commons in the future. In the case of these commons, the key was to study what had been done, what had worked, and what had not, and by that to reveal a much wider range of possibilities. Hardin got people's attention; Ostrom et al. pointed toward realistic solutions and possibilities.

Ostrom et al. could point to possibilities because of research, and by extension, if we intend to manage, or at least make strategic choices about the temporal commons, much more research is necessary. For example, disturbing data were presented in Chapter 7 questioning the wisdom of shifting into and out of daylight saving time annually. When governments decided to start making these changes early in the twentieth century, those decisions were made based on certain presumed positive outcomes. No systematic research was done until late in the century about the costs of those changes. But only such research can help us make an informed strategic choice about whether we want to retain the practice of daylight saving time as part of our temporal commons. I am persuaded that we should not retain it, albeit I could be persuaded otherwise by research yet to be conducted.

As concluded in Chapter 5, the minimum of humanity's goals is to develop futures in which humanity can live, but the minimum would be unacceptable to most of us because it would certainly allow a Hobbesian world in which life was "solitary, poore, nasty, brutish, and short" (Hobbes 1968, p. 186). We can do better than that, and we should aspire to do so. We should do so for ourselves as well as for our children, for the descendants of the Iroquois seven generations hence, for Matsushita's employees in the twenty-third century, for all who will labor, love, and live under the same sun that has nourished humanity and its hominid ancestors for millions of years. And perhaps even for humanity's successors, who may someday labor, love, and live under the warmth of more than a single proximate star.

We cannot guarantee any future to subsequent generations; indeed, we cannot even guarantee our own future. For those futures will always be determined in part by their residents. But they will also be *partly* determined by us. And this is why the concept of the future as a temporal commons is so important. It is also a major reason why studying time is so important: We are the current stewards of that temporal commons. To conduct that stewardship intelligently and wisely, we need to know more about time than we know currently. Failing to develop a sound knowledge base about one of humanity's, indeed, the cosmos's most fundamental phenomena limits our ability, individually and collectively, to make informed decisions about our stewardship.

In Chapter 8 the discussion addressed the issue of regret, which seems to have been associated with the things not done. So not to study time has the potential to become a major collective regret for future generations, regret for the knowledge that will not be available as well as for the way such knowledge might have influenced stewardship decisions. Although we know some things, we are really just now learning how much there is to learn about how times differ and how those differences affect our lives.

So given what we know to date, it is hard to disagree with a general prescription to pay attention to all three temporal foci, but is an absolutely even balance among the three universally best, a balance apportioning equal importance and attention to each component of this temporal trinity? This is harder to say, but one suspects not (e.g., Brown and Eisenhardt 1998). As individuals and groups alike proceed through their activities, through the course of their life cycles, one suspects that what makes up the optimal balance changes.

Some balances are better than others, and some are outright disasters, as Peter Drucker recognized: "The all too common case of the great man in management who produces startling economic results as long as he runs the company but leaves behind nothing but a sinking hulk is an example of irresponsible managerial action and of failure to balance present and future" (1974, p. 43).

James March argued for a balance of basically the same two foci in noting that exploitation and exploration need to be balanced.

Others have called for changes in the balance regarding specific issues. For example, the president of Mexico, Vicente Fox, noted soon after his election, "The United States has always tried to resolve migration, drug trafficking or trading problems on a day-to-day basis, and we will never solve it in that way. We need to think long-term, 20–30 years from now, where we want to be" (Vicente Fox quoted in Price 2000).

So how can the temporal foci be balanced? It seems unlikely that a satisfactory balance can be achieved if the past, present, and future are not connected, for as has been argued in several places (i.e., Chapters 5, 7, and 8), the foci influence each other. Paule Marshall provided yet one more statement of this point in her novel *The Chosen Place, The Timeless People*: "But sometimes it's necessary to go back before you can go forward, really forward" (1969, p. 359). And Lewis Mumford was even more forceful: "But the notion that the past, instead of being respected, must be liquidated is a peculiar mark of the megatechnic power system" (1970, p. 282).

Yet recognizing the need for connections to work toward a reasonable balance, how can one develop such connections? An image I saw time and again while traveling across the Serengeti may help. One way to describe the ecology of the Serengeti is that some animals are always looking for food while others are always trying to avoid becoming it. What is so remarkable about the Serengeti is that so much of this occurs in plain view so much of the time. And part of this perpetual drama is the behavior of a well-known grazer, the zebra.

Zebras are social and are normally seen in herds, which provides survival value for them. While a herd of zebras grazes, a few individuals in the herd usually pair up and stand side by side. But they face opposite directions. So they are side by side, but one is pointed east, the other west. And by this they serve the herd as lookouts for the big cats, especially the lions. By standing as they do, they can view most of the horizon, and if they see a potential threat getting too close, the lookouts start running and their running alerts the rest of the herd, which starts to run along with them.[6]

But how effective would this system be if both members of a zebra lookout pair looked in the same direction? They would see to the horizon in one direction in greater detail, but nothing in the other direction. Clearly the *balanced* approach of being oriented in opposite directions is a better solution. If it did not work passably well, the lions would have caused the zebras to become extinct long ago.

To me the zebras are a metaphor for balancing the past, present, and future. The zebras who graze while a few stand lookout represent the present, a present that is possible only because some of them take their turn looking into the distance in two different directions, the two different directions representing the past and the future. And perhaps the metaphor extends further than that, because most of us most of the time probably do not devote much attention to matters beyond shallow pasts and futures. Deeper times tend to be left to temporal specialists like historians, archaeologists, futurists, and strategic planners. But time is too important to be left to such temporalists alone.[7] We may have abdicated too much of our temporal responsibility to the strategic planners. So a reasonable reform might be to reassume a sporadic interest in the things that were and the things that may yet be, and take our turns doing so just as the zebras do. Who knows what some of us might see if we occasionally looked outside of now?

And getting oneself to do so may not require a huge change effort. Kathy Marko and Mark Savickas (1998) found that a modest counseling intervention (i.e., about two hours' worth of activity) produced a significant increase in participants' orientation to the future and an increased sense of continuity between the past, present, and future. Although their data did not allow them to estimate how long these changes would last, the results are intriguing. Of course, El Sawy's (1983) experiment also showed that future temporal depth could be modified easily and quickly—at least in the short term (see Chapter 5).

All in all, I suspect the balance in temporal focus would improve if from time to time people would think mindfully about things that happened before they were born and about what things may be like after they die. We do not like to think about death, but thinking about death is not the point. The point is what the world will be like after that. For instance, will there be any trees in it that we planted and in whose shade our great-great-grandchildren may sit? To me this is a pleasing image, not a morbid one. In fact, the image creates a kind of connection to that time, a connection to what may be and how we may make it be. What is imagined may never happen, but we will never know. The habit of thought is key. And perhaps to encourage that habit of thought, in June 2000 the Norwegian government asked the citizens of Norway to devote one hour of a working day to thinking about time and how they use it (Kahn 2000).

As important as it is to balance the three temporal foci, the other temporal forms must be balanced as well. Polychronicity is fundamental, for example, and as described in Chapters 3 and 6, when people diverge from each other too much in terms of their polychronicity, especially when they deal with each other in close quarters, it is challenging for them to get along. What, then, would be the best polychronicity strategy from the standpoint of balance?

There seem to be at least two viable possibilities. One is flexibility; the other is moderation. Edward Hall (Bluedorn 1998, p. 114) noted that some people seem able to adjust to different times daily, if not more frequently, thus to fit the times as Seneca advised. But Hall also noted that other people either could not do this or could do so only to a modest extent. Being able to be this flexible is thus one way to balance polychronicity by, in effect, adopting an in-phase entrainment strategy; that is, staying in phase by adjusting to the predominant polychronicity patterns one encounters.

But as Hall noted, people vary in this flexibility, which brings us to the second answer. If one had to choose just one point along the polychronicity continuum as one's general polychronicity pattern, from the standpoint of balance, the midpoint on this continuum, or better yet, the average of the patterns one will typically encounter would seem most balanced. Why? Because at the average or midpoint people will, almost by definition, be closer to the patterns they will encounter more of the time than will those at any other position along the continuum. This also means that people who can be flexible and adjust to other polychronicity patterns will have smaller adjustments to make, on average, than will those at any other position on the continuum. And making smaller adjustments would presumably be easier and less stressful overall than making larger ones. But as Hall noted, some people are more flexible than others; hence they have more choices about polychronicity and other temporal factors such as speed.

For speed too should be balanced. Problems arising from doing things too fast were discussed in Chapters 4 and 7, and the existence of problems suggests that a proper balance would improve things (see Freedman and Edwards 1988). Perhaps this is a reason the syndrome of behaviors surrounding doing so many things too rapidly has been dubbed "hurry sickness" (Gleick 1999, p. 9). This sickness has deep roots, for de Tocqueville detected a major strain of it during his visit in the 1830s: "He is so hasty in grasping at all [goods] within his reach that one would suppose he was constantly afraid of not living long enough to enjoy them" (1945b, p. 144), and "He who has set his heart exclusively upon the pursuit of worldly welfare is always in a hurry, for he has but a limited time at his disposal to reach, to grasp, and to enjoy it" (1945b, p. 145).

De Tocqueville also observed a corollary to these points about gratifications: "the means to reach that object must be prompt and easy or the trouble of acquiring the gratification would be greater than the gratification itself" (1945b, p. 145). It also seems that aspiring to the prompt and easy extends beyond the economic to the moral sphere, where it is called cheap grace.

Dietrich Bonhoeffer coined the phrase and wrote of cheap grace as "grace without price; grace without cost!" (1959, p. 35). In the words of a contemporary writer, cheap grace is "salvation, or balm for the spirit, that requires little work and absolutely no sacrifice" (Oppenheimer 2000). It is the redemptive version of fast food, although that is unfair to fast food, because fast food is real food. Cheap grace, conversely, is not even real. It is a facade, a melange of kitschy artifacts and sentimentality. It is buying a bracelet rather than taking the time to learn what the bracelet means and then behave accordingly, behavior that traditionally involves "a measure of sacrifice" such as "charity, self-examination and abstention from some worldly pleasures" (Oppenheimer 2000). As Bonhoeffer described it, cheap grace is "forgiveness without requiring repentance" (1959, p. 36).[8]

The issue of cheap grace connects most directly to the theories of justice in the study of ethics. According to the theories of justice, if one person injures or harms another, the person causing the injury must make restitution to make good the injury (Cavanagh, Moberg, and Velasquez 1981, p. 366). For example, my son John once pulled out of a parking space and backed into another parked car. He got out of his car and saw that the collision had damaged the other car significantly. The owner of the damaged car was not in the car, nor was anyone close by who seemed to be walking toward the car. John could have driven away, and no one would have been the wiser. Instead, he wrote a note that explained what had happened and included his name, address, and phone number for the owner of the damaged car; he placed the note under one of the damaged car's windshield wipers. He did this even though he realized his insurance rates would go up as a result of the claim the other driver would file (and they did, by several hundred dollars annually). John's grace was not cheap, but it was real.

Cheap grace lessens the quality of the moral environment, just as the desire for quick fixes in general often lessens the quality of the results obtained. The quest to do too many things too quickly is a sickness indeed. And slowing down may not only help restore balance to the pace of life, but also help restore balance with the temporal foci by aiding in their reconnections. For example, Neil Altman described his experience in the much slower paced society (compared to the United States) of southern India as leading to "a sense of continuity with other times produced in me by the slower pace" (Neil Altman quoted in Levine 1997, p. 204). Somehow slowing down led Altman to connections with other times. So if slower paces lead to more connections, do faster paces sever them?

These considerations should help us make better choices. For in the United States and other parts of the world the realistic choice seems to be between two options described in Chapter 7, options about which my preference should have been clear. I repeat the two options here as a reminder that a choice is possible: "Do everything faster!" (Cottrell and Layton 2000, p. 34); or, "Knowing when to think and act quickly, and knowing when to think and act slowly" (Robert Sternberg as quoted in Gleick 1999, p. 114). The choice is yours.

CREATION

John Hassard noted that *"time is a basic element of human organization"* (Hassard's emphasis; 1989, p. 80). Moreover, *"It is the very fact of man's biological and thus ultimately finite existence that compels him to 'organize' time* [Hassard's emphasis]. As time cannot be conserved nor cultivated, it must be organized. The finite nature of human-time means that it must be sub-divided and prioritized. Because of this, social as well as biological agencies must be *created* in order to harness temporal potential and make it productive [emphasis added]" (p. 80).

Times must be created "to harness temporal potential," to develop temporal possibilities. Some times, some temporal forms, have already been created. For example, Heschel wrote of the Sabbath, "The seventh day is a *palace in time* which we build" (Heschel's emphasis; 1951, pp. 14–15). Note the phrase "which we build." Building is creating, and as the material about temporal estuaries suggests, other times are likely to be created still. Yet to what ends will they be created? Just because something can be done does not mean it should be done, which is one of the most important principles of all.[9] So for what reasons should times be created? Perhaps one answer can be found in child's play.

Many years ago I called home from my office at the university, and my son Nick answered the phone. It was around 4:00 P.M. on a fall afternoon a few weeks after school had started—for both of us—and when I asked what he was doing, he replied, "I was playing." After the conversation I experienced two emotions. The first was joy over the transcendent condition of simple, unstructured play, play with the freedom to go wherever one wants, an activity that is clearly one of the best of times.

But juxtaposed with this was another emotion, one I felt just as powerfully. That reaction was a melancholy sorrow, because I realized that soon in his life such activities and statements would cease. Indeed, I never heard him say "I was playing" again. This wistful reaction reflected my knowledge that progressing into adult life in late-twentieth-century America meant leaving play behind, that at best American adults are allowed to play only within narrowly bounded conditions. After work we are allowed to play sports recreationally, but compare memories of pick-up games in your youth, sports or games of any kind, with the bureaucratically templated—location, hour, and rules—play of adults. Which was more intrinsically satisfying? In which did you experience more joy? So my melancholy reaction was really a form of mourning for both myself and my son; for myself, because Nick's statement led me to realize that the phase of my life in which pure play was permitted, even encouraged, had passed long ago; and for Nick, because I realized that phase was soon to pass. This is one of the worst of times because it leads one to realize that some of the best of times were ending and would no longer be a part of one's life. Bid this time return indeed.

And what is the goal of play? Peter Berger identified it: "Joy is play's intention" (1969, p. 58). Play is the antithesis of alienation, for it is done for its own sake. The joy does not happen at the end of play, it happens *during* play. And one suspects that spontaneous, nonbureaucratic play may be the most joyful of all. For when one's play is organized by others, the others tend to attach to it goals other than joy, goals such as learning or winning. To the extent that satisfaction, the shadow of joy, is contingent on outcomes rather than the activity itself, by that margin the activity differs from true play. As Sebastian de Grazia noted, "Play ceases when at the player's shoulder pallid necessity appears. If starvation or death is the outcome of a contest, then it is neither game nor play" (1962, p. 374). Yet true play by adult hominids is not quite legitimate.

Perhaps if Ecclesiastes 3 had included a verse indicating a time to work and a time to play—and notice how that order seems so natural, work first, then play, a time for play and a time for work seeming to elevate play to too prominent a position—play would have greater legitimacy. Play can certainly be justified instrumentally as an inexpensive way to practice for the future, an interpretation that gives play profound evolutionary significance (Alexander 1990, p. 7; also see Chapter 8). But the very act of giving play this type of instrumental justification distracts from the point that play is its own justification.

Ecclesiastes may not have mandated a time for play, but another prescription from the Hebrew Scriptures comes close, at least mandating a form of time that might be play's first cousin. The prescription is for a sabbatical *year*, and it is presented in Leviticus 25:3–4: "Six years thou shalt sow thy field, and six years thou shalt prune thy vineyard, and gather in the fruit thereof; But in the seventh year shall be a sabbath of rest unto the land, a sabbath for the Lord: thou shalt neither sow thy field, nor prune thy vineyard." Thus was ordained a sabbath year of rest for both the land and the people—though what they were supposed to eat is not addressed. The word *sabbatical* in the phrase "sabbatical year," which is mainly an academic practice in modern times, is etymologically derived from the word *sabbath*. And both of them refer to distinct times created by human beings, times created and constructed to be different from other times. Although not really prescribing play, the idea of both a sabbath day and a sabbatical year is that they are times for doing something other than one's regular work. And perhaps something like a sabbatical is the closest to play that the bureaucracies adults inhabit can deliberately structure, a practice Theodore Zeldin (1994, pp. 355–56) suggests may become a more important institution in organizational life. And as regards pure play, adults will apparently have to do that on their own. But having permission to do so would help.

All in all, perhaps the cumulative moral to be taken from all of these principles is that times should differ, but those differences need to be balanced; and to achieve a better balance, new times will need to be created. But as always the questions are, which times, which balance? Emily Dickinson noted, "I dwell in Possibility" (1960, p. 327), as do we all, because today was one of yesterday's possibilities, only some of which were desirable. The findings presented throughout this book can provide some guidance about what can happen, about the possibilities, and even about what may happen if certain choices are made, but they cannot tell us what we should want to happen. Those criteria come from other sources.

Four hundred years ago Shakespeare wrote, "You waste the treasure of your time" (*Twelfth Night*, ii, 5). But treasuring time is not the point, as important as that point may be. The point is the very meaning of life itself, for the meaning of life is in striving to create times worth living, times worth revering, times worth treasuring. This is the point of studying time.

Appendix

The Temporal Depth Index
and Its Development

The Temporal Depth Index (TDI) is a measure of an individual's future, past, and total temporal depths (i.e., the temporal distances into the past and future that individuals and collectivities typically consider when contemplating events that have happened, may have happened, or may happen; see Chapter 5). The TDI is presented in this appendix, as are instructions for scoring it. An account of the TDI's development, including pertinent psychometric data, is then presented.

BACKGROUND

When this development effort began circa 1990, my motivation for doing so was to develop a scale or scales that would (1) consist of multiple items (to promote psychometric quality) that could be combined into a scale or index score; (2) assess the temporal distances into both the future and the past directions (i.e., temporal depths) individuals typically considered; (3) provide the respondent with a format that could be responded to easily and quickly; and (4) generate responses in a form that could be easily entered into computer databases (i.e., statistical program or spreadsheet files) with little or no coding by the researcher. In my view, when I began this development effort no measure existed that met all four of these criteria. Most studies of temporal depth (usually labeled time horizon) dealt exclusively with the future direction and measured the depth of that horizon with a single item or question (e.g., Javidan 1984; Lindsay and Rue 1980; Tung 1979). A rare exception was El Sawy's (1983) Vista scale, which dealt with both past and future directions, and it did so with a multiple-item approach (i.e., asking respondents to list ten events in each direction along with the dates for each event). T. K. Das (1986) took a similar approach to measuring future horizons. So an approach such as the Vista scale would have satisfied my first two criteria.

However, my experience with psychometric scales led me to believe that the structured format provided by Likert-style and similar scales might be easier and quicker for respondents to complete (criterion 3), and also easier to input during the data-entry process (criterion 4). This seemed to me to favor a Likert-style format (assuming validity and reliability) rather than a more open-ended format, especially in research projects involving lengthy self-administered questionnaires given to large samples (i.e., hundreds of respondents).[1] So I proceeded with the development work with the goal of creating a measure that would meet all four of the criteria already mentioned, and I assumed that to satisfy these criteria, the measure would need to be either a Likert scale or a measure that was similar to the Likert format.

Over several years these efforts produced a six-item scale—three about the future and three about the past—each of which respondents answered by selecting a response from a common set of fifteen temporal depths (i.e., time spans ranging from one day to more than twenty-five years) that indicated the varying lengths of time they used to think about matters in the future and the past. To provide comparability between future and past depths, the items asked respondents about their short-term, mid-term, and long-term future depths as well as their counterparts in the past (recent, middling, and long ago, respectively). This scale is the Temporal Depth Index, a measure of an *individual's* temporal depths, not the temporal depths of organizations or other groups (although it can be adapted for use at the group level, as will be described later).

The Temporal Depth Index is presented in the following section, beginning with the instructions respondents receive about how to complete it.

THE TEMPORAL DEPTH INDEX

This set of questions concerns how you typically consider the past and the future when you make plans or decisions. Please use the following choices to respond to items 1–6 by writing the appropriate number on the blank line in front of each statement. If you use choice 15 to respond to any of the items, also write the specific number of years on the blank after the item.

1=One day	6=Six months	11=Ten years
2=One week	7=Nine months	12=Fifteen years
3=Two weeks	8=One year	13=Twenty years
4=One month	9=Three years	14=Twenty-five years
5=Three months	10=Five years	15=More than twenty-five years

_____ 1. When I think about the *short-term future*, I usually think about things this far ahead.

(If response 15, please write the specific number of years: _____ years)

_____ 2. When I think about the *mid-term future*, I usually think about things this far ahead.

(If response 15, please write the specific number of years: _____ years)

_____ 3. When I think about the *long-term future*, I usually think about things this far ahead.

(If response 15, please write the specific number of years: _____ years)

_____ 4. When I think about things that happened *recently*, I usually think about things that happened this long ago.

(If response 15, please write the specific number of years: _____ years)

_____ 5. When I think about things that happened a *middling time ago*, I usually think about things that happened this long ago.

(If response 15, please write the specific number of years: _____ years)

_____ 6. When I think about things that happened a *long time ago*, I usually think about things that happened this long ago.

(If response 15, please write the specific number of years: _____ years)

Scoring Instructions

Future Temporal Depth. Items 1, 2, and 3 make up the measure of *future temporal depth*, so to calculate the future temporal depth score, add these three items together (i.e., add the numbers from one to fifteen that respondents select as their responses to items 1, 2, and 3) and divide the sum by three.

Past Temporal Depth. Items 4, 5, and 6 make up the measure of *past temporal depth*, so to calculate the past temporal depth score, add these three items together (i.e., add the numbers from one to fifteen that respondents select as their responses to items 4, 5, and 6) and divide the sum by three.

Total Temporal Depth. If both *future temporal depth* and *past temporal depth* have already been calculated following the instructions just presented, those two scores can simply be added together to obtain the total temporal depth score. If the past and future depth scores have not been calculated, to calculate the total temporal depth score, add all six items together (i.e., add the numbers from one to fifteen that respondents select as their responses to items 1, 2, 3, 4, 5, and 6) and divide the sum by three. (Divide by three rather than six so the score obtained will be the same as if the past and future temporal scores had been calculated and then added together.) Note that combining the past and future temporal depth scores into a measure of total temporal depth is consistent with El Sawy's research in which he calculated similar measures ("total Vista spans") by adding together past and future components from the Vista scale (El Sawy 1983, pp. 122–23).

Options

Perceptions of Group Depths. If the research questions involve temporal depth as an attribute of *group* culture (e.g., the culture of a department or organization), the Temporal Depth Index can be modified to obtain perceptions of the group's temporal depths. The modifications are as follows: (1) replace each "I" with "we" in items 1–6, and (2) replace each "you" in the instruction paragraph to the appropriate group referent such as "people in your department." The TDI was modified this way for research on a national sample of publicly traded American companies, and the results produced alpha coefficients for the future, past, and total scales comparable to those obtained for the individual version presented in this appendix (Bluedorn and Ferris 2000).

Saving Questionnaire Space. Each item is followed by the statement "If response 15, please write the specific number of years: _____ years." These statements are included when there is an interest in calculating means precisely in terms of fungible time units such as days or years. If there is no interest in such calculations, these statements may be eliminated to save space on the questionnaire.

THE DEVELOPMENT OF THE TEMPORAL DEPTH INDEX

I developed the Temporal Depth Index over a period of several years spanning the 1990s, a development effect that involved several combinations of different items about the past and future as well as different temporal intervals with which individuals could respond to them. The instrument that resulted from these efforts is the Temporal Depth Index presented at the beginning of the previous section. The TDI's psychometric properties are described in the following section.

Principle components analyses of the Temporal Depth Index.

Temporal Depth Item	Orthogonal (Varimax) Rotation		Oblique Rotation	
	Past Component	Future Component	Past Component	Future Component
Recent past	.20	**.77**	.06	**.77**
Middling past	.14	**.90**	−.02	**.92**
Long-ago past	.11	**.76**	−.03	**.78**
Short-term future	**.91**	.08	**.94**	−.09
Mid-term future	**.94**	.19	**.95**	.02
Long-term future	**.83**	.24	**.83**	.10
Eigenvalue	3.11	1.45	3.11	1.45

Factor Structure

I conducted a principle components analysis on TDI responses from a large sample (listwise N = 362) of University of Missouri undergraduate students and used an orthogonal (varimax) rotation. This analysis produced two components with eigenvalues greater than 1.00 that together explained 75.94 percent of the variance. Knowing that the past and future components might be correlated, I repeated this analysis with an oblique rotation to allow the two components to be correlated. Although the correlation between the two components was .35, the substantive interpretation of the results did not change. The factor loading matrices for both rotations are presented in Table A.1.

These results reveal two factors that are obviously interpretable as future and past temporal depths, and they account for a large proportion of the variance (75.94 percent). As for total temporal depth, a one-factor solution accounted for 51.83 percent of the variance and produced the following six factor loadings: .63, .66, .56, .76, .86, and .80 for the items "Recent past" through "Long-term future," respectively.

Descriptive Statistics and Alphas

The following descriptive statistics were produced in the same sample of undergraduate students that provided the data for the factor-analysis results just presented (listwise N = 362).

Means and standard deviations. The mean for future temporal depth was 7.61, and its standard deviation was 2.13; the mean for past temporal depth was 5.84, and its standard deviation was 1.52; and the mean for total temporal depth was 13.46, and its standard deviation was 3.02. (Note that the mean is 13.46 rather than 13.45 —7.61 plus 5.84—because of rounding differences produced by adding all six items and dividing by three. I calculated the mean this way because the alpha reported in the next section is based on all six items.)

Alphas. The alpha coefficients (Cronbach 1951) were as follows: for future temporal depth, .90; for past temporal depth, .76; and for total temporal depth, .81.

Correlations Between Future and Past Temporal Depth

In the same sample of undergraduate students (listwise $N = 362$) that provided the data for the factor analyses and descriptive statistics just presented, the correlation between the respondents' future temporal depths and their past temporal depths was $r = .35$, $p \leq .001$ (two-tailed test).

Because the relationship between future and past temporal depths is so important, I am also going to report correlations from four other samples that were involved in the development of the Temporal Depth Index. Although the scales used in these samples were not identical to the Temporal Depth Index, they were very similar to it, often differing in just one or two adjectives in items 1 through 6 (e.g., "near future" rather than "short-term future") and having different temporal intervals in the response scale. All of these scales used the same general approach: three items about the future and three items about the past, and all four samples were made up of students at the University of Missouri-Columbia, nearly all of whom were undergraduate students. The correlations and other data about them are presented in Table A.2.

So consistent with El Sawy's (1983) experimental results, which I discussed in Chapter 5, I found that the further into the past individuals reported thinking

TABLE A.2

Correlations between future and past temporal depths in four samples.

Sample	Future–Past Temporal Depth Correlation	Statistical Significance (Two-Tailed Tests)	Sample Size
A	$r = .43$	$p \leq .001$	$N = 184$
B	$r = .31$	$p \leq .001$	$N = 137$
C	$r = .31$	$p \leq .001$	$N = 474$
D	$r = .38$	$p \leq .001$	$N = 381$

about things, the further into the future they reported thinking about things too. And since most of this development work took place with college students at the University of Missouri-Columbia (mainly undergraduates), the positive correlations between past and future temporal depths obtained in this development work make this relationship more general because they were obtained in samples drawn from different populations (college students) than El Sawy's CEOs.

Independence from Temporal Focus

Temporal focus is the degree of emphasis on the past, present, and future (Bluedorn 2000e, p. 124), and following an uncritical intuition it would be easy to assume that the more a person focuses on the future, the longer that person's future temporal depth will be. So to show that temporal depth is different from temporal focus, that the two concepts are indeed conceptually and empirically distinct, I examined their relationships in the sample of student data that produced the factor analyses presented earlier in the appendix. (The listwise N was 361 for the analyses that follow.)

I calculated the correlation between each of the three future temporal depth items and their respective counterparts among the following three questions: "For planning and decision making, how important do you consider the (1) Short-Term Future, (2) Mid-Term Future, and (3) Long-Term Future?" Respondents could answer each of these questions on a seven-point scale ranging from 1 = Unimportant to 7 = Extremely important. The three correlations were as follows: short-term future, $r = -.04$, ns (two-tailed test); mid-term future, $r = -.03$, ns (two-tailed test); and long-term future, $r = .08$, ns (two-tailed test). Since how important a person rates each of these regions of the future indicates the degree of emphasis the person gives to that region, each of the three questions taps the respondent's temporal focus about the respective region of the future. The three correlations, none of which were significantly different from zero, thus support the proposition that temporal depth and focus are distinct concepts.

This conclusion was also supported when I repeated this same type of analysis on the three past temporal depth items, which I correlated with their respective counterparts from the following three questions: "For planning and decision making, how important do you consider the: (1) Recent Past, (2) Middling Past, and (3) Long-Term Past?" The three correlations were as follows: recent past, $r = .11$, $p \leq .05$ (two-tailed test); middling past, $r = .00$, ns (two-tailed test); and long-term past, $r = -.04$, ns (two-tailed test). Although one correlation was statistically significant (for the recent past), it was very small ($r = .11$), and the other two correlations were not statistically significant. So overall the temporal depth items were generally unrelated to the corresponding questions that indicated temporal focus, which indicates that temporal depth and focus are distinct concepts.

But there is more evidence that supports this same conclusion. In data from another large student sample (listwise N = 380) from the University of Missouri-Columbia, I correlated a version of the future Temporal Depth Index from the development process (the differences from the TDI were that the adjective *near* was used in the first future-depth item rather than the word *short-term*, and the intervals in the response categories were different, alpha = .83) with Jean-Claude Usunier and Pierre Valette-Florence's (1994) Orientation Towards the Future scale (M = 6.09, SD = .80, alpha = .88). I also correlated a version of the past Temporal Depth Index from the development process (the three items were the same, but the intervals in the response categories were different, alpha = .76) with Usunier and Valette-Florence's Orientation Towards the Past scale (M = 4.27, SD = 1.26, alpha = .80). The correlation between the two future scales was r = .15, $p \leq$ .01 (two-tailed test), and the correlation between the two past scales was r = .06, ns (two-tailed test). The Orientation Towards the Future and Orientation Towards the Past scales indicate temporal focus on the past and future. Although the two future scales were correlated, the correlation, r = .15, is far too small to argue that the concepts they measure are the same. These results parallel those I found in another sample (Bluedorn 2000e, p. 124), again using an earlier version of the TDI. They are also similar to El Sawy's (1983, pp. 126–27) results, which were discussed in Chapter 5, albeit El Sawy's results, if anything, support the independence (i.e., noncorrelation) of the two variables even more strongly. (Note that El Sawy's 1983 work did not use the term *temporal depth*, even though much of that work is basically about this phenomenon.)

Taken together, these tests in three different large samples, combined with similar results from other research (i.e., Bluedorn 2000e; El Sawy 1983) all reveal little or no association between the measures of temporal depth and temporal focus, thereby supporting the view that temporal depth and temporal focus are distinct concepts, both of which should be investigated. These results also indicate that what is discovered about one variable (i.e., temporal depth or temporal focus) cannot be assumed to be true about the other.

Notes

CHAPTER 1   *All Times Are Not the Same*

1. Although it is often assumed that days and nights are equally long on the equinoxes (e.g., Gimpel 1976, pp. 167–68), they are not. For example, at about 40 degrees latitude daytime and nighttime are equal about four days before the equinox in the northern hemisphere and about four days after the equinox in the southern hemisphere (Steel 2000, pp. 372–73).

2. A picture of a watch from the French Revolutionary period like the one described in the text is presented in both Aveni (1989, p. 145) and Barnett (1998, p. 145). Less imaginative, or at least less efficient, clock designs employed separate clock faces, each with one dial, to tell time in both systems (see Lippincott 1999, p. 149).

3. The term *escapement* is used here to refer to the entire mechanism of controller and escapement, which, as David Landes noted, "were linked and influenced (controlled) each other," so the term *escapement* was applied "understandably" to "the whole device" (1983, p. 11). According to Landes, the great breakthrough was not just the escapement but the "use of oscillatory motion to divide time into countable beats" (Landes 1983, p. 11). The pendulum in a pendulum clock provides an easily observable example of such oscillatory motion, although the pendulum clock was not invented until the seventeenth century (Andrewes 1994a, pp. 124–25).

4. In this translation and many others made in the twentieth century, the Italian word *orologio* was given in English as *clock*. But in the nineteenth century, many translators chose *horologe* instead. *Horologe* is an English word that refers to the general category of timekeeping devices that can tell the hour (i.e., clocks, sundials, hourglasses, etc.); it does not refer solely to clocks. Indeed, the word's etymology traces back via Old French to the Latin *hōrologium*, and from there to the Greek *hōrológion* (Barnhart 1988, p. 491; Weekley 1967, p. 727), both meaning basically the same thing as the English *horologe*. The earliest record of the word in English is its use as the surname *Orloge* in 1266 (Barnhart 1988, p. 491), which may suggest the name-bearer's occupation had something to do with making or tending to devices that could tell the hours. The *Oxford English Dictionary* (2nd ed.,

under *horologe*) cited Wyclif as having used this word with the spelling "oriloge" in 1382, which is very similar to the Italian version (*orologio*). So as literal translations, *horologe* and *horologes* can certainly be justified. Examples of these choices are found in Henry Wadsworth Longfellow's (Dante 1879) translations of a line from Canto 10 of *Paradiso* and of the same passage quoted in the chapter, respectively: "Then, as a horologe that calleth us" (p. 50), and "And as the wheels in works of horologes / Revolve so that the first to the beholder / Motionless seems, and the last one to fly" (p. 116).

Which is the better choice? Although *horologe* may be the more literal choice, because instruments such as sundials and hourglasses do not contain wheels, let alone wheels that move at strikingly different speeds, the mechanical clock seems likely to be the device Dante had in mind in the simile about wheels in horologes. But the uncertainty about his reference illustrates one of the problems of identifying early appearances of mechanical clocks. The use of the generic term *horologe* makes its specific referent ambiguous, so the appearance of the word *horologe* by itself does not necessarily mean the writer was referring to a mechanical clock (see Barnett 1998, p. 68). To make this determination, especially in what appears to be the transitional period, the record must provide other clues such as a physical description of the device itself, just as Dante provided the clues about slow and rapidly moving wheels.

5. Otto Mayr (1986, p. 38) cited the following passage: "For if someone should construct a material clock would he not make all the motions and wheels as nearly commensurable as possible? How much more [then] ought we to think [in this way] about that architect who, it is said, has made all things in number, weight, and measure?" (Mayr's bracketed words). He attributed this passage to a treatise written by Oresme entitled *De commensurabilitate vel incommensurabilitate motuum celi*, which he believed was written in the 1350s.

6. Bede (1969, p. 44) also used the form *anno Dominicae incarnationis*, which has been similarly translated as "in the year of our Lord."

7. A further illustration of the socially constructed nature of time is the issue of how to designate the years. In the West the Dionysius Exiguus-inspired Bedan *distinction* between B.C. and A.D. has been used extensively for a thousand years, albeit the use of B.C. itself was not common until after "another Dionysius," Dionysius Petavius, argued for the use of *ante Christum* (before Christ) in 1627 (Steel 2000, p. 114). Recently some chroniclers have begun using C.E. and B.C.E. to designate the years (Common Era and Before the Common Era, respectively) because they feel the references to Christ in the A.D. and B.C. designations are culturally biased. Gregory Benford (1999, p. 209) has argued that this approach only

camouflages the issue because when addressing the issue of "what is common about our Common Era," the answer is year reckoning from the birth of Christ, thereby returning one to the original issue. Further, Benford advocated a "Deep Time Perspective" (see Chapter 5 for a discussion of this concept), which he believed would be inhibited by the confusion that would be generated by a change to the Common Era nomenclature. Duncan Steel took a different tack. He noted that most of the labels in the West's dating system (e.g., names of the days, names of the months, etc.) take their origins from various sectarian sources, and as such, all should be changed if one objects to the religious sources of the labels. To single out the year designation is, in his words, a "selective argument" (Steel 2000, p. 111).

Let me propose a third take on this issue. If one really wanted to secularize the system for designating the years but avoid major work and confusion, there is a simple solution based on the philosophy of science principle of parsimony (roughly, the simpler solution is the better solution). That solution would be to simply report the year A.D. 2000 (or 2000 C.E.) as 2000. Drop any designation at all for the years that have previously been designated A.D. or C.E., and let the convention develop that a year number without a designation is in the A.D. or C.E. category. I have done this several times in the chapter already, and I suspect that no one even noticed. Take a look at the years 1266 and 1382 in note 4 to this chapter for examples. Similarly, for the years in the B.C. or B.C.E. category, rather then appending such designations, refer to the year 4 B.C. as the year −4. I am not sure that I advocate this change, being sympathetic as I am to Steel's and Benford's point that the B.C. and A.D. designations are widely used and, more important, widely understood, and have been for a long time. But if a change appears necessary, let's make the system simpler rather than more complex. Please note that I will generally follow my own proposed system—using "−" rather than "B.C." and giving a year without further designation rather than using "A.D."—throughout the rest of the book; that way we can all see how it works.

Readers familiar with the system astronomers use to assign dates (Steel 2000, p. 113) will recognize the principle I am proposing, except I am not suggesting the addition of the year 0, which the astronomers' system includes. The astronomers' system makes 1 B.C. in the Dionysian system the year 0.

8. This phrase, *que no son todos los tiempos unos* (Cervantes 1831, p. 541), has been rendered into English with amazing variety. For example, unlike the translation I have used, Alexander Duffield's version does not include the word *for* (Cervantes 1881, p. 301), which is likely the most modest difference among the set of this statement's translations into English. The English version I have used is from the

translation by Henry Edward Watts (Cervantes 1898, p. 270), and it may be found on page 270 of volume 2 of that translation, spoken by Sancho.

9. The fundamental importance of possibilities was suggested to me in a statement the character Q made to Captain Picard near the end of the final episode of *Star Trek: The Next Generation*. Q said, "That is the exploration that awaits you . . . charting the unknown possibilities of existence" (Moore and Bragga 1995). Then I encountered the following line from an Emily Dickinson poem, "I dwell in Possibility" (Dickinson 1960, p. 327), and realized that so do we all, which reinforced the point about the importance of possibilities. Although this entire book can be seen as a treatise about possibilities, Chapter 9 is particularly devoted to them and deals with them most explicitly.

CHAPTER 2   *Temporal Realities*

1. J. T. Fraser has presented the hierarchical model of time in several books and articles (1975, 1978, 1981, 1987, 1990, 1994, 1999); however, in my opinion its presentation in *Time, Conflict, and Human Values* (1999) is both the most explicitly formalized and the clearest of the several presentations. Thus for readers interested in exploring this model in greater detail, I recommend starting with that work.

2. I have quoted this line from "Setting Sail," which appeared in Emily Dickinson's *Poems* (1890, p. 116). A version of this poem published many years later replaced the exclamation point at the end of the line with a dash (Dickinson 1960, p. 39). If poetry about time and temporal matters is of interest, Dickinson (1960) is a good place to start, because she wrote many poems that deal with such themes.

3. This statement is widely attributed to Benjamin Franklin. In my attempts to verify Franklin as the source, I found citations to it as occurring in a publication titled "Advice to a Young Tradesman." The two microform versions of this brief essay from the eighteenth century—one was from 1762, the other from 1792—that I located in the University of Missouri-Columbia's library were indeed listed in the electronic card catalog under Franklin's name, but when I examined the essays themselves, the author was given as "Written by an Old One." The essay does begin, "Remember that Time is Money." And given Franklin's listing as author in the electronic catalog, I cite Franklin as its author even though his name was not given on the documents I examined.

4. This quotation has been attributed to Mark Twain (Least Heat Moon 1982, p. 10), and many other writers and speakers have attributed it and similar versions to Twain as well. However, none that I have encountered have actually cited a source for it in Twain's work, and I have been unable to locate it there myself—and I have been searching for years. So its origin in Twain's work or that of others re-

mains in question—which is unfortunate, because whoever said or wrote it first deserves credit for having produced a noteworthy insight.

5. As a graduate student I studied and taught hominid evolution, and the summary of hominid evolution just presented in the text is based on that learning and teaching, supplemented with some updating provided by physical anthropologist Carol Ward, who is engaged in an ongoing program of research on hominid evolution (e.g., Leakey, Feibel, McDougall, Ward, and Walker 1998; Brown, Walker, Ward, and Leakey 1993; Ward, Leakey, Brown, Brown, Harris, and Walker 1999; Ward, Leakey, and Walker 1999). The summary is consistent with the interpretation presented in Coppens (1994).

6. A major alternative explanation for increasing brain size called the social brain hypothesis proposes that the information-processing demands of increasingly complex social systems was the selective force that led to increased brain size (Dunbar 1998, p. 178). Robin Dunbar's (1998) research using group size as the measure of social system complexity has produced an impressive set of statistical results that support this hypothesis over the ecological information-processing perspective. Johanson and Edgar's suggestion (1996, p. 92) about fruit eaters and leaf eaters is an example of the ecological information-processing perspective. Whether either or both of these propositions withstand the test of time and additional research, the basic point remains that the ability to use temporal information likely provided significant survival value to our ancestors, just as it does for us today. And certain types of temporal information may have played an especially important role in human evolution, as will be discussed in Chapter 8.

7. Although I suspect she overstated the case a bit, in doing so she effectively made the point about what is probably the principle evolutionary role of language.

8. The tracks were actually made and preserved in volcanic ash (Leakey 1979).

9. Even most digital clocks and watches preserve a portion of the clockwise heritage by being set to count up rather than down (i.e., they start from zero).

CHAPTER 3   *Polychronicity*

1. Carol Kaufman, Paul Lane, and Jay Lindquist (1991a) developed the Polychronic Attitude Index, a measure of an individual's polychronicity. See Kaufman-Scarborough and Lindquist (1999) for additional work on this measure as well as important findings about the relationship between individual polychronicity and time management behaviors.

2. On two occasions nine days apart, 230 undergraduate students took the individual version of the Inventory of Polychronic Values (see note 3 to this chapter for more details about this scale). The students were from a course in fundamen-

tals of management at the University of Missouri-Columbia, and the correlation between their two scores on this measure of polychronicity was a statistically significant $r = .73$ (Bluedorn 2000c).

3. I developed the Inventory of Polychronic Values (IPV) to measure perceptions of polychronicity as one component of a group's culture (e.g., as a part of a department's or organization's culture). The IPV consists of ten items, some of which are adapted and significantly modified from Kaufman, Lane, and Lindquist's (1991a) Polychronic Attitude Index, with the majority of the IPV's items being original items developed specifically for the scale. A full report of the IPV's development and psychometric validation is presented in Bluedorn et al. 1999, as well as the IPV itself. Although the IPV is a scale that measures perceptions of polychronicity at the level of group culture, as note 2 to this chapter indicates, it can be modified to measure individual polychronicity as well. Instructions for making these modifications are presented in Bluedorn et al. (1999). When scored as the sum of its ten items divided by ten, both the group and the individual versions of the IPV have a range from one (the lowest level of polychronicity) to seven (the highest level of polychronicity). The IPV is scored this way for all original data analyses I have conducted with it that are presented in this book.

4. Onken (1999) used the Inventory of Polychronic Values (see note 3 to this chapter) to measure the polychronicity of twenty firms in the telecommunications and publishing industries, and she did so by having several people in each firm complete the IPV about their firm. Onken then calculated the mean of the responses about a firm as that firm's polychronicity score. To justify aggregating these responses by firm, Onken conducted a within-and-between analysis (WABA), which strongly supported the conclusion that the perceptions were indeed perceptions of *firm* polychronicity, hence supporting the approach of aggregating the respondents' perceptions by firm (Onken 1999, pp. 237–38).

To further demonstrate the existence of polychronicity as a component of organizational culture, I will now present the results from a study of organizational culture in fifty dental practices that Gregg Martin and I conducted. The following results are original analyses conducted for this book, and I am very grateful for Gregg Martin's enthusiastic permission to use these data to conduct these and other analyses and to present the results here and elsewhere in this chapter.

Dentists and their employees in fifty dental practices from Florida and Missouri completed a questionnaire about themselves and their organizations, and the questionnaire included the Inventory of Polychronic Values. (Please note that the number of respondents for this sample will vary in the subsequent notes because of missing values on different variables and my use of listwise deletion options.)

For the analyses reported in this note, 208 respondents from the fifty dental practices returned questionnaires usable for the analyses, with at least two respondents providing data about each practice. To demonstrate that the responses to the IPV represent perceptions of a group-level phenomenon (a component of organizational culture in this case), I conducted three analyses described in Klein et al. (2000) that can justify aggregation. First, I calculated the intraclass correlation coefficient ICC(1) (Bartko 1976) for the practices and obtained a statistically significant ICC(1) of .20, $F(49, 158) = 2.005$, $p \leq .001$. Next, I conducted a WABA I analysis (Dansereau, Alutto, and Yammarino 1984), which produced an E ratio of .788. Finally, I calculated $r_{wg}(I)$ (James, Demaree, and Wolf 1984) and obtained a mean $r_{wg}(I)$ of .81.

All three of these findings support the conclusion that polychronicity is a group-level (cultural) phenomenon in these dental practices. The ICC(1) of .20 is not only statistically significant but also compares favorably with the .05–.20 range of ICC(1) values Paul Bliese reports as typical for the many ICC(1)s he has calculated with data collected from numerous military units (Bliese 2000, p. 361). The E ratio of .788 is a result Fred Dansereau, Joseph Alutto, and Francis Yammarino (1984, pp. 169–71) classify as equivocal, indicating that the group has an effect, but it does not totally determine the perceptions or responses. And following a suggestion from Fran Yammarino (personal communication, 2001), I used the overall IPV score to calculate the r_{wg} designed for a single item, $r_{wg}(I)$. An r_{wg} of .7 or greater is considered justification for aggregating data for a group (Klein et al. 2000, p. 517), and the mean $r_{wg}(I)$ of .81 is clearly above this standard. In fact, 41 of the 50 dental practices, 82 percent, had $r_{wg}(I)$s above .70. Overall, all three tests support the conclusion that these responses to the IPV about the dental practices represent perceptions of a genuinely group-level phenomenon, that in these dental practices polychronicity was an attribute of the practices' culture.

5. I thank Stephen Ferris for giving me permission, enthusiastically, to perform original analyses of these data for this book. The correlation between size and polychronicity was a nonsignificant $r = .13$, ns (two-tailed test), but the correlation between the logarithm (base 10) of size and polychronicity was a statistically significant $r = .20$, $p \leq .02$. The reason for this difference is likely the difference in skewness for the two size variables. For size (number of employees), the skewness index was nearly eight, indicating the distribution of size scores is greatly skewed as distributions of organizational size often are. For the logarithm (base 10) of size (number of employees), the skewness index was reduced almost to zero, which indicates no significant skewing of the distribution. The listwise N for these analyses was 154 firms, and the variables' means and standard deviations were as follows:

size (M = 3973.71, SD = 13308.78), logarithm (base 10) of size (M = 2.78, SD = .87), and polychronicity (M = 5.05, SD = .97). Cronbach's (1951) alpha coefficient for the Inventory of Polychronic Values (group version) from this set of 154 responding executives was .83.

6. Onken reported a statistically significant correlation between speed values and polychronicity (measured with the Inventory of Polychronic Values, see note 3 to this chapter) of r = .44 (1999, p. 238), and the statistically significant correlation I obtained in the national sample data (see note 5 to this chapter) between speed values and polychronicity was r = .19 (N = 183, $p \leq$ .01, two-tailed test). However, because the alpha coefficient (Cronbach 1951) for the speed-values scale in the national sample was only .55 (the alpha coefficient for polychronicity was .82), I recalculated this correlation coefficient to correct it for attenuation (Cohen and Cohen 1983, p. 69). The correlation corrected for attenuation was r = .28.

Speed values were measured in the national sample with Jacquelyn Schriber and Barbara Gutek's (1987) work-pace scale, and they were scored so that higher scores represented a greater emphasis on doing things fast (i.e., more highly valuing speed). The means and standard deviations follow for the two variables in the polychronicity-speed values correlation obtained from the national sample: polychronicity, measured with the group version of the Inventory of Polychronic Values (M = 5.05, SD = .93), and speed values measured with Schriber and Gutek's (1987) work-pace scale scored on a 1-to-7 response scale so that higher scores indicate speed is more highly valued (M = 4.32, SD = .84).

7. All four samples reported in Bluedorn (2000c) were large samples (ranging from 248 to 493 respondents) of students at the University of Missouri-Columbia, nearly all of whom were undergraduates. The fifth sample, in which women were more polychronic than men, was a sample of undergraduate students at the University of Missouri-Columbia (N = 165), and the correlation was r = -.15 ($p \leq$ .056, two-tailed test; women were coded as 1, men as 2, as they were coded in the other samples too). Although this correlation does not quite reach the customary .05 significance level, I applied the more conservative two-tailed test and feel that between the choices of declaring this correlation significant or nonsignificant, classifying it as significant more accurately represents the finding.

In the small sample of college food service managers (N = 34), the correlation between gender and polychronicity was r = .33 ($p \leq$ .057, two-tailed test). And just as with the other correlation that did not quite achieve the customary .05 significance level, I believe classifying this correlation as significant more accurately represents the finding than would a nonsignificant classification. The correlation between gender and polychronicity in the sample of dental practice personnel (N = 204) was

$r = -.13$, ns (two-tailed test), and in the student sample (N = 136), the correlation was $r = .07$, ns (two-tailed test).

In all of the original data analyses, polychronicity was measured with the Inventory of Polychronic Values (individual version, see note 3 to this chapter), whose means, standard deviations, and alpha coefficients (Cronbach 1951) are presented next, along with the means and standard deviations for gender: first University of Missouri undergraduate student sample (polychronicity: M = 3.48, SD = 1.02, alpha = .86; gender: M = 1.52, SD = .50); college food service managers (polychronicity: M = 4.50, SD = .91, alpha = .84; gender: M = 1.29, SD = .46); dental practices (polychronicity: M = 3.18, SD = 1.09, alpha = .82; gender: M = 1.75, SD = .43); and the second University of Missouri undergraduate student sample (polychronicity: M = 3.68, SD = .98, alpha = .85; gender: M = 1.67, SD = .47).

8. The correlation between age and individual polychronicity in the sample of dental practice personnel (listwise N = 204) was a nonsignificant $r = .08$, ns (two-tailed test). The mean and standard deviation for age were M = 39.14 years and SD = 10.42. See note 7 to this chapter for the descriptive statistics about polychronicity in this sample. In the sample of college food service managers (N = 34), the correlation between age and polychronicity was a nonsignificant $r = .02$, ns (two-tailed test). The mean and standard deviation for age were M = 39.62 years and SD = 8.70. See note 7 to this chapter for the descriptive statistics about polychronicity in this sample.

9. Stress was measured with Thomas Dougherty and Robert Pritchard's tension-anxiety scale (1985), and polychronicity was measured with the individual version of the Inventory of Polychronic Values (see note 3). The listwise N was 207 (employees and dentists), and the means, standard deviations, and scale alpha coefficients (Cronbach 1951) for the variables in this sample were as follows: stress (M = 3.33, SD = 1.36, alpha = .74) and polychronicity (M = 3.17, SD = 1.09, alpha = .82). The correlation between polychronicity and stress in this *entire* sample was a nonsignificant $r = -.09$, ns (two-tailed test).

A standard test was conducted for the interaction between polychronicity and category of dental practice employee (dentists versus all other employees) by first regressing polychronicity and employee status on stress and then adding the multiplicative interaction term to the equation (Cohen and Cohen 1983). Adding the interaction term produced a significant increase in R^2 with the increment-in-R^2 test giving $F(1, 203) = 4.74$, $p \le .04$, two-tailed test. For the dentists (N = 44), the zero-order correlation between polychronicity and stress was $r = -.38$ ($p \le .02$, two-tailed test), and for the other practice employees (N = 163) it was $r = -.04$, ns. After controlling for age, gender, and years with the practice, the partial correlation between polychronicity and stress for the dentists was $pr = -.46$ ($p \leq .01$, two-tailed test); and for the other practice employees it was $pr = -.01$, ns (two-tailed test). Note: For the partial correlation analysis, the number of dentists was unchanged, but missing values reduced the number of other practice employees to 157.

10. The formal test of the model is the path analysis presented in the following path diagram. As indicated in the diagram, the path model is overidentified, because the direct path from polychronicity to job satisfaction is not significantly different from zero (see Pedhazur 1982, p. 617). To test how well this overidentified path model fits the data, I calculated Q and its associated inferential statistic, W. Q varies from 0 to 1, and the closer to 1, the better the fit (Pedhazur 1982, pp. 619–20). In the case of the path model in the diagram, the value of Q was 1, which indicates a perfect fit to the data, and W was 0, which also indicates a perfect fit (Pedhazur 1982, p. 619). In mediation terms, this is an example of complete mediation (James and Brett 1984, p. 307), because all of polychronicity's effect on job satisfaction is indirect, through stress.

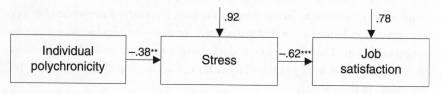

NOTE: The path coefficient between polychronicity and job satisfaction was a nonsignificant .01, so it is not included in the diagram. N = 44, $^{**}p \leq .01$, $^{***}p \leq .001$, two-tailed tests.

In this test of the model, job satisfaction was measured with the three-item version of Richard Hackman and Greg Oldham's (1975) general satisfaction scale, stress with Dougherty and Pritchard's (1985) tension-anxiety scale, and polychronicity with the individual version of the Inventory of Polychronic Values (IPV) (see note 3 to this chapter). The listwise N was forty-four dentists, and the means and standard deviations of the variables are as follows: job satisfaction (M = 5.98, SD = 1.08), stress (M = 3.70, SD = 1.43), and the IPV (individual version) (M = 3.49, SD = 1.06). The alpha coefficients (Cronbach 1951) for the forty-four dentists on these three scales are .75 for job satisfaction, .75 for stress, and .84 for polychronicity. The R^2 for stress was .15, and for job satisfaction it was .39.

11. The samples are the same two samples of undergraduate students at the University of Missouri-Columbia about which data were presented in note 7 to this chapter. In the first of those two samples (N = 165), the correlation was $r = .10$, ns (two-tailed test) between polychronicity and grade point average (GPA: M = 3.05,

SD = .52); in the second sample, r = .06, ns (two-tailed test) between polychronicity and grade point average (GPA: M = 3.04, SD = .46). See note 7 to this chapter for the means, standard deviations, and alpha coefficients about the polychronicity measure in these two samples.

12. Frei, Racicot, and Travagline employed a separate measure of "monochronic behavior." However, it was operationalized in terms of things faculty did to minimize interruptions while working (1999, p. 380). Although this operationalization could be interpreted as the inverse of polychronicity as measured by one of the psychometric polychronicity scales (i.e., the Polychronic Attitude Index, see note 1 to this chapter; the Inventory of Polychronic Values, see note 3 to this chapter), this measure is sufficiently narrow to make such an interpretation ambiguous. Thus Frei, Racicot, and Travagline's findings for "projects in progress" (1999, pp. 380–81), which were operationalized separately from the "monochronic work behaviors" measure, have been used in the text discussion instead. "Projects in progress" is also more directly comparable to Taylor et al.'s (1984) measures.

13. The listwise N was 173 companies for this analysis from the national sample of American publicly traded firms (see note 5 to this chapter), and the specific correlations with the firms' polychronicity follow: with return on assets r = .04, ns (two-tailed test); with return on equity r = .17, $p \leq$ .03 (two-tailed test); and with return on sales r = −.01, ns (two-tailed test). The means and standard deviations for the variables in this analysis follow: polychronicity, measured with the group version of the Inventory of Polychronic Values, see note 3 to this chapter (M = 5.05, SD = .95, alpha = .82); return on assets (M = .01, SD = .19); return on equity (M = .18, SD = .96); and return on sales (M = −.06, SD = .68). Data to calculate the three measures of financial performance using standard formulas were taken from the CompuStat Annual Data Tapes for the year immediately preceding the year in which the top executives responded to the questionnaires. All the top executives completed the questionnaires in a two-month period during the same year.

CHAPTER 4 *Seldom Early, Never Late*

1. Prizes for developing a practical method for determining the longitude at sea were offered as early as 1567 by Philip II of Spain, followed by prize offers in France and the Dutch states (Barnett 1998, p. 107).

2. The ship's longitude would be 30 degrees *west* of Greenwich because the earth rotates to the east, which means that places to the east must have times "*more advanced*" (Abell's emphasis) than those to the west (Abell 1969, p. 143). In the example, 2:00 P.M. is "more advanced" than noon, the local time at the ship's position, so the ship is 30 degrees of longitude *west* of Greenwich.

3. A sobering story has become associated with this tragedy. The story has it that a sailor on Sir Clowdisley's ship had kept his own reckoning of the fleet's position. Believing the official statement of the fleet's position to be in error, he voiced his concern, and Sir Clowdisley promptly had the sailor hanged for mutiny (Sobel and Andrewes 1998, pp. 16–17). Versions of this story have been given in other sources (e.g., Brown 1949, pp. 225–26; Gould 1960, p. 2). Note that the spelling of the admiral's name was given as Clowdisley by Sobel and Andrewes, but both Brown and Gould gave it as Clowdesley.

4. Landes's explanation in terms of a belief in the enhanced power of actions performed simultaneously included a statement that "the whole was greater than the sum of the parts" (1983, p. 63), which anticipates the discussion of entrainment in Chapter 6.

5. Plautus was given as the author of these lines; however, the scholar who translated them, Henry Thomas Riley, cited a source that indicates Plautus may have translated them from a play written in Greek by another author (Plautus 1902, p. 517). David Landes quoted a portion of these lines, but from a different translation (Landes 1983, pp. 15–16). Landes's description of them as being "attributed by Roman authors to Plautus" (1983, p. 393) also indicates a lack of certainty about their author's identity.

The translation of Plautus's works that I have quoted was published in 1902, so the actual volume I read was a century old. When I located the page containing the quoted material, I was surprised also to find a small piece of paper that had been neatly torn from what looked like a lined sheet of notebook or tablet paper, a piece of paper just a little larger than a standard business card. What surprised me was what was written on this piece of paper: "Sundial into Rome." Someone many years ago (judging by the condition of this piece of paper) had been interested in this same passage and the information the translator provided about it. When I saw it, I felt a little bit like Mary Leakey did when she interpreted the behavior of one of the australopithecines whose footprints had been preserved in volcanic ash (see Chapter 2). Across what may have been two or three generations, I shared a common bond with someone who was interested in the sundials of Rome and how the Romans may have reacted to them. Who knows whether that person's reasons for being interested were the same as mine, but we were both interested nevertheless. And then I realized that I faced a decision: Should I keep the small piece of paper or should I leave it in the book to mark this page, so the next person to track down this passage could experience the same connection I did? I decided to make a photocopy of the note as a memento and to leave the original as I had found it, keeping pages 516 and 517 company in the stacks between visits from seekers of information about ancient time reckoners. I hope the future seekers who find their way to these two pages will notice the note on the faded piece of paper as I did and experience a connection with those who had visited before them, perhaps many years before.

6. The prohibition against workers having watches at work seems to parallel the British navy's prohibition against sailors keeping their own reckoning of their ships' positions (Sobel and Andrewes 1998, p. 17), a key element in the story described in note 2 to this chapter.

7. Levine and Norenzayan (1999, pp. 195–96) reported an adjusted R^2 of .38 for the equation that regressed clock accuracy on climate, productivity of the economy, and individualism/collectivism. Levine and Bartlett (1984, p. 244) reported statistically significant positive correlations between punctuality (as measured by clock accuracy) and both speeds of completion of postal requests and walking speeds of $r = .71$ and $r = .82$, respectively.

8. The national sample of publicly traded companies is described in Chapter 3 and Bluedorn and Ferris (2000). The correlation between punctuality and speed values was $r = .28$ ($p \leq .001$, two-tailed test). Controlling for schedules-and-deadlines values, a variable likely to be related to both punctuality and speed values, the partial correlation was still statistically significant, $pr = .21$ ($p \leq .01$, two-tailed test). All three of these variables were measured with scales developed by Jacquelyn Schriber and Barbara Gutek (1987), their work-pace, punctuality, and schedules and deadlines scales. All three scales were scored so that higher scores indicated a greater value being placed on speed, punctuality, and schedules and deadlines, respectively. The listwise N for these analyses was 181, and the means, standard deviations, and alpha coefficients (Cronbach 1951) for these three variables were as follows: speed values (M = 4.32, SD = .84, alpha = .54), punctuality values (M = 4.55, SD = 1.21, alpha = .76), and schedules and deadlines values (M = 5.35, SD = .98, alpha = .84). Because the alpha for speed values was relatively low, I recalculated the correlation between speed values and punctuality values to correct it for attenuation (Cohen and Cohen 1983, p. 69), and the resulting correlation corrected for attenuation was $r = .44$.

9. I calculated these statistics from data presented in Table 1 in Levine, West, and Reis 1980 (p. 544).

10. I calculated these averages (means) and their differences from data presented in Table 2 in Levine, West, and Reis (1980, p. 546). In this table, data were presented about what constitutes being early and late for several other events in both cultures too.

11. Bluedorn et al. (1999, p. 225) reported a statistically significant correlation of

r = .39 between punctuality and schedule-and-deadlines values, and Benabou's correlation was a statistically significant r = .29 (1999, p. 263). In the data from the national sample of American publicly traded companies, the correlation between punctuality values and schedules-and-deadlines values was r = .42 ($p \leq$ .001, two-tailed test). After controlling for speed values, the partial correlation between punctuality and speed-and-deadlines values was pr = .39 ($p \leq$ .001, two-tailed test). See note 8 to this chapter for how these variables were measured in this study and for their means and standard deviations.

12. As part of Deloitte Consulting's recruiting efforts, I received an e-mail that provided a job description of its systems analyst position. This description included the following point: "Multi-tasking is essential for success."

13. In their sample of thirty-six American cities, Levine et al. (1989, p. 517) found a statistically significant correlation of r = .50 between pace of life and coronary heart disease death rates.

14. In their sample of thirty-one countries, Levine and Norenzayan (1999, p. 191) found a statistically significant correlation of r = .35 between pace of life and coronary heart disease death rates.

CHAPTER 5 *Eternal Horizons*

1. I used listwise deletion of missing cases in order to conduct all t-tests on the same respondents. Fortunately, this procedure reduced the sample by only one respondent, resulting in a listwise N of 362 people whose responses are reported in Tables 5.1 and 5.2 and Figure 5.1 and are the basis for the t-tests reported here. I used dependent sample t-tests (Glass and Stanley 1970, pp. 297–300) because the same respondents provided judgments about themselves on all six temporal depth items. All of the t-tests were two-tailed tests with 361 degrees of freedom, and all were significant at the $p \leq$ .001 level. The t statistics for the past items are as follows: recent versus middling, t = 10.18; middling versus long ago, t = 17.09; and recent versus long ago, t = 18.55. For the future items the t statistics are as follows: short-term versus mid-term, t = 15.20; mid-term versus long-term, t = 20.21; and short-term versus long-term, t = 21.46. And for the three pairs of parallel items, the t statistics are as follows: recent past versus short-term future, t = 8.97; middling past versus mid-term future, t = 10.23; and long-ago past versus long-term future, t = 10.54. The means for all six temporal depth items are presented in Figure 5.1, and the standard deviations for all six items are as follows: recent past, 145.13; middling past, 505.79; long-ago past, 1,585.57; short-term future, 358.39; mid-term future, 1,080.62; and long-term future, 2,722.92.

2. A few time capsules have been buried in the United States with instructions

to wait this long before opening, such as one interred at the New York World's Fair in 1939, which specified an opening date five thousand years in the future (Benford 1999, p. 23). A huge room-sized capsule, "The Crypt of Civilization," at Oglethorpe University in Atlanta, Georgia, was sealed in 1940 and is scheduled to be opened in 8113, which is even longer (Benford 1999, p. 24). But as Gregory Benford concluded, "Roughly ten thousand time capsules already await future historians. Notably, their time horizon is quite short, usually a century" (p. 23).

3. See note 4 to Chapter 2 concerning the uncertain origins of this quotation.

4. The approach to budgeting known as *incrementalism* is consistent with both examples and the correlations too. The incremental approach bases one year's budget on the previous year's, "with special attention given to a narrow range of increases and decreases" (Wildavsky 2001, p. 182). So this approach to budgeting seems to base itself on the assumption that the next year will be about the same as the preceding year, a specific example of using the past as a metaphor for the future (the future will be like the past). Making this assumption and budgeting incrementally has its political advantages too: "By encapsulating the past in the present through the base, budgeters limit future disputes" (Wildavsky 2001, p. 182). And limiting disputes peacefully is no mean accomplishment.

5. One Westerner who seemed to agree with Makeba was F. Scott Fitzgerald, who referred to "the contradiction between the dead hand of the past and the high intentions of the future" (1945, p. 70). The "dead hand of the past" certainly seems like part of a "dead animal," a "carcass" as Makeba described the West's attitude about the past.

6. Students from the University of Missouri-Columbia made up both samples. For this analysis, one sample contained 481 respondents, whose mean and standard deviation on the orientation toward the future scale were 5.94 and .89, respectively; on the orientation toward the past scale, 4.29 and 1.29, respectively. The dependent sample t-test (Glass and Stanley 1970, pp. 297–300) for the difference in these two means was $t(480) = .23.92$, $p \leq .001$, two-tailed test. The second sample was composed of 383 respondents for this analysis, whose mean and standard deviation on the orientation toward the future scale were 6.10 and .80, respectively; for the orientation toward the past scale, 4.27 and 1.25, respectively. The dependent sample t-test for the difference in these two means was $t(382) = 24.20$, $p \leq .001$, two-tailed test.

7. As in the analyses reported in note 1, the listwise N is 362 for the t-tests reported here. And as in the analyses reported in note 1, I used dependent sample t-tests (Glass and Stanley 1970, pp. 297–300) because the same respondents provided judgments about themselves on all six temporal depth items. All of the t-tests were two-tailed tests with 361 degrees of freedom. The t statistics for the

past items are as follows: recent versus middling, $t = 12.34$, $p \leq .001$; middling versus long ago, $t = 8.93$, $p \leq .001$; and recent versus long ago, $t = 12.96$, $p \leq .001$. For the future items the t statistics are as follows: short-term versus mid-term, $t = 1.87$, ns; mid-term versus long-term, $t = 3.18$, $p \leq .01$; and short-term versus long-term, $t = 1.09$, ns. And for the three pairs of parallel items, the t statistics are as follows: recent past versus short-term future, $t = 5.02$, $p \leq .001$; middling past versus mid-term future, $t = 13.95$, $p \leq .001$; and long-ago past versus long-term future, $t = 17.61$, $p \leq .001$. The means for all six importance (focus) items are presented in Figure 5.2, and the standard deviations for all six items are as follows: recent past, 1.71; middling past, 1.53; long-ago past, 1.88; short-term future, 1.40; mid-term future, 1.02; and long-term future, 1.39.

CHAPTER 6 *Convergence*

1. Jürgen Aschoff described a phase as an "instantaneous state of an oscillation" that is defined by "the value of the variable and all its time derivatives" (1979, p. 5). Further, the phase angle is the point on the abscissa (the X value in X-Y Cartesian coordinates) that corresponds to the phase, and "it is measured in fractions of the entire period from an arbitrary zero point on the abscissa and is expressed in units of time or in angular degrees (one period = 360°)" (p. 5). A period is the time between successive events in a series of regularly occurring events (*Longman Dictionary of Scientific Usage* 1979, p. 459).

2. The mechanical clock was probably invented in or shortly after the 1270s. See Chapter 1 for a more complete discussion of this issue.

3. The canonical hours were mandated in the Rule of St. Benedict (Rothwell 1959, pp. 140–41; Zerubavel 1981, pp. 34–35) and came to be designated as follows: matins, lauds, prime, terce, sext, none, vespers, and compline (Zerubavel 1981, p. 35; see also Landes 1983, p. 61; Macey 1994, p. 441; and Richards 1998, p. 44). As noted in the chapter, these were temporal hours (see Chapter 1), because they were part of a system that divided the day into twelve hours, and the night into twelve hours as well, all of which varied with the seasons (Rothwell 1959, p. 241).

4. Whiggish history's emphasis on historical progress may also foster the belief that because things are progressing historically, things today are therefore better than they were yesterday. This translates into what I call *chronocentrism*, the tendency to attribute more positive attributes (e.g., morality, technology, sophistication) to one's own times than to those of others, especially the times of those who lived in bygone eras. For example, Nigel Calder noted, "Radiocarbon rescued archaeology from the circular reasoning that assigned relative dates according to how advanced a culture seemed to be" (1983, p. 30). This suggests that radiocarbon

dating helped prevent purely chronocentric interpretations of archaeological materials (i.e., "the circular reasoning"), and one can only wish that such prophylactic effects were more widespread. For I fear that chronocentric attitudes lead us to believe that *our times* are the best that have ever been, and by extension, that *we* are the best that have ever been, a dangerous cultural hubris (see Chapter 9). (*Note*: At press time a chance Web finding led to a Web search that located many similar uses of "chronocentrism" by postmodernists and historians. Thus my usage is not the first, although I developed it independently several years ago.)

5. One point in Dohrn-van Rossum's analysis appears to be debatable, and that is where he located the Horloge du Palais—in the Louvre—which I believe is off by several city blocks. In at least three instances Dohrn-van Rossum described the clock at the center of this temporal legend as being located in the Louvre (1996, pp. 142, 188, 217), and twice indicated that it was located in a *corner tower* of the Louvre (1996, pp. 142 and 188). Yet no other source I have examined mentioned the Louvre as the location of this storied clock. (Alexander Waugh [1999, p. 58] mentioned that the king had clocks installed in several of his palaces, including the Louvre, but Waugh's statement of this point was too brief and too general to inform this discussion one way or the other.) Some simply indicated the clock's location by referring to the "clock of the royal palace" (Whitrow 1988, p. 110), the "Palais Royal" (Crombie 1959, p. 212), "the Palais-Royal" (Le Goff 1980, p. 50—note that this source was originally published in French), "one of the towers of the royal palace" (Cipolla 1978, p. 41), or "in his palace" (Borst 1993, p. 97). David Landes (1983, p. 75) identified the location more precisely as "in his palace on the Ile de la Cité." The Île de la Cité (Island of the City) is an island in the Seine River on which the original settlement began that eventually grew into Paris. The Louvre is not on the Île de la Cité; it is on the right bank of the Seine.

Alfred Crosby also located the clock on the Île de la Cité: "the clock he [Charles V] was installing in his palace on the Ile de la Cité" (1997, p. 82). And he added an additional clue: "The quai de l'Horloge, with a clock, is still there" (p. 82), a location seconded by Jean Gimpel, "on one of the towers of the Royal Palace in Paris, now on the corner of the boulevard du Palais and the quai de l'Horloge" (1976, p. 168—another source that was originally published in French). So exactly where on the Île de la Cité did Charles V have Henry de Vic build this famous clock? I believe it was in a royal palace; however, that palace was not the Louvre but the building that is known today as the Conciergerie.

The Conciergerie began as a fortress during the days of the Roman Empire, and by the end of the tenth century it had been enlarged and improved, earning it the name Palais de la Cité. It also became the residence of the French Capetian kings. In 1187 the French king Philip Augustus received England's Richard Lion Heart in this palace. By the beginning of the fourteenth century the royal palace had become "one of the most admirable residences of the Middle Ages" (Bouvet, Malécot, and Sallé 2000, pp. 13–15). When Charles V attained the throne in 1364, he decided to live elsewhere, including residence in the Château du Louvre. Nevertheless, he and subsequent monarchs continued to use the Palais for state functions and receptions, including a banquet with eight hundred place settings that Charles V held in honor of his uncle Charles IV of Luxembourg on January 26, 1378. (This brief summary of the Conciergerie is based on information in Bouvet, Malécot, and Sallé 2000.)

So even though Charles lived in both places, the Château du Louvre and the Palais de la Cité, the evidence seems to support the Palais de la Cité as the location of the famous clock. Evidence in favor of this conclusion includes its location on the Île de la Cité by a number of scholars, *and* the official French statement that the clock was located there: "In 1370, Charles V had the first public clock in Paris placed there, the work of Henri de Vic" (Bouvet, Malécot, and Sallé 2000, p. 41). The "there" is a reference to "the famous Tour de l'Horloge," which stands "at the corner of the Quai de l'Horloge and the Boulevard du Palais" (p. 41).

While writing this book, I stood one day "at the corner of the Quai de l'Horloge and the Boulevard du Palais," looking at "the famous Tour de l'Horloge." And I saw that the Tour is a *corner tower* of the Conciergerie, née the Palais de la Cité. Although at the time unaware of Dohrn-van Rossum's claim about a location in the Louvre, it seemed notable to me that the specific corner tower was named the Tour de l'Horloge and that the street atop the quay immediately in front of the Tour was named the Quai de l'Horloge. It seemed notable because these names pay considerable homage to the location of a single clock, and this suggested that this clock was important, very important. And so it was, in both fact and legend.

Charles V made at least one other major contribution to the history of human time, for it was Charles V who commissioned Nicole Oresme to translate scientific works by Aristotle, which resulted in Oresme's *Le Livre du ciel et du monde* (*The Book of the Heavens and the World*) (Menut 1968, pp. 3–9). Within Oresme's commentaries in this volume is the great metaphor comparing God to a clockmaker and the universe to a clock (see Chapter 1 for my discussion of this metaphor and its impact).

6. Ramon Aldag suggested this pun to me after reading a draft of this chapter, and I am grateful that he did so, because it helps reinforce the point so well.

7. The tropical year can be thought of as "the time it takes for the seasons to

cycle through and start again" (Duncan 1998, p. 182), but it is more precisely defined as "the period of one complete revolution of the mean longitude of the Sun with respect to the dynamical equinox" (*Explanatory Supplement to the Astronomical Almanac* as quoted in Steel 2000, p. 382).

8. At the time Gregory XIII signed the papal bull in 1582, it would have been possible to reduce future calendar-tropical year discrepancies even further. However, issues of practicality were involved, a crucial one being the desire to have rules involving leap years that would be simple enough that they could be understood "by any parish priest" (Richards 1998, p. 250).

9. The meridian was defined more precisely as "the one passing through the center of the transit instrument at the Royal Observatory at Greenwich" (Bartky 2000, p. 151).

10. Even scientists took several transient periods to come into phase. One of the conference's other temporal recommendations was that "as soon as may be practicable the nautical and astronomical days will be arranged everywhere to begin at midnight" (quoted in Bartky 2000, p. 152). But as Bartky noted drolly, "Among the world's astronomers, 'as soon as may be practicable' took forty-four years" (p. 152). One wonders whether this leisurely pace might have been inherited from the Board of Longitude (see Chapter 4).

11. I came across the advertisement for the Atomic Watch while on a transatlantic flight in October 2000.

12. Institutionalization of such entrained rhythms and the organizational adjustments that develop around them are the sociological equivalents of the complexes of genetic instructions that presumably underlie free-running circadian rhythms, such rhythms presumably having been acted upon by natural selection over periods of deep time (as geologists would define it).

13. I tried to follow the principles of this theory of writing as I wrote this book. For details about positions in a sentence and reader expectations for them, see Gopen and Swan (1990) and Williams (2000, pp. 41–135).

14. Mary Austin also understood the principle of a zeitgeber: "Let any one of them [distinctive rhythms] attain a position of dominant stress, and the whole organism is brought into subjective obedience to it, or ruptures itself in the attempt" (1970, p. 5).

15. Details about measuring the time span of discretion are presented in the *Time-Span Handbook* (Jaques 1998b).

16. A noteworthy attribute associated with our results is the statistical methodology we used to obtain them (Slocombe and Bluedorn 1999). Our research was one of the few studies of congruence effects in general to use the polynomial re-

gression and response surface analysis procedures Jeffrey Edwards proposed in a series of groundbreaking articles (Edwards 1993, 1994; Edwards and Parry 1993; Edwards and Harrison 1993). In these articles, Edwards critiques traditional statistical approaches used to study congruence (e.g., difference scores, squared differences scores, profile similarity indexes) and explains why they may lead to ambiguous or incorrect conclusions, problems eliminated with the statistical approach he presented as an alternative (i.e., polynomial regression and response surface analysis). We used that approach in our analysis of the data collected from the business school graduates, thereby avoiding the problems often found with other statistical methods used to study congruence.

CHAPTER 7   *The Best of Times and the Worst of Times*

1. The *New York Times* published an article about the research in its Science Times section (Berger 1999). Other print coverage included *Psychology Today* (Wynne 1999), *Newsweek* (McGinn 2000), an Associated Press story carried by many newspapers, and a boxed feature in at least one management textbook (Kreitner 2001, p. 384). Electronic media coverage included two appearances on BBC radio, one in July 1999 and the other in February 2000.

2. The task was the famous "lost-on-the moon" exercise developed by Hall and Watson (1970; see also Hall 1971). This exercise presents decision makers with a scenario in which people have traveled to the moon and their sight-seeing vehicle has crashed. The people have survived along with some of their equipment, and the task is to rank the pieces of equipment in the order of their value for helping the people return to the main spaceship, which will return them to earth. The consensus answer a group of NASA astronauts and space scientists reached about this problem is used as the "right answer," against which the rank-orderings produced in this exercise are judged.

3. I was invited to discuss this issue as part of a symposium on small-group research at the 2000 Academy of Management Meeting (Bluedorn 2000d), and I presented some of these ideas at that symposium.

4. Jean Bartunek suggested using more of the full quotation after she read an earlier draft of this chapter. She felt providing more of Lewin's complete statement would reveal more of the thinking behind it and result in a more powerful message. I thank Jean for the suggestion, and also note that doing so provides more of a connection to other ideas, which makes it a more meaningful statement, the matter of connections and meaning being an issue explored extensively in this chapter.

Fans of this quotation should note that Lewin used a slightly different version of it in an article he published in *Sociometry* in 1945, and he placed it in quotation marks when he did so as follows: "nothing is as practical as a good theory" (p. 129). Also, a note appears at the front of the chapter from which I quoted the passage in the text, and the note indicates that the last part of the chapter comes from "Kurt Lewin: Constructs in Psychology and Psychological Ecology" (*Univ. Iowa Stud. Child Welf.*, 1944, 20, 23–27), so the passage I quoted, which is part of the next-to-last paragraph in the chapter, may also have been published in the 1944 article.

5. The "shorter time" information was that the work was produced either immediately after it was assigned (well before the deadline) or immediately before the deadline. These two conditions were contrasted with subjects who performed the work steadily over the entire period from when it was assigned up to the deadline, a much longer interval. (See Chapter 6 and Persing 1992.)

6. In one of those incredible coincidences (i.e., connections!) one encounters in life, the astrophysicist husband of my new colleague Cathleen Burns turned out to be a long-time friend and collaborator with Gregory Benford. So Jack Burns was kind enough to provide the introduction, and I am most grateful that he did.

7. I presented several of these in a symposium about time at the 2000 Academy of Management Meeting (Bluedorn 2000b). I interviewed Greg on March 28, 2000.

8. As noted, this story was related to me by a resident of Missouri. I am sure residents of Kansas will have no trouble modifying the story to fit their tastes.

9. The ability to deal with the situation is an important part of flow theory too, because an important quality of the flow experience is the feeling that one's abilities "are well matched to the opportunities for action" (Csikszentmihalyi 1996, p. 111).

10. Certain situations in police work can generate a similar experience, as Flaherty describes: "the vivid perceptions, thoughts, and emotions of a police officer who is desperately trying to make sense of what is happening to him" (1999, p. 15).

11. Except for Lopata (1986), all of these sources explicitly addressed the perceived passage of time. However, Helena Lopata's statement that "at first, many widows feel in a 'limbo' location" (1986, p. 705) strongly suggests that movement toward an experience of more protracted duration is likely during the early stages of widowhood. Ken Starkey's descriptions of unemployment—"impairs the time sense" and "the individual becomes trapped in an endless present"—suggest a similar possibility for the unemployed (1989, p. 53).

12. The correct phrase appears to be the singular, daylight *saving* time. For example, see the usage in O'Malley (1990, pp. 256–308), especially the reference to the "original Daylight Saving Act of 1918" (p. 290). Also see the usage in Stephens (1994, p. 576) and the title of a U.S. Department of Transportation document given in Macey, which includes the phrase "Daylight Saving Time" (1994, p. 444). I must

confess that I, like many, have referred to it as daylight *savings* time most of my life; nevertheless, the formally given designation seems to be daylight saving time.

13. Actually, most of the United States makes these changes. Arizona stays on standard time throughout the year, and Indiana copes with being split by two time zones by observing a single time during the daylight-saving period (Bartky 2000, p. 261).

14. The changes into and out of daylight saving time have also been grouped with flying "across several time zones" as potential disrupters (Monk and Folkard 1976, p. 688), a classification with which anyone who has ever experienced jet lag will agree.

15. I calculated this percentage from data in Table 2 in Hicks, Lindseth, and Hawkins (1983, p. 66): $(2095/59008) \times 100\% = 3.6\%$.

16. The three items that measured the difficulty of the fall shift out of daylight saving time were (1) "I find the change out of daylight savings time each fall very disruptive"; (2) "I find it harder to deal with the change out of daylight savings time in the fall than the change into daylight savings time in the spring"; and (3) "I find the change out of daylight savings time each fall easy to deal with." Item 3 was reverse scored. The three items that measured the difficulty of the spring shift into daylight saving time were (4) "I find it harder to deal with the change into daylight savings time in the spring than the change out of daylight savings time in the fall;" (5) "I find the change into daylight savings time each spring easy to deal with;" and (6) "I find the change into daylight savings time each spring very disruptive." Item 5 was reverse scored. The same seven-item response scale was used for all six items: 1 = strongly disagree; 2 = moderately disagree; 3 = slightly agree; 4 = neither agree nor disagree; 5 = slightly agree; 6 = moderately agree; 7 = strongly agree. Note that the more common usage, "daylight *savings* time," is used in each item because it was more familiar to respondents. See note 12 to this chapter.

In the fall of 1997, on the Tuesday immediately following the shift out of daylight saving time, I administered both scales to 415 students in an undergraduate management course at the University of Missouri-Columbia. Similarly, in the spring of 1998, on the Tuesday immediately following the shift into daylight saving time, I administered both scales to 316 students in that semester's version of the same undergraduate management course. Missing values in the fall sample questionnaires reduced that sample size to 406 for analysis, and three missing values on these six items reduced the spring sample to 313 for this analysis.

I performed principle components analysis of the six daylight-saving-time-difficulty items in both samples using an orthogonal (varimax) rotation on these six items in each sample. The results are presented in the following table:

| Scale Item | Fall Sample (N = 406) | | Spring Sample (N = 313) | |
	Fall Difficulty Factor	Spring Difficulty Factor	Fall Difficulty Factor	Spring Difficulty Factor
1	.11	**.89**	.18	**.83**
2	−.25	**.82**	−.32	**.80**
3*	.03	**.87**	−.02	**.80**
4	**.85**	−.09	**.87**	−.14
5*	**.91**	−.03	**.87**	−.07
6	**.90**	.06	**.88**	.11
Eigenvalue	2.50	2.17	2.50	1.91

* Reversed scored

The analysis in each sample produced two eigenvalues greater than 1.00, with two factors in the fall data that explained 77.83 percent of the variance and two corresponding factors that explained 73.50 percent of the variance in the spring data. Cronbach's (1951) alpha coefficient for the scale created by the three spring items was .86 in the fall data and .84 in the spring data. For the scale created by the three fall items, the alpha coefficient was .83 in the fall data and .74 in the spring data. The means for the two scales were given in the text in Figure 7.2, and the standard deviations were 1.47 for the fall sample and .86 for the spring sample. The independent-sample t-test for the difference in the sample means revealed a statistically significant difference: $t(717)$ = 10.65, $p \leq .001$ (two-tailed test). The means compared in this t-test and in Figure 7.2 are the mean for the *fall scale* for the fall 1997 respondents and the mean for the *spring scale* for the spring 1998 respondents.

CHAPTER 8 *Carpe Diem*

1. Buehler, Griffin, and Ross found that connecting properly with past events counteracted the planning fallacy, but doing so did not improve the accuracy of predicted completion times. Rather, the errors became balanced between pessimistic and optimistic estimates (i.e., "subjects' predictions were as likely to be too long as they were to be too short," [1994, p. 376]).

2. See note 1 to this chapter.

3. The mean correlation between planning and growth was .17, and the mean correlation between planning and profitability was .12 (Miller and Cardinal 1994, p. 1656).

In further analysis, Miller and Cardinal treated the planning-growth and planning-profitability correlations from each sample as data points, hence as variables whose variance was to be explained, and they conducted separate multiple regression analyses on each set of correlations. Neither organizational size nor capital intensity was significantly related to either set of correlations, meaning that the planning-growth and planning-profitability relationships were about the same (positive) for both small and large companies, and for capital and labor-intensive companies.

Miller and Cardinal did find that the positive correlations between planning and organizational growth were larger when the studies controlled for industry effects, when performance data were obtained from people in the companies rather than from archival sources, when the measure of planning was all of the strategic planning process rather than just written guidelines for planning, and when the researchers used a high-quality strategy to assess the organization's strategy, one that assessed planning practices at the appropriate time vis-à-vis performance (1994, p. 1658). For the planning-profitability correlations, three of the four methodological variables—all but controlling for industry—had similar effects, and environmental turbulence was significantly related to the correlations too in that stronger planning-profitability correlations occur in more turbulent environments (pp. 1658–60).

4. This process of intragroup cooperation and intergroup competition was illustrated well in "The Dawn of Man" segment at the beginning of Stanley Kubrick's film *2001: A Space Odyssey*. Further, the end of that segment illustrated clearly how technological differences can confer major advantages on one group vis-à-vis other groups.

5. For examples of additional theory and research about frames and framing, see Bazerman (1984), Paese (1995), and Tversky and Kahneman (1981).

6. Ricardo Quinones's (1972) entire book is about the changing attitudes and beliefs concerning time that occurred during the Renaissance. His discussion of Petrarch's views, especially on pp. 135–52, are particularly relevant to beliefs and attitudes that can be related to time management, and elsewhere he uses and interprets Alberti's ideas as important evidence for the new temporal attitudes and beliefs (pp. 190–91).

7. As follows in the text, Alberti (1971, p. 180) anticipated this overall approach, and if Helen Reynolds and Mary Tramel quoted Benjamin Franklin correctly— "Take enough time to think and plan things in the order of their importance" (1979, p. 1)—Charles Schwab could have gotten this advice free from him too. To put the twenty-five thousand dollars Schwab was said to have paid Ivy Lee into perspective, note that these were early-twentieth-century dollars. Schwab was president of Bethlehem Steel Corporation from 1904 to 1916 (Bethlehem Steel

2001), and using 1913 as a year to convert a dollar value from (the first year available for such conversions on the Federal Reserve Bank of Minneapolis's Web site), $25,000 in 1913 would be worth $450,757.58 in 2001 (Federal Reserve Bank of Minneapolis 2001). If only Schwab had read Franklin or Alberti!

8. Alan Lakein actually distinguished "internal prime time" from "external prime time" (1973, pp. 48–50), with the internal form referring to when a person works best and the external form referring to when it is best to deal with other people. As I am using the phrase, prime time refers more or less to what Lakein identified as "internal prime time."

9. Although quiet time was clearly less polychronic in terms of interruptions from other engineers, Perlow's accounts do not directly reveal how monochronically or polychronically the engineers behaved while working by themselves during quiet time (some variation between engineers would be expected). Nevertheless, given the time log provided by one of the engineers (Perlow 1997, pp. 73–75), which reveals a day punctuated with interruptions from other engineers, it seems reasonable to conclude that by removing these interruptions, quiet time represented relatively less polychronic work periods for the engineers.

10. After reading a draft of this chapter, Ramon Aldag made the interesting observation that a therapy for what appears to be the excessive regret people can feel for actions forgone might be to have them write scenarios in which they identify some possible negative consequences if they had taken the action. Because the regret may stem in part from unrealistic, perhaps Pollyannaish images of what would have happened if the forgone action had been taken, Ray's idea seems worth exploring.

CHAPTER 9   *New Times*

1. These eight verses are taken from the King James Version of the Bible, of which Daniel Boorstin (1983) wrote, "It is perhaps the only literary masterpiece ever written by a committee" (p. 523). These eight verses in particular were especially well written, as the frequency with which they are still used today gives testimony.

2. Thanksgiving also seems to have been spared the Christmas cloning that is transforming Halloween in the United States. For one now sees trees and homes decorated with Halloween lights, just as one sees them bedecked with lights during the Christmas season. The leapfrogging of Thanksgiving in this regard is again noteworthy, as is the mimetic imitation of Christmas celebration practices.

3. The underlying principle is similar to the one followed by the famous discount retailer Sam Walton. Walton made it his personal mission to "ensure that constant change" was "a vital part of the Wal-Mart culture," so he sometimes

forced change "for change's sake alone" (Walton 1992, p. 169). The idea is that the more experience one has with change, the more successfully one can deal with it.

4. Coren's summary of the research on these shifts rings true for me. As an undergraduate, I preferred to write during the evening, as an assistant professor I preferred to write during the early afternoon, and today I prefer to write during the morning.

5. The Washington State University Foundation (2001) *Catalyst* stated this proverb the way I have presented it in the text, and the *Catalyst* identified it as a Greek proverb. CARE's Annual Report for 1997 (2001) identified a similar proverb as African in origin: "A society grows great when old men plant trees under whose shade they know they will never sit."

6. I learned from our guide why the zebras stood side by side and have had this explanation confirmed by others familiar with them. The explanation is also consistent with my own observations while on the Serengeti. And after reading about the zebra's behavior in a draft of this chapter, Ramon Aldag suggested that the zebras were the "Janus of the Serengeti." (Janus is the Roman god of beginnings, who was often portrayed as looking both forward and backward [Adkins and Adkins 1996, pp. 111–12]).

7. I coined the word *temporalist* and presented it at a workshop at the Academy of Management (Bluedorn 2000a) as a designation for those who study time. I distributed a handout at the workshop containing the following lexical formalization: "tem-por-al-ist (tem' prɛ list, tem' pɐr ɛ list), n. 1. One who explicitly studies time and temporal phenomena. 2. One who believes time and temporal topics are important and worthy of investigation, esp., in a scholarly manner. The eleven temporalists met at the Academy of Management to help themselves and others study time [TEMPORAL + -IST]."

8. I am indebted to Jean Bartunek, who directed me to Bonhoeffer as the source of the cheap grace concept. Dietrich Bonhoeffer was a Lutheran theologian with an ecumenical orientation. Thus it is not surprising that I first heard the phrase *cheap grace* in a sermon given by a Methodist minister, who said he had heard about the concept from a Catholic priest.

In addition to his writing, Bonhoeffer lived what he wrote and preached. While he could have spent World War II living safely in the United States, Bonhoeffer felt he was morally obligated to return to Germany and share what happened there if he were to have a justifiable role in helping rebuild the country after the war: "I cannot make this choice [to favor Germany's defeat in the war] in security" (Leibholz 1959, p. 13). Bonhoeffer played an active role in German resistance to the Nazis, was arrested by the Gestapo on April 5, 1943, and spent the rest of his life in that demonic regime's prisons and concentration camps, including Büchenwald. Death came near the end of the war at Flossenburg concentration camp on April 9, 1945, the result of special orders from Himmler himself (details from Leibholz 1959, pp. 11–17).

9. Carl Rogers addressed this point so very well in his discussion of science: "In sharp contradiction to some views that have been advanced, I would like to propose a two-pronged thesis: (i) In any scientific endeavor—whether 'pure' or applied science—there is a prior subjective choice of the purpose or value which that scientific work is perceived as serving. (ii) This subjective value choice which brings the scientific endeavor into being must always lie outside of that endeavor and can never become a part of the science involved in that endeavor" (Rogers and Skinner 1956, p. 1061).

When Rogers wrote these words, Frederick Taylor had long been in the grave, but if Taylor had followed these principles, he likely would have avoided making some fundamentally flawed claims about his system of scientific management. For Taylor made explicit claims that his methods could determine how much work a person "should" do in a day: "and the more modern and scientific management based on an accurate knowledge of how long it *should* take to do the work" (emphasis added; 1947b, p. 61), a claim he repeated later: "What constitutes a fair day's work will be a question for scientific investigation, instead of a subject to be bargained and haggled over" (Taylor 1947a, pp. 142–43).

The phrase a "fair day's work" also has implications for a fair day's pay, as yet one more passage from Taylor's work makes very clear: "The amount of work which a man should do in a day, what constitutes proper pay for this work, and the maximum number of hours per day which a man should work, together form the most important elements which are discussed between workmen and their employers. The writer has attempted to show that these matters can be much better determined by the expert time student than by either the union or a board of directors, and he firmly believes that in the future scientific time study will establish standards which will be accepted as fair by both sides" (1947b, pp. 186–87).

Taylor mixed "should" and "can," the value choice of the purpose of the experiment or study with the study and its findings themselves, although the results of Taylor's work can themselves be questioned (e.g., Wrege and Hodgetts 2000). Scientific management might be able to determine how fast someone *could* work; it can never determine how fast someone *should* work. Thus Robert Kanigel concluded about this issue, "The tools of scientific management alone could supply no certain answer, even in principle. The truth they discovered was only a limited truth that must be balanced against others" (1997, p. 518).

Taylor may have been oblivious to this distinction, or he may have simply re-fused to acknowledge it, but neither problem seems to have affected workers who were subjected to early scientific management practices: "The people of the United States have a right to say we want to work only so fast. We don't want to work as fast as we are able to. We want to work as fast as we think it's comfortable for us to work. We haven't come into existence for the purpose of seeing how great a task we can perform through a lifetime." And continuing, "Most people walk to work in the morning, if it isn't too far. If somebody should discover that they could run to work in one third the time, they might have no objection to have that fact as-certained, but if the man who ascertained it had the power to make them run, they might object to having him find it out" (both from N. P. Alifas as quoted in Com-mons 1921, pp. 148–49). These statements make clear that American labor was very aware of the distinction between *can* and *should*, even if Frederick Taylor mixed or ignored them. To me, because it is so fundamental, this is the most serious flaw in Taylor's work, yet it is seldom noted.

APPENDIX   *The Temporal Depth Index and Its Development*

1. Arguably, T. K. Das (1986, 2001) used the open-ended format successfully in two large survey research projects. However, the point is not whether one approach or the other can be used, but whether in a specific research context (e.g., large-sample survey research that uses self-administered questionnaires) a particular ap-proach might be easier for respondents to answer and would produce data easier for researchers to process. The Temporal Depth Index is not a general replacement for any other method of measuring temporal depth; instead, it is an instrument de-liberately designed to facilitate the collection of data about temporal depths in re-search projects that will employ self-administered questionnaires to collect data from large samples of respondents. It is also designed to facilitate the data-entry task during the data-processing and analysis phases of such research projects.

References

Abell, George. (1969). *Exploration of the Universe* (2nd ed.). New York: Holt, Rinehart and Winston.

Adam, Barbara. (1990). *Time and Social Theory*. Philadelphia: Temple University Press.

Adkins, Lesley, and Roy A. Adkins. (1996). *Dictionary of Roman Religion*. Oxford: Oxford University Press.

Albert, Stuart. (1995). Towards a Theory of Timing: An Archival Study of Timing Decisions in the Persian Gulf War. *Research in Organizational Behavior*, 17, 1–70.

Alberti, Leon Battista. (1969). *The Family in Renaissance Florence: A Translation by Renée Neu Watkins of I Libri Della Famiglia by Leon Battista Alberti*. Columbia: University of South Carolina Press. (Original work published circa 1434–43.)

Alberti, Leon Battista. (1971). *The Albertis of Florence: Leon Battista Alberti's Della Famiglia* (Guido A. Guarino, trans.). Lewisburg, Pa.: Bucknell University Press. (Original work published circa 1434–43.)

Alexander, Richard D. (1990). How Did Humans Evolve? Reflections on the Uniquely Unique Species. Ann Arbor, Mich.: Museum of Zoology Special Publication no. 1, University of Michigan.

Allen, Frederick. (2000). Technology at the End of the Century: A Look at Where We've Been and Where We May Be Going. *American Heritage of Invention and Technology*, 15(3), 10–16.

Ambrose, Stephen E. (1997). *Citizen Soldiers: The U.S. Army from the Normandy Beaches to the Bulge to the Surrender of Germany, June 7, 1944–May 7, 1945*. New York: Simon & Schuster.

Ancona, Deborah, and Chee-Leong Chong. (1996). Entrainment: Pace, Cycle, and Rhythm in Organizational Behavior. *Research in Organizational Behavior*, 18, 251–84.

Ancona, Deborah, and Chee-Leong Chong. (1999). Cycles and Synchrony: The Temporal Role of Context in Team Behavior. *Research on Managing Groups and Teams*, 2, 33–48.

Andrewes, William J. H. (1994a). Clocks and Watches: The Leap to Precision. In Samuel L. Macey (ed.), *Encyclopedia of Time* (pp. 123–27). New York: Garland.

Andrewes, William J. H. (1994b). Longitude. In Samuel L. Macy (ed.), *Encyclopedia of Time* (pp. 346–50). New York: Garland.

Aristotle. (1911). *The Poetics* (S. H. Butcher, trans.). In S. H. Butcher, *Aristotle's Theory of Poetry and Fine Art: With a Critical Text and Translation of the Poetics*. London: Macmillan. (Original work published circa –340.)

Aristotle. (1961). *Poetics* (Kenneth A. Telford, trans.). In Kenneth A. Telford, *Aristotle's Poetics: Translation and Analysis*. Chicago: Henry Regnery. (Original work published circa –340.)

Aschoff, Jürgen. (1965). The Phase-Angle Difference in Circadian Periodicity. In Jürgen Aschoff (ed.), *Circadian Clocks: Proceedings of the Feldafind Summer School, 7–18 Sept. 1964* (pp. 262–76). Amsterdam: North-Holland Publishing.

Aschoff, Jürgen. (1979). Circadian Rhythms: General Features and Endocrinological Aspects. In Dorothy T. Krieger (ed.), *Endocrine Rhythms* (pp. 1–61). New York: Raven Press.

Ashcraft, Mark H. (1989). *Human Memory and Cognition*. New York: HarperCollins.

Asimov, Isaac. (1972). *Asimov's Biographical Encyclopedia of Science and Technology* (new rev. ed.). Garden City, N.Y.: Doubleday.

Augustine, Saint. (1912). *St. Augustine's Confessions* (vol. 2) (William Watts, trans.). New York: Macmillan. (Original work published circa 400.)

Austin, Mary. (1970). *The American Rhythm: Studies and Reëxpressions of Amerindian Songs* (new and enlarged ed.). New York: Cooper Square. (Original work published in 1923 and 1930.)

Aveni, Anthony F. (1989). *Empires of Time: Calendars, Clocks, and Cultures*. New York: Kodansha International.

Axelrod, Robert. (1984). *The Evolution of Cooperation*. New York: Basic Books.

Backoff, Robert W. (1999, August 6–11). Presentation in "Theories and Research About Concepts of Time in Organizations: Three Conversations," a professional development workshop at the Annual Meeting of the Academy of Management, Chicago, Ill.

Bailyn, Lotte. (1993). *Breaking the Mold: Women, Men, and Time in the New Corporate World*. New York: Free Press.

Baldwin, James (1985). Every Good-Bye Ain't Gone. In James Baldwin (ed.), *The Price of the Ticket: Collected Nonfiction: 1948–1985* (pp. 641–47). New York: St. Martin's/Marek.

Baltes, Boris B., Thomas E. Briggs, Joseph W. Huff, Julie A. Wright, and George A. Neuman. (1999). Flexible and Compressed Workweek Schedules: A Meta-Analysis of Their Effects on Work-Related Criteria. *Journal of Applied Psychology*, 84, 496–513.

Barker, Joel A. (1992). *Paradigms: The Business of Discovering the Future.* New York: HarperBusiness.

Barley, Stephen R. (1988). On Technology, Time, and Social Order: Technically Induced Change in the Temporal Organization of Radiological Work. In Frank A. Dubinskas (ed.), *Making Time: Ethnographies of High-Technology Organizations* (pp. 123–69). Philadelphia: Temple University Press.

Barnett, Jo Ellen. (1998). *Time's Pendulum: From Sundials to Atomic Clocks, the Fascinating History of Timekeeping and How Our Discoveries Changed the World.* San Diego, Calif.: Harcourt, Brace.

Barnhart, Robert K. (ed.). (1988). *The Barnhart Dictionary of Etymology.* New York: The H. W. Wilson Company.

Bartko, John J. (1976). On Various Intraclass Correlation Reliability Coefficients. *Psychological Bulletin*, 83, 762–65.

Bartky, Ian R. (2000). *Selling the True Time: Nineteenth-Century Timekeeping in America.* Stanford, Calif.: Stanford University Press.

Baum, J. Robert, Edwin A. Locke, and Shelley A. Kirkpatrick. (1998). A Longitudinal Study of the Relation of Vision and Vision Communication to Venture Growth in Entrepreneurial Firms. *Journal of Applied Psychology*, 83, 43–54.

Bavelas, Janet B. (1973). Effects of the Temporal Context of Information. *Psychological Reports*, 32, 695–98.

Bazerman, Max H. (1984). The Relevance of Kahneman and Tversky's Concept of Framing to Organizational Behavior. *Journal of Management*, 10, 333–43.

Bede, The Venerable. (1969). *Bede's Ecclesiastical History of the English People* (Bertram Colgrave, trans., Bertram Colgrave and R. A. B. Mynors, eds.). Oxford, U.K.: Clarendon Press, Oxford University Press. (Original work published in 731.)

Benabou, Charles. (1999). Polychronicity and Temporal Dimensions of Work in Learning Organizations. *Journal of Managerial Psychology*, 14, 257–68.

Benford, Gregory. (1992). *Timescape.* New York: Bantam Books. (Original work published in 1980.)

Benford, Gregory. (1999). *Deep Time: How Humanity Communicates Across Millennia.* New York: Bard.

Bennett, Joel B. (2000). *Time and Intimacy: A New Science of Personal Relation-ships*. Mahway, N.J.: Lawrence Erlbaum.

Berger, Alisha. (1999, June 22). The All-Rise Method for Faster Meetings. *New York Times*, D7.

Berger, Peter L. (1969). *A Rumor of Angels: Modern Society and the Rediscovery of the Supernatural*. Garden City, N.Y.: Anchor Books.

Bergin, Thomas G. (1965). *Dante*. New York: The Orion Press.

Bergson, Henri. (1959). *Time and Free Will* (F. L. Pogson, trans.). New York: Macmillan. (Original work published in 1889.)

Berkman, Lisa F., and S. Leonard Syme. (1979). Social Networks, Host Resistance, and Mortality: A Nine-Year Follow-up Study of Alameda County Residents. *American Journal of Epidemiology*, 109, 186–204.

Berliner, Joseph S. (1970). A Problem in Soviet Business Administration. In Oscar Grusky and George A. Miller (eds.), *The Sociology of Organizations: Basic Studies* (pp. 557–64). New York: Free Press.

Berry, William B. N. (1968). *Growth of a Prehistoric Time Scale: Based on Organic Evolution*. San Francisco: W. H. Freeman.

Bethlehem Steel. (2001, March 27). A Brief Chronology of Bethlehem Steel. http://www.bethsteel.com/about/history/index.shtml.

Blackburn, Jospeh D. (ed.). (1991). *Time-Based Competition: The Next Battle-ground in American Manufacturing*. Homewood, Ill.: Business One Irwin.

Bliese, Paul D. (2000). Within-Group Agreement, Non-Independence, and Reliability: Implications for Data Aggregation and Analysis. In Katherine J. Klein and Steve W. J. Kozlowski (eds.), *Multilevel Theory, Research, and Methods in Organizations: Foundations, Extensions, and New Directions* (pp. 349–81). San Francisco: Jossey-Bass.

Bloomfield, Louis A. (1997). *How Things Work: The Physics of Everyday Life*. New York: Wiley.

Blount, Sally, and Gregory Janicik. (In press). Getting and Staying In-Pace: The "In-Synch" Preference and Its Implications for Work Groups. *Research on Managing Groups and Teams* (a JAI Press series).

Bluedorn, Allen C. (1980). Cutting the Gordian Knot: A Critique of the Effectiveness Tradition in Organizational Research. *Sociology and Social Research*, 64, 477–96.

Bluedorn, Allen C. (1993). Pilgrim's Progress: Trends and Convergence in Research on Organizational Size and Environments. *Journal of Management*, 19, 163–91.

Bluedorn, Allen C. (1997). Primary Rhythms, Information Processing, and

Planning: Toward a Strategic Temporal Technology. *Technology Studies*, 4, 1–36.

Bluedorn, Allen C. (1998). An Interview with Anthropologist Edward T. Hall. *Journal of Management Inquiry*, 7, 109–15.

Bluedorn, Allen C. (2000a, August 4–9). All Times Are Not the Same: A Workshop on Temporal Questions in Organizational Research and Methods for Studying Them. Workshop at the Annual Meeting of the Academy of Management, Toronto, Ontario, Canada.

Bluedorn, Allen C. (2000b, August 4–9). Management and the Challenge of Deep Time. Presentation in the symposium "The Role of Time in Organizational Life," at the Annual Meeting of the Academy of Management, Toronto, Ontario, Canada.

Bluedorn, Allen C. (2000c, April 14–16). Polychronicity, Change Orientation, and Organizational Attractiveness. Paper presented at the Annual Meeting of the Society for Industrial and Organizational Psychology, New Orleans, La.

Bluedorn, Allen C. (2000d, August 4–9). Stand-Up Meetings and the Quest for Speed. Presentation in the symposium "The Impact of Time Pressure and Timing on Work Group Effectiveness," at the Annual Meeting of the Academy of Management, Toronto, Ontario, Canada.

Bluedorn, Allen C. (2000e). Time and Organizational Culture. In Neal M. Ashkanasy, Celeste P. M. Wilderom, and Mark F. Peterson (eds.), *Handbook of Organizational Culture and Climate* (pp. 117–28). Thousand Oaks, Calif.: Sage Publications.

Bluedorn, Allen C., and Robert B. Denhardt. (1988). Time and Organizations. *Journal of Management*, 14, 299–320.

Bluedorn, Allen C., and Stephen P. Ferris. (2000, October 13). Temporal Depth, Age, and Organizational Performance. Paper presented in the INSEAD Strategic Management Seminar Series, Fontainebleau, France.

Bluedorn, Allen C., Thomas J. Kalliath, Michael J. Strube, and Gregg D. Martin. (1999). Polychronicity and the Inventory of Polychronic Values (IPV): The Development of an Instrument to Measure a Fundamental Dimension of Organizational Culture. *Journal of Managerial Psychology*, 14, 205–30.

Bluedorn, Allen C., Carol F. Kaufman, and Paul M. Lane. (1992). How Many Things Do You Like to Do at Once? An Introduction to Monochronic and Polychronic Time. *Academy of Management Executive*, 6(4), 17–26.

Bluedorn, Allen C., and Earl F. Lundgren. (1993). A Culture-Match Perspective for Strategic Change. *Research in Organizational Change and Development*, 7, 137–79.

Bluedorn, Allen C., Daniel B. Turban, and Mary Sue Love. (1999). The Effects of Stand-Up and Sit-Down Meeting Formats on Meeting Outcomes. *Journal of Applied Psychology*, 84, 277–85.

Bluedorn, John C. (2001). Can Democracy Help? Growth and Ethnic Divisions. *Economics Letters*, 70, 121–26.

Bolton, L. (1924). *Time Measurement: An Introduction to Means and Ways of Reckoning Physical and Civil Time.* London: G. Bell and Sons.

Bonhoeffer, Dietrich. (1959). *The Cost of Discipleship* (rev. and unabridged ed.) (R. H. Fuller and Irmgard Booth, trans.). New York: Macmillan. (Original work published in 1937.)

Boorstin, Daniel J. (1983). *The Discoverers.* New York: Vintage Books.

Borst, Arno. (1993). *The Ordering of Time: From the Ancient Computus to the Modern Computer* (Andrew Winnard, trans.). Chicago: University of Chicago Press.

Bourgeois, L. J., III, and Kathleen M. Eisenhardt. (1988). Strategic Decision Processes in High Velocity Environments: Four Cases in the Microcomputer Industry. *Management Science*, 34, 816–35.

Bouvet, Vincent, Claude Malécot, and Alix Sallé (eds.). (2000). *The Conciergerie: Palais de la Cité.* Paris: Caisse Nationale des Monuments Historiques et des Sites/Éditions du Patrimoine.

Boyd, John N., and Philip G. Zimbardo. (1997). Constructing Time After Death: The Transcendental-Future Time Perspective. *Time & Society*, 6, 35–54.

Brand, Stewart. (1999). *The Clock of the Long Now: Time and Responsibility.* New York: Basic Books.

Bridger, Jeffery C. (1994). Time and Narrative. In Samuel L. Macey (ed.), *Encyclopedia of Time* (pp. 605–6). New York: Garland.

Britton, Bruce K., and Abraham Tesser. (1991). Effects of Time-Management Practices on College Grades. *Journal of Educational Psychology*, 83, 405–10.

Brown, Alyson. (1998). 'Doing Time': The Extended Present of the Long-Term Prisoner. *Time & Society*, 7, 93–103.

Brown, B., A. Walker, C. V. Ward, and R. E. Leakey (1993). New *Australopithecus boisei* Calvaria from East Lake Turkana, Kenya. *American Journal of Physical Anthropology*, 91, 137–59.

Brown, Lloyd A. (1949). *The Story of MAPS.* New York: Bonanza Books.

Brown, Shona L., and Kathleen M. Eisenhardt. (1997). The Art of Continuous Change: Linking Complexity Theory and Time-Paced Evolution in Relentlessly Shifting Organizations. *Administrative Science Quarterly*, 42, 1–34.

Brown, Shona L., and Kathleen M. Eisenhardt. (1998). *Competing on the Edge: Strategy as Structured Chaos*. Boston: Harvard Business School Press.

Buehler, Roger, Dale Griffin, and Michael Ross. (1994). Exploring the "Planning Fallacy": Why People Underestimate Their Task Completion Times. *Journal of Personality and Social Psychology*, 67, 366–81.

Bunge, Mario. (1986). Book review of *Rational Thermodynamics* by C. Truesdell. *Philosophy of Science*, 53, 305–6.

Bunker, Barbara B., and Billie T. Alban. (1997). *Large Group Interventions: Engaging the Whole System for Rapid Change*. San Francisco: Jossey-Bass.

Burns, Tom, and G. M. Stalker. (1961). *The Management of Innovation*. London: Tavistock.

Butler, Richard. (1995). Time in Organizations: Its Experience, Explanations, and Effects. *Organization Studies*, 16, 925–50.

Calder, Nigel. (1983). *Timescale: An Atlas of the Fourth Dimension*. New York: Viking Press.

Calo, Bob (producer). (1996, August 25). Labor of Love, *Dateline NBC*. New York: National Broadcasting Company.

Canetti, Elias. (1989). *The Secret Heart of the Clock: Notes, Aphorisms, Fragments, 1973–1985* (Joel Agee, trans.). New York: Farrar, Strauss & Giroux.

CARE. (2001, February 14). *Annual Report 1997*. Http://www.care.org/publications/annualreport97/report_ea-me.html.

Carroll, Paul. (1993). *Big Blues: The Unmaking of IBM*. New York: Crown.

Carton, Barbara. (2001, July 6). In 24-Hour Workplace, Day Care Is Moving to the Night Shift. *Wall Street Journal*, A1, A4.

Cartwright, John. (2000). *Evolution and Human Behavior: Darwinian Perspectives on Human Nature*. Cambridge: A Bradford Book of MIT Press.

Carver, Charles S. (1989). How Should Multifaceted Personality Constructs Be Tested? Issues Illustrated by Self-Monitoring, Attributional Style, and Hardiness. *Journal of Personality and Social Psychology*, 56, 577–85.

Cavanagh, Gerald F., Dennis J. Moberg, and Manuel Velasquez. (1981). The Ethics of Organizational Politics. *Academy of Management Review*, 6, 363–74.

Cervantes Saavedra, Miguel de. (1831). *El Ingenioso Hidalgo Don Quijote de la Mancha* (parte segunda, tomo III). Madrid: D. J. Espinosa. (Original work published in 1605 and 1615.)

Cervantes Saavedra, Miguel de. (1881). *The Second Part of the Ingenious Knight, Don Quixote de la Mancha* (vol. 3) (Alexander J. Duffield, trans.). London: C. Kegan Paul. (Original work published in 1615.)

Cervantes Saavedra, Miguel de. (1898). *Don Quixote de la Mancha* (vols. 1 and 2)

(Henry E. Watts, trans.). New York: D. Appleton. (Original work published in 1605 and 1615.)

Chandler, Alfred D., Jr. (1962). *Strategy and Structure: Chapters in the History of the Industrial Enterprise*. Cambridge: MIT Press.

Chapple, Eliot D. (1970). *Culture and Biological Man: Explorations in Behavioral Anthropology*. New York: Holt, Rinehart and Winston.

Chapple, Eliot D. (1971). Toward a Mathematical Model of Interaction: Some Preliminary Considerations. In Paul Kay (ed.), *Explorations in Mathematical Anthropology* (pp. 141–78). Cambridge: MIT Press.

Charmaz, Kathy. (1991). *Good Days, Bad Days: The Self in Chronic Illness and Time*. New Brunswick, N.J.: Rutgers University Press.

Chatfield, Michael. (1996). Periodicity. In Michael Chatfield and Richard Vangermeersch (eds.), *The History of Accounting: An International Encyclopedia* (pp. 456–59). New York: Garland.

Churchill, Winston S. (1943). *The End of the Beginning: War Speeches by the Right Hon. Winston S. Churchill, C.H., M.P.* (Charles Eade, compiler). Boston: Little, Brown.

Churchill, Winston S. (1974). *Winston S. Churchill: His Complete Speeches, 1897–1963* (vol. 7: *1943–1949*) (Robert R. James, ed.). New York: Chelsea House.

Cicero, Marcus Tullius. (1962). *Orator*, with an English translation by H. M. Hubbell. In *Brutus* (G. L. Hendrickson, trans.); *Orator* (H. M. Hubbell, trans.) (pp. 295–509). Cambridge: Harvard University Press. (Original work published in –46.)

Cipolla, Carlo M. (1978). *Clocks and Culture: 1300–1700*. New York: W. W. Norton.

Clark, Peter. (1978). Temporal Inventories and Time Structuring in Large Organizations. In J. T. Fraser, N. Lawrence, and D. Park (eds.), *The Study of Time III* (pp. 391–416). New York: Springer-Verlag.

Clark, Peter. (1985). A Review of the Theories of Time and Structure for Organizational Sociology. *Research in the Sociology of Organizations*, 4, 35–80.

Cohen, Jacob, and Patricia Cohen. (1983). *Applied Multiple Regression/Correlation Analysis for the Behavioral Sciences* (2nd ed.). Hillsdale, N.J.: Lawrence Erlbaum.

Cohen, Sheldon, William J. Doyle, David P. Skoner, Bruce S. Rabin, and Jack M. Gwaltney. (1997). Social Ties and Susceptibility to the Common Cold. *Journal of the American Medical Association*, 277, 1940–44.

Cohen, Sheldon, and Thomas A. Wills. (1985). Stress, Social Support, and the Buffering Hypothesis. *Psychological Bulletin*, 98, 310–57.

Collins, James C., and Jerry I. Porras. (1997). *Built to Last: Successful Habits of Visionary Companies*. New York: HarperBusiness.

Commons, John R. (ed.). (1921). *Trade Unionism and Labor Problems* (second series). Boston: Ginn.

Comte, Auguste. (1970). *Introduction to Positive Philosophy* (Frederick Ferré, trans.). Indianapolis, Ind.: Bobbs-Merrill. (Original work published in 1830.)

Considine, Douglas M., and Glenn D. Considine (eds.). (1989). *Van Nostrand's Scientific Encyclopedia* (7th ed.). New York: Van Nostrand Reinhold.

Conte, Jeffrey M. (2000, April 14–16). Examining Relationships Among Polychronicity, Big Five Personality Dimensions, Absence, and Lateness. Paper presented at the Annual Meeting of the Society for Industrial and Organizational Psychology, New Orleans, La.

Conte, Jeffrey M., Tracey E. Rizzuto, and Dirk D. Steiner. (1999). A Construct-Oriented Analysis of Individual-Level Polychronicity. *Journal of Managerial Psychology*, 14, 269–87.

Coppens, Yves. (1994). East Side Story: The Origin of Humankind. *Scientific American*, 270(5), 88–95.

Coren, Stanley. (1996a). Daylight Savings Time and Traffic Accidents. *New England Journal of Medicine*, 334, 924.

Coren, Stanley. (1996b). *Sleep Thieves: An Eye-Opening Exploration into the Science and Mysteries of Sleep*. New York: Free Press.

Cotte, June, and S. Ratneshwar. (1999). Juggling and Hopping: What Does It Mean to Work Polychronically? *Journal of Managerial Psychology*, 14, 184–204.

Cottle, Thomas J. (1976). *Perceiving Time: A Psychological Investigation with Men and Women*. New York: Wiley.

Cottrell, David, and Mark C. Layton. (2000). *175 Ways to Get More Done in Less Time!* Dallas: CornerStone Leadership Institute.

Coveney, Peter, and Roger Highfield. (1990). *The Arrow of Time: A Voyage Through Science to Solve Time's Greatest Mystery*. New York: Fawcett, Columbine.

Covey, Stephen R. (1989). *The Seven Habits of Highly Effective People: Restoring the Character Ethic*. New York: Simon & Schuster.

Crichton, Michael. (1999). *Timeline*. New York: Ballantine.

Crombie, A. C. (1959). *Medieval and Early Modern Science (vol 1: Science in the Middle Ages)*. New York: Doubleday Anchor.

Cronbach, Lee J. (1951). Coefficient Alpha and the Internal Structure of Tests. *Psychometrika*, 16, 297–334.

Crosby, Alfred W. (1997). *The Measure of Reality: Quantification and Western Society, 1250–1600*. Cambridge: Cambridge University Press.

Csikszentmihalyi, Mihaly. (1975). *Beyond Boredom and Anxiety: The Experience of Play in Work and Games*. San Francisco: Jossey-Bass.

Csikszentmihalyi, Mihaly. (1990). *Flow: The Psychology of Optimal Experience*. New York: HarperPerennial.

Csikszentmihalyi, Mihaly. (1996). *Creativity: Flow and the Psychology of Discovery and Invention*. New York: HarperCollins.

Cushman, John H., Jr. (2000, June 15). Frederic Cassidy, 92, Expert on American Folk Language, Dies. *New York Times*, Arts/Cultural Desk, section B, 15.

Dansereau, Fred, Joseph A. Alutto, and Francis J. Yammarino. (1984). *Theory Testing in Organizational Behavior: The Varient Approach*. Englewood Cliffs, N.J.: Prentice-Hall.

Dante Alighieri. (1879). *The Divine Comedy of Dante Alighieri* (vol. 3: *Paradiso*) (Henry W. Longfellow, trans.). Boston: Houghton, Osgood. (Original work published in 1321.)

Dante Alighieri. (1921). *The Divine Comedy of Dante Alighieri* (vol. 3: *Paradiso*). Translated in English blank verse and with a commentary by Courtney Langdon; includes the Italian text. Cambridge: Harvard University Press. (Original work published in 1321.)

Dante Alighieri. (1980). *The Divine Comedy of Dante Alighieri: A Verse Translation* (Allen Mandelbaum, trans.). Berkeley, Calif.: University of California Press. (Original work published in 1321.)

Darwin, Charles. (1859). *On the Origin of Species by Means of Natural Selection, or the Preservation of Favoured Races in the Struggle for Life*. London: John Murray.

Das, T. K. (1986). *The Subjective Side of Strategy Making: Future Orientations and Perceptions of Executives*. New York: Praeger.

Das, T. K. (1987). Strategic Planning and Individual Temporal Orientation. *Strategic Management Journal*, 8, 203–9.

Das, T. K. (2001). Time and Ethics: An Empirical Study of the Temporal Orientations and Ethical Valences of Senior Business Executives. Working paper, Department of Management, Baruch College, City University of New York.

Davies, Paul. (1995). *About Time: Einstein's Unfinished Revolution*. New York: Simon & Schuster.

Deacon, Terrence W. (1997). *The Symbolic Species: The Co-Evolution of Language and the Brain*. New York: W. W. Norton.

Deal, Terrence E., and Allan A. Kennedy. (1999). *The New Corporate Cultures: Revitalizing the Workplace After Downsizing, Mergers, and Reengineering.* Reading, Mass.: Perseus Books.

de Grazia, Sebastian. (1962). *Of Time, Work, and Leisure.* New York: Twentieth Century Fund.

Denbigh, Kenneth G. (1994). Thermodynamics. In Samuel L. Macey (ed.), *Encyclopedia of Time* (pp. 600–601). New York: Garland.

DePalma, Anthony. (1994, June 26). It Takes More Than a Visa to Do Business in Mexico. *New York Times*, F5.

de Tocqueville, Alexis. (1945a). *Democracy in America* (vol. 1) (Henry Reeve, trans., Phillips Bradley, ed.). New York: Vintage Books. (Original work published in 1835.)

de Tocqueville, Alexis. (1945b). *Democracy in America* (vol. 2) (Henry Reeve, trans., Phillips Bradley, ed.). New York: Vintage Books. (Original work published in 1840.)

Dickens, Charles. (1984). *A Christmas Carol and Other Christmas Stories.* New York: Signet Classic from New American Library. (Original work published in 1843.)

Dickinson, Emily. (1890). *Poems* (8th ed.) (Mabel Loomis Todd and T. W. Higginson, eds.). Boston: Roberts Brothers.

Dickinson, Emily. (1960). *The Complete Poems of Emily Dickinson* (Thomas H. Johnson, ed.). Boston: Little, Brown.

Digman, John M. (1990). Personality Structure: Emergence of the Five-Factor model. *Annual Review of Psychology*, 41, 417–40.

DiMaggio, Paul J., and Walter W. Powell. (1983). The Iron Cage Revisited: Institutional Isomorphism and Collective Rationality in Organizational Fields. *American Sociological Review*, 48, 147–60.

Dohrn-van Rossum, Gerhard. (1996). *History of the Hour: Clocks and Modern Temporal Orders* (Thomas Dunlap, trans.). Chicago: University of Chicago Press.

Donaldson, Lex. (2001). *The Contingency Theory of Organizations.* Thousand Oaks, Calif.: Sage Publications.

Doob, Leonard W. (1971). *Patterning of Time.* New Haven, Conn.: Yale University Press.

Dougherty, Thomas W., and Robert D. Pritchard. (1985). The Measurement of Role Variables: Exploratory Examination of a New Approach. *Organizational Behavior and Human Decision Processes*, 35, 141–55.

Drucker, Peter F. (1967). *The Effective Executive.* New York: Harper & Row.

Drucker, Peter F. (1974). *Management: Tasks, Responsibilities, Practices.* New York: Harper & Row.

Dubinskas, Frank A. (1988). Janus Organizations: Scientists and Managers in Genetic Engineering Firms. In Frank A. Dubinskas (ed.), *Making Time: Ethnographies of High-Technology Organizations* (pp. 170–232). Philadelphia: Temple University Press.

Dunbar, Robin I. M. (1998). The Social Brain Hypothesis. *Evolutionary Anthropology*, 6, 178–90.

Duncan, David E. (1998). *Calendar: Humanity's Epic Struggle to Determine a True and Accurate Year.* New York: Bard.

Dunham, Randall B. (1977). Shift Work: A Review and Theoretical Analysis. *Academy of Management Review*, 2, 624–34.

Durkheim, Émile. (1915). *The Elementary Forms of the Religious Life: A Study in Religious Sociology* (Joseph W. Swain, trans.). London: George Allen & Unwin.

Earley, P. Christopher, and Elaine Mosakowski. (2000). Creating Hybrid Team Cultures: An Empirical Test of Transnational Team Functioning. *Academy of Management Journal*, 43, 26–49.

Ebert, Ronald J., and DeWayne Piehl. (1973). Time Horizon: A Concept for Management. *California Management Review*, 15(4), 35–41.

Eddington, A. S. (1928). *The Nature of the Physical World.* New York: Macmillan.

Edwards, Jeffrey R. (1993). Problems with the Use of Profile Similarity Indices in the Study of Congruence in Organizational Research. *Personnel Psychology*, 46, 641–65.

Edwards, Jeffrey R. (1994). The Study of Congruence in Organizational Behavior Research: Critique and a Proposed Alternative. *Organizational Behavior and Human Decision Processes*, 58, 51–100. Erratum 58, 323–25.

Edwards, Jeffrey R., and R. V. Harrison. (1993). Job Demands and Worker Health: Three-Dimensional Reexamination of the Relationship Between Person-Environment Fit and Strain. *Journal of Applied Psychology*, 78, 628–48.

Edwards, Jeffrey R., and Mark E. Parry. (1993). On the Use of Polynomial Regression Equations as an Alternative to Difference Scores in Organizational Research. *Academy of Management Journal*, 36, 1577–613.

Einstein, Albert. (1949). Autobiographical Notes. In Paul A. Schilpp (ed.), *Albert Einstein: Philosopher-Scientist* (pp. 1–95). Evanston, Ill.: Library of Living Philosophers.

Einstein, Albert. (1961). *Relativity: The Special and the General Theory* (includes

"Note to the Fifteenth Edition"). (Robert W. Lawson, trans.). New York: Crown. (Original work published in 1920.)

Eiseley, Loren. (1958). *Darwin's Century: Evolution and the Men Who Discovered It*. Garden City, N.Y.: Anchor Books.

Eiseley, Loren. (1960). *The Firmament of Time*. New York: Atheneum.

Eisenhardt, Kathleen M. (1989). Making Fast Strategic Decisions in High-Velocity Environments. *Academy of Management Journal*, 32, 543–76.

Eldredge, Niles, and Stephen J. Gould. (1972). Punctuated Equilibria: An Alternative to Phyletic Gradualism. In Thomas J. M. Schopf (ed.), *Models in Paleobiology* (pp. 82–115). San Francisco: Freeman, Cooper.

Eliade, Mircea. (1954). *The Myth of the Eternal Return or, Cosmos and History* (Williard R. Trask, trans.). Princeton, N.J.: Princeton University Press. (Original work published in 1949.)

El Sawy, Omar A. (1983). *Temporal Perspective and Managerial Attention: A Study of Chief Executive Strategic Behavior. Dissertation Abstracts International*, 44(05A), 1556–57.

Emerson, Ralph Waldo. (1983). *The Collected Works of Ralph Waldo Emerson* (vol. 3: *Essays: Second Series*) (text established by Alfred R. Ferguson and Jean Ferguson Carr). Cambridge: Harvard University Press, Belknap Press. (Original work published in 1844.)

Emery, Douglas R., John D. Finnerty, and John D. Stowe. (1998). *Principles of Financial Management*. Upper Saddle River, N.J.: Prentice-Hall.

Evans-Pritchard, E. E. (1940). *The Nuer: A Description of the Modes of Livelihood and Political Institutions of a Nilotic People*. London: Oxford University Press.

Fayol, Henri. (1949). *General and Industrial Management* (Constance Storrs, trans.). London: Sir Isaac Pitman and Sons. (Original work published in 1916.)

Federal Reserve Bank of Minneapolis. (2001, March 27). What is a Dollar Worth? http://woodrow.mpls.frb.fed.us/economy/calc/cpihome.html.

Feldman, David. (1987). *Why Do Clocks Run Clockwise? And Other Imponderables*. New York: Hallmark Books and HarperPerennial.

Ferner, Jack D. (1980). *Successful Time Management*. New York: Wiley.

Fine, Charles H. (1998). *Clockspeed: Winning Industry Control in the Age of Temporary Advantage*. Reading, Mass.: Perseus Books.

Fishbein, Martin, and Icek Ajzen. (1975). *Belief, Attitude, Intention, and Behavior: An Introduction to Theory and Research*. Reading, Mass.: Addison-Wesley.

Fitzgerald, F. Scott. (1945). *The Crack-Up* (Edmund Wilson, ed.). New York: New Directions Books.

Flaherty, Michael G. (1999). *A Watched Pot: How We Experience Time*. New York: New York University Press.

Ford, Henry. (1922). *My Life and Work* (in collaboration with Samuel Crowther). Garden City, N.Y.: Doubleday, Page.

Fraisse, Paul. (1963). *The Psychology of Time* (Jennifer Leith, trans.). New York: Harper & Row.

Fraisse, Paul. (1984). Perception and Estimation of Time. *Annual Review of Psychology*, 35, 1–36.

Fraser, J. T. (1975). *Of Time, Passion, and Knowledge: Reflections on the Strategy of Existence*. New York: George Braziller.

Fraser, J. T. (1978). Temporal Levels: Sociobiological Aspects of a Fundamental Synthesis. *Journal of Social and Biological Structures*, 1, 339–55.

Fraser, J. T. (1981). Temporal Levels and Reality Testing. *International Journal of Psycho-Analysis*, 62, 3–26.

Fraser, J. T. (1987). *Time, the Familiar Stranger*. Redmond, Wash.: Tempus Books of Microsoft Press.

Fraser, J. T. (1990). *Of Time, Passion, and Knowledge: Reflections on the Strategy of Existence* (2nd ed.). Princeton, N.J.: Princeton University Press.

Fraser, J. T. (1994). Hierarchical Theory of Time. In Samuel L. Macey (ed.), *Encyclopedia of Time* (pp. 262–64). New York: Garland.

Fraser, J. T. (1999). *Time, Conflict, and Human Values*. Urbana, Ill.: University of Illinois Press.

Freedman, Jonathan L., and Donald R. Edwards. (1988). Time Pressure, Task Performance, and Enjoyment. In Joseph E. McGrath (ed.), *The Social Psychology Time: New Perspectives* (pp. 113–33). Newbury Park, Calif.: Sage Publications.

Frei, Richard L., Bernadette Racicot, and Angela Travagline. (1999). The Impact of Monochronic and Type A Behavior Patterns on Research Productivity and Stress. *Journal of Managerial Psychology*, 14, 374–87.

Friedman, Meyer, and Ray H. Rosenman. (1974). *Type A Behavior and Your Heart*. New York: Alfred A. Knopf.

Ganey, Terry. (1983, June 21). Bond Signs Post-Labor Day School Bill. *St. Louis Post-Dispatch*, 1A, 4A.

Ganitsky, Joseph, and Gerhard E. Watzke. (1990). Implications of Different Time Perspectives for Human Resource Management in International Joint Ventures. *Management International Review*, 30 (special issue), 37–51.

Gee, Henry. (1999). *In Search of Deep Time: Beyond the Fossil Record to a New History of Life*. New York: Free Press.

George, Jennifer M., and Gareth R. Jones. (1999). *Understanding and Managing Organizational Behavior* (2nd ed.). Reading, Mass.: Addison-Wesley.

George, Jennifer M., and Gareth R. Jones. (2000). The Role of Time in Theory and Theory Building. *Journal of Management*, 26, 657–84.

Gersick, Connie J. G. (1988). Time and Transition in Work Teams: Toward a New Model of Group Development. *Academy of Management Journal*, 31, 9–41.

Gersick, Connie J. G. (1989). Marking Time: Predictable Transitions in Task Groups. *Academy of Management Journal*, 32, 274–309.

Gersick, Connie J. G. (1991). Revolutionary Change Theories: A Multilevel Exploration of the Punctuated Equilibrium Paradigm. *Academy of Management Review*, 16, 10–36.

Gersick, Connie J. G. (1994). Pacing Strategic Change: The Case of a New Venture. *Academy of Management Journal*, 37, 9–45.

Gesteland, Richard R. (1999). *Cross-Cultural Business Behavior: Marketing, Negotiating, and Managing Across Cultures*. Copenhagen: Handelshøjskolens Forlag, Copenhaygen Business School Press.

Giddens, Anthony. (1984). *The Constitution of Society: Outline of the Theory of Structuration*. Cambridge, U.K.: Polity Press.

Giddens, Anthony. (1995). *A Contemporary Critique of Historical Materialism* (2nd ed.). Stanford, Calif.: Stanford University Press.

Gilbert, Martin. (1994). *The First World War: A Complete History*. New York: Henry Holt.

Gilbreth, Frank B., Jr., and Ernestine G. Carey. (1948). *Cheaper by the Dozen*. New York: Thomas Y. Crowell.

Gilovich, Thomas, and Victoria H. Medvec. (1994). The Temporal Pattern to the Experience of Regret. *Journal of Personality and Social Psychology*, 67, 357–65.

Gilovich, Thomas, and Victoria H. Medvec. (1995). The Experience of Regret: What, When, and Why. *Psychological Review*, 102, 379–95.

Gilovich, Thomas, Victoria H. Medvec, and Daniel Kahneman. (1998). Varieties of Regret: A Debate and Partial Resolution. *Psychological Review*, 105, 602–5.

Gimpel, Jean. (1976). *The Medieval Machine: The Industrial Revolution of the Middle Ages*. New York: Holt, Rinehart and Winston.

Glass, Gene V., and Julian C. Stanley. (1970). *Statistical Methods in Education and Psychology*. Englewood Cliffs, N.J.: Prentice-Hall.

Gleick, James. (1999). *Faster: The Acceleration of Just About Everything*. New York: Pantheon Books.

Glennie, Paul, and Nigel Thrift. (1996). Reworking E. P. Thompson's 'Time, Work-Discipline and Industrial Capitalism.' *Time & Society*, 5, 275–99.

Goffman, Erving. (1974). *Frame Analysis: An Essay on the Organization of Experience*. Cambridge: Harvard University Press.

Goodman, Ellen. (1979). *Close to Home*. New York: Fawcett Crest.

Gopen, George D., and Judith A. Swan. (1990). The Science of Scientific Writing. *American Scientist*, 78, 550–58.

Gore, Rick. (1997). The First Steps. *National Geographic*, 191(2), 72–99.

Gould, Rupert T. (1960). *The Marine Chronometer: Its History and Development*. London: Holland Press. (Original work published in 1923.)

Gould, Stephen Jay. (1987). *Time's Arrow, Time's Cycle: Myth and Metaphor in the Discovery of Geological Time*. Cambridge: Harvard University Press.

Gourley, Catherine. (1997). *Wheels of Time: A Biography of Henry Ford*. Brookfield, Conn.: Millbrook Press.

Grove, Andy. (1996, May 13). Taking on Prostate Cancer. *Fortune*, 54–72.

Guest, Robert H. (1956). Of Time and the Foreman. *Personnel*, 32, 478–86.

Gulick, Luther. (1987). Time and Public Administration. *Public Administration Review*, 47, 115–19.

Gulick, Luther, and L. Urwick (eds.). (1937). *Papers on the Science of Administration*. New York: Institute of Public Administration, Columbia University.

Gupta, Shalini, and L. L. Cummings. (1986). Perceived Speed of Time and Task Affect. *Perceptual and Motor Skills*, 63, 971–80.

Gurvitch, Georges. (1964). *The Spectrum of Social Time* (Myrtle Korenbaum, trans.). Dordrecht-Holland, Netherlands: D. Reidel Publishing.

Guthrie, James P., Ronald A. Ash, and Venkat Bendapudi. (1995). Additional Validity Evidence for a Measure of Morningness. *Journal of Applied Psychology*, 80, 186–90.

Haase, Richard F., Dong Y. Lee, and Donald L. Banks. (1979). Cognitive Correlates of Polychronicity. *Perceptual and Motor Skills*, 49, 271–82.

Hackman, J. Richard, and Greg R. Oldham. (1975). Development of the Job Diagnostic Survey. *Journal of Applied Psychology*, 60, 159–70.

Hackman, J. Richard, and Greg R. Oldham. (1976). Motivation Through the Design of Work: Test of a Theory. *Organizational Behavior and Human Performance*, 16, 250–79.

Hall, Edward T. (1981a). *Beyond Culture*. New York: Anchor Books. (Original work published in 1976.)

Hall, Edward T. (1981b). *The Silent Language*. New York: Anchor Books. (Original work published in 1959.)

Hall, Edward T. (1982). *The Hidden Dimension*. New York: Anchor Books. (Original work published in 1966.)

Hall, Edward T. (1983). *The Dance of Life: The Other Dimension of Time*. Garden City, N.Y.: Anchor Press.

Hall, Edward T., and Mildred R. Hall. (1987). *Hidden Differences: Doing Business with the Japanese*. New York: Anchor Books.

Hall, Edward T., and Mildred R. Hall. (1990). *Understanding Cultural Differences: Germans, French, and Americans*. Yarmouth, Maine: Intercultural Press.

Hall, Jay. (1971). Decisions, Decisions, Decisions. *Psychology Today*, 5(6), 51–54, 86, 88.

Hall, Jay, and W. H. Watson. (1970). The Effects of a Normative Intervention on Group Decision-Making Performance. *Human Relations*, 23, 299–317.

Hannan, Michael T., and John Freeman. (1989). *Organizational Ecology*. Cambridge: Harvard University Press.

Hardin, Garrett. (1968). The Tragedy of the Commons. *Science*, 162, 1243–48.

Harrison, Lionel G. (1988). Kinetic Theory of Living Pattern and Form and Its Possible Relationship to Evolution. In Bruce H. Weber, David J. Depew, and James D. Smith (eds.), *Entropy, Information, and Evolution: New Perspectives on Physical and Biological Evolution* (pp. 53–74). Cambridge: MIT Press.

Hassard, John. (1989). Time and Organization. In Paul Blyton, John Hassard, Stephen Hill, and Ken Starkey (eds.), *Time, Work and Organization* (pp. 79–104). London: Routledge.

Hassard, John. (1996). Images of Time in Work and Organization. In Stewart R. Clegg, Cynthia Hardy, and Walter R. Nord (eds.), *Handbook of Organization Studies* (pp. 581–98). London: Sage Publications.

Hastings, Donald F. (1999). Lincoln Electric's Harsh Lessons from International Expansion. *Harvard Business Review*, 77(3), 162–78.

Hawking, Stephen W. (1988). *A Brief History of Time: From the Big Bang to Black Holes*. New York: Bantam Books.

Hay, Michael, and Jean-Claude Usunier. (1993). Time and Strategic Actions: A Cross-Cultural View. *Time & Society*, 2, 313–33.

Heath, Louise R. (1936). *The Concept of Time*. Chicago: The University of Chicago Press.

Hedrick, Lucy H. (1992). *365 Ways to Save Time*. New York: Hearst Books.

Heiken, Grant H., David T. Vaniman, and Bevan M. French. (1991). *Lunar Sourcebook: A User's Guide to the Moon*. Cambridge: Cambridge University Press.

Heinlein, Robert A. (1973). *Time Enough for Love*. New York: Berkeley Medallion.

Heschel, Abraham J. (1951). *The Sabbath: Its Meaning for Modern Man*. New York: Farrar, Straus and Giroux.

Hicks, Robert A., Kristin Lindseth, and James Hawkins. (1983). Daylight Saving-Time Changes Increase Traffic Accidents. *Perceptual and Motor Skills*, 56, 64–66.

Hirsch, James S. (2000). *Hurricane: The Miraculous Journey of Rubin Carter*. Boston: Houghton Mifflin.

Hobbes, Thomas. (1968). *Leviathan* (C. B. MacPherson, ed.). London: Penguin Books. (Original work published in 1651.)

Hochschild, Arlie R. (1989). *The Second Shift: Working Parents and the Revolution at Home*. New York: Viking.

Hochschild, Arlie R. (1997). *The Time Bind: When Work Becomes Home and Home Becomes Work*. New York: Metropolitan Books.

Hofstede, Geert, and Michael H. Bond. (1988). The Confucius Connection: From Cultural Roots to Economic Growth. *Organizational Dynamics*, 16(4), 4–21.

Hogan, H. Wayne. (1975). Time Perception and Stimulus Preference as a Function of Stimulus Complexity. *Journal of Personality and Social Psychology*, 31, 32–35.

House, James S., Karl R. Landis, and Debra Umberson. (1988). Social Relationships and Health. *Science*, 241, 540–45.

Husserl, Edmund. (1964). *The Phenomenology of Internal Time-Consciousness*. (Martin Heidegger, ed.; James S. Churchill, trans.) Bloomington, Ind.: Indiana University Press.

Hutton, James. (1959). *Theory of the Earth with Proofs and Illustrations*. Weinheim, Germany: H. R. Engelmann (J. Cramer) and Wheldon & Wesley. (Original work published in 1795.)

Huy, Quy Nguyen. (2001). Time, Temporal Capability, and Planned Change. *Academy of Management Review*, 26, 601–23.

Hymowitz, Carol. (2001, February 27). Taking Time to Focus on the Big Picture Despite Flood of Data. *Wall Street Journal*, B1.

Izraeli, Dafna M., and Todd D. Jick. (1986). The Art of Saying No: Linking Power to Culture. *Organization Studies*, 7, 171–92.

Jamal, Muhammad. (1981). Shift Work Related to Job Attitudes, Social Participation and Withdrawal Behavior: A Study of Nurses and Industrial Workers. *Personnel Psychology*, 34, 535–47.

James, Lawrence R., and Jeanne M. Brett. (1984). Mediators, Moderators, and Tests for Mediation. *Journal of Applied Psychology*, 69, 307–21.

James, Lawrence R., Robert G. Demaree, and Gerrit Wolf. (1984). Estimating Within-Group Interrater Reliability With and Without Response Bias. *Journal of Applied Psychology*, 69, 85–98.

James, William. (1918). *The Principles of Psychology* (vol. 1). New York: Henry Holt. (Original work published in 1890.)

Janicik, Gregory A., and Caroline A. Bartel. (2001, August 3–8). Managing Time and Pace in Project Groups: Effects of Temporal Planning on Coordination. Paper presented at the Annual Meeting of the Academy of Management, Washington, D.C.

Janis, Irving L. (1972). *Victims of Groupthink: A Psychological Study of Foreign-Policy Decisions and Fiascoes*. Boston: Houghton Mifflin.

Jaques, Elliott. (1982). *The Form of Time*. New York: Crane Russak.

Jaques, Elliott. (1998a). *Requisite Organization: A Total System for Effective Managerial Organization and Managerial Leadership for the 21st Century* (rev. 2nd ed.). Arlington, Va.: Cason Hall.

Jaques, Elliott. (1998b). *Time-Span Handbook*. Arlington, Va.: Cason Hall.

Javidan, Mansour. (1984). The Impact of Environmental Uncertainty on Long-Range Planning Practices of the U.S. Savings and Loan Industry. *Strategic Management Journal*, 5, 381–92.

Johanson, Donald, and Blake Edgar. (1996). *From Lucy to Language*. New York: Simon & Schuster.

Judge, William Q., and Alex Miller. (1991). Antecedents and Outcomes of Decision Speed in Different Environmental Contexts. *Academy of Management Journal*, 34, 449–63.

Judge, William Q., and Mark Spitzfaden. (1995). The Management of Strategic Time Horizons Within Biotechnology Firms: The Impact of Cognitive Complexity on Time Horizon Diversity. *Journal of Management Inquiry*, 4, 179–96.

Kahn, Joseph P. (2000, June 20). Till the End of Time. *Boston Globe*, D11.

Kahneman, Daniel, and Amos Tversky. (1979). Intuitive Prediction: Biases and Corrective Procedures. *TIMS Studies in the Management Sciences*, 12 (S. Makridakis and S. C. Wheelwright, eds.), 313–27. Amsterdam: North-Holland Publishing.

Kamstra, Mark J., Lisa A. Kramer, and Maurice D. Levi. (2000). Losing Sleep at the Market: The Daylight Saving Anomaly. *American Economic Review*, 90, 1005–11.

Kanigel, Robert. (1997). *The One Best Way: Frederick Winslow Taylor and the Enigma of Efficiency*. New York: Viking.

Kaplan, Robert. (1999). *The Nothing That Is: A Natural History of Zero*. New York: Oxford University Press.

Kasof, Joseph. (1997). Creativity and Breadth of Attention. *Creativity Research Journal*, 10, 303–15.

Kaufman, Carol F., and Paul M. Lane. (1996). Time and Technology: The Growing Nexus. In Ruby R. Dholakia, Norbert Mundorf, and Nikhilesh Dholakia (eds.), *New Infotainment Technologies in the Home: Demand-Side Perspectives* (pp. 134–56). Mahwah, N.J.: Lawrence Erlbaum.

Kaufman, Carol F., Paul M. Lane, and Jay D. Lindquist. (1991a). Exploring More Than 24 Hours a Day: A Preliminary Investigation of Polychronic Time Use. *Journal of Consumer Research*, 18, 392–401.

Kaufman, Carol F., Paul M. Lane, and Jay D. Lindquist. (1991b). Time Congruity in the Organization: A Proposed Quality-of-Life Framework. *Journal of Business and Psychology*, 6, 79–106.

Kaufman-Scarborough, Carol, and Jay D. Lindquist. (1999). Time Management and Polychronicity: Comparisons, Contrasts, and Insights for the Workplace. *Journal of Managerial Psychology*, 14, 288–312.

Kern, Stephen. (1983). *The Culture of Time and Space: 1880–1918*. Cambridge: Harvard University Press.

Klein, Katherine J., Paul D. Bliese, Steve W. J. Kozlowski, Fred Dansereau, Mark B. Gavin, Mark A. Griffin, David A. Hofmann, Lawrence R. James, Francis J. Yammarino, and Michelle C. Bligh. (2000). Multilevel Analytical Techniques: Commonalities, Differences, and Continuing Questions. In Katherine J. Klein and Steve W. J. Kozlowski (eds.), *Multilevel Theory, Research, and Methods in Organizations: Foundations, Extensions, and New Directions* (pp. 512–53). San Francisco: Jossey-Bass.

Kluckhohn, Florence R., and Fred. L. Strodtbeck. (1961). *Variations in Value Orientations*. Evanston, Ill.: Row, Peterson.

Knight, Charles F. (1992). Emerson Electric: Consistent Profits, Consistently. *Harvard Business Review*, 70(1), 57–69.

Koestler, Arthur. (1959). *The Sleep Walkers: A History of Man's Changing Vision of the Universe*. New York: Macmillan.

Kondratieff, N. D. (1935). The Long Waves in Economic Life. *Review of Economic Statistics*, 17, 105–15. (Original work published in 1926; 1935 translation by W. F. Stolper.)

Kotter, John P. (1982). *The General Managers*. New York: Free Press.

Kreitner, Robert. (2001). *Management* (8th ed.). Boston: Houghton Mifflin.

Kroeber, A. L. (1917). The Superorganic. *American Anthropologist*, 19, 163–213.

Kuhn, Thomas S. (1970). *The Structure of Scientific Revolutions* (2nd ed., enlarged). Chicago: University of Chicago Press.

Lakein, Alan. (1973). *How to Get Control of Your Time and Your Life*. New York: Signet, New American Library.

Landes, David S. (1983). *Revolution in Time: Clocks and the Making of the Modern World*. Cambridge: Harvard University Press, Belknap Press.

Landes, David S. (1998). *The Wealth and Poverty of Nations: Why Some Are So Rich and Some So Poor*. New York: W. W. Norton.

Larwood, Laurie, Cecilia M. Falbe, Mark P. Kriger, and Paul Miesing. (1995). Structure and Meaning of Organizational Vision. *Academy of Management Journal*, 38, 740–69.

Lauer, Robert H. (1981). *Temporal Man: The Meaning and Uses of Social Time*. New York: Praeger.

Lawrence, Barbara S. (1984). Age Grading: The Implicit Organizational Timetable. *Journal of Occupational Behaviour*, 5, 23–35.

Lawrence, Paul R., and Jay W. Lorsch. (1967). *Organization and Environment: Managing Differentiation and Integration*. Boston: Harvard University, Graduate School of Business Administration.

Lay, Clarry H. (1986). At Last, My Research Article on Procrastination. *Journal of Research in Personality*, 20, 474–95.

Leakey, Mary D. (1979). Footprints in the Ashes of Time. *National Geographic*, 155(4), 446–57.

Leakey, Meave G., Craig S. Feibel, Ian McDougall, Carol Ward, and Alan Walker. (1998). New Specimens and Confirmation of an Early Age for *Australopithecus anamensis*. *Nature*, 393, 62–66.

Least Heat Moon, William. (1982). *Blue Highways: A Journey into America*. New York: Fawcett Crest.

LeBoeuf, Michael. (1979). *Working Smart: How to Accomplish More in Half the Time*. New York: Warner Books.

Lee, H. (1999). Time and Information Technology: Monochronicity, Polychronicity, and Temporal Symmetry. *European Journal of Information Systems*, 8, 16–26.

Lee, Heejin, and Jonathan Liebenau. (1999). Time in Organizational Studies: Towards a New Research Direction. *Organization Studies*, 20, 1035–58.

Le Goff, Jacques. (1980). *Time, Work, and Culture in the Middle Ages* (Arthur Goldhammer, trans.). Chicago: University of Chicago Press.

Leibholz, G. (1959). Memoir. In Dietrich Bonhoeffer, *The Cost of Discipleship* (rev. and unabridged ed.) (pp. 9–27). New York: Macmillan.

Levine, Robert. (1989, October). The Pace of Life. *Psychology Today*, 23, 42–46.

Levine, Robert. (1997). *A Geography of Time: The Temporal Misadventures of a Social Psychologist, or How Every Culture Keeps Time Just a Little Bit Differently*. New York: Basic Books.

Levine, Robert V., and Kathy Bartlett. (1984). Pace of Life, Punctuality, and Coronary Heart Disease in Six Countries. *Journal of Cross-Cultural Psychology*, 15, 233–55.

Levine, Robert V., Karen Lynch, Kunitate Miyake, and Marty Lucia. (1989). The Type A City: Coronary Heart Disease and the Pace of Life. *Journal of Behavioral Medicine*, 12, 509–24.

Levine, Robert V., Todd S. Martinez, Gary Brase, and Kerry Sorenson. (1994). Helping in 36 U.S. Cities. *Journal of Personality and Social Psychology*, 67, 69–82.

Levine, Robert V., and Ara Norenzayan. (1999). The Pace of Life in 31 Countries. *Journal of Cross-Cultural Psychology*, 30, 178–205.

Levine, Robert V., Laurie J. West, and Harry T. Reis. (1980). Perceptions of Time and Punctuality in the United States and Brazil. *Journal of Personality and Social Psychology*, 38, 541–50.

Levine, Robert, and Ellen Wolff. (1985). Social Time: The Heartbeat of Culture. *Psychology Today*, 19(3), 28–35.

Lewin, Kurt. (1945). The Research Center for Group Dynamics at Massachusetts Institute of Technology. *Sociometry*, 8, 126–36.

Lewin, Kurt. (1951). *Field Theory in Social Science: Selected Theoretical Papers* (Dorwin Cartwright, ed.). New York: Harper & Brothers.

Lightfoot, Robert W., and Christopher A. Bartlett. (1995). Philips and Matsushita: A Portrait of Two Evolving Companies. In Christopher A. Bartlett and Sumantra Ghoshal, *Transnational Management: Text, Cases, and Readings in Cross-Border Management* (2nd ed.) (pp. 73–91). Chicago: Richard D. Irwin.

Lightman, Alan. (1993). *Einstein's Dreams*. New York: Pantheon Books.

Lim, Stephen G.-S., and J. Keith Murnighan. (1994). Phases, Deadlines, and the Bargaining Process. *Organizational Behavior and Human Decision Processes*, 58, 153–71.

Lindecke, Fred W. (1983, June 2). Bill for Later Start of Schools Approved. *St. Louis Post-Dispatch*, 1A, 4A.

Lindsay, William M., and Leslie W. Rue. (1980). Impact of the Organization

Environment on the Long-Range Planning Process: A Contingency View. *Academy of Management Journal*, 23, 385–404.

Lippincott, Kristen. (1999). *The Story of Time*. London: Merrell Holberton.

Locke, Edwin A., and Gary P. Latham. (1990). *A Theory of Goal Setting and Task Performance*. Englewood Cliffs, N.J.: Prentice-Hall.

Locke, John. (1959). *An Essay Concerning Human Understanding* (Alexander C. Fraser, collater and annotater). New York: Dover. (Original work published in 1690.)

Longfellow, Henry Wadsworth. (1883). *The Complete Poetical Works of Henry Wadsworth Longfellow*. Boston: Houghton Mifflin.

Longman Dictionary of Scientific Usage. (1979). Burnt Mill, U.K.: Longman Group.

Lopata, Helena Z. (1986). Time in Anticipated Future and Events in Memory. *American Behavioral Scientist*, 29, 695–709.

Lucas, George R. (1994). Whitehead, Alfred North (1861–1947). In Samuel L. Macey (ed.), *Encyclopedia of Time* (pp. 670–71). New York: Garland.

Lyell, Charles. (1868). *Principles of Geology; or, The Modern Changes of the Earth and Its Inhabitants* (new and entirely rev. ed.). New York: D. Appleton.

Lyons, Oren. (1980). An Iroquois Perspective. In Christopher Vecsey and Robert W. Venables (eds.), *American Indian Environments: Ecological Issues in Native America History* (pp. 171–74). Syracuse, N.Y.: Syracuse University Press.

Macan, Therese H. (1994). Time Management: Test of a Process Model. *Journal of Applied Psychology*, 79, 381–91.

Macan, Therese H., Comila Shahani, Robert L. Dipboye, and Amanda P. Phillips. (1990). College Students' Time Management: Correlations with Academic Performance and Stress. *Journal of Educational Psychology*, 82, 760–68.

Macey, Samuel L. (1994). Partitioning the Day. In Samuel L. Macey (ed.), *Encyclopedia of Time* (pp. 441–44). New York: Garland.

Mackenzie, R. Alec. (1972). *The Time Trap: How to Get More Done in Less Time*. New York: McGraw-Hill.

Mackenzie, R. Alec. (1997). *The Time Trap* (3rd ed.). New York: AMACOM.

Maines, David R. (1993). Narrative's Moment and Sociology's Phenomena: Toward a Narrative Sociology. *Sociological Quarterly*, 34, 17–38.

Makeba, Miriam. (1987). *Makeba: My Story* (written with James Hall). New York: Nal Books.

Malinowski, Bronislaw. (1990). Time-Reckoning in the Trobriands. In John

Hassard (ed.), *The Sociology of Time* (pp. 203–18). New York: St. Martin's Press. (Original work published in 1926–27.)

Mannix, Elizabeth A., and George F. Loewenstein. (1993). Managerial Time Horizons and Interfirm Mobility: An Experimental Investigation. *Organizational Behavior and Human Decision Processes*, 56, 266–84.

Manrai, Lalita A., and Ajay K. Manrai. (1995). Effects of Cultural-Context, Gender, and Acculturation on Perceptions of Work Versus Social/Leisure Time Usage. *Journal of Business Research*, 32, 115–28.

March, James G. (1991). Exploration and Exploitation in Organizational Learning. *Organization Science*, 2, 71–87.

March, James G. (1999). Research on Organizations: Hopes for the Past and Lessons from the Future. *Nordiske Organisasjonsstudier*, 1, 69–83.

March, James G., and Herbert A. Simon. (1958). *Organizations*. New York: Wiley.

Marion, Russ. (1999). *The Edge of Organization: Chaos and Complexity Theories of Formal Social Systems*. Thousand Oaks, Calif.: Sage Publications.

Marko, Kathy W., and Mark L. Savickas. (1998). Effectiveness of a Career Time Perspective Intervention. *Journal of Vocational Behavior*, 52, 106–19.

Marshack, Alexander. (1964). Lunar notation on upper paleolithic remains. *Science*, 146, 743–45.

Marshack, Alexander. (1972). *The Roots of Civilization: The Cognitive Beginnings of Man's First Art Symbol and Notation*. New York: McGraw-Hill.

Marshall, Paule. (1969). *The Chosen Place, The Timeless People*. New York: Vintage Contemporaries.

Martens, M. F. J., F. J. N. Nijhuis, M. P. J. Van Boxtell, and J. A. Knottnerus. (1999). Flexible Work Schedules and Mental and Physical Health. A Study of a Working Population with Non-traditional Working Hours. *Journal of Organizational Behavior*, 20, 35–46.

Martin, Joanne. (1992). *Cultures in Organizations: Three Perspectives*. New York: Oxford University Press.

Mayr, Otto. (1986). *Authority, Liberty, and Automatic Machinery in Early Modern Europe*. Baltimore: Johns Hopkins University Press.

Mazzotta, Giuseppe. (1993). Life of Dante. In Rachel Jacoff (ed.), *The Cambridge Companion to Dante* (pp. 1–13). Cambridge: Cambridge University Press.

McCartney, Scott, and Jonathan Friedland. (1995, June 29). Computer Sales Sizzle as Developing Nations Try to Shrink PC Gap. *Wall Street Journal*, A1, A8.

McCollum, James K., and J. Daniel Sherman. (1991). The Effects of Matrix

Organization Size and Number of Project Assignments on Performance. *IEEE Transactions on Engineering Management*, 38, 75–78.

McGinn, Daniel. (2000, October 16). Mired in Meetings. *Newsweek*, 52–54.

McGrath, Joseph E., and Janice R. Kelly. (1986). *Time and Human Interaction: Toward a Social Psychology of Time*. New York: Guilford Press.

McGrath, Joseph E., and Nancy L. Rotchford. (1983). Time and Behavior in Organizations. *Research in Organizational Behavior*, 5, 57–101.

McKenna, Regis. (1997). *Real Time: Preparing for the Age of the Never Satisfied Customer*. Boston: Harvard Business School Press.

McLain, David L. (1993). The MSTAT-I: A New Measure of an Individual's Tolerance for Ambiguity. *Educational and Psychological Measurement*, 53, 183–89.

McPhee, John. (1981). *Basin and Range*. New York: Farrar, Straus and Giroux.

McTaggart, J. M. E. (1927). *The Nature of Existence* (vol. 2) (C. D. Broad, ed.). Cambridge: Cambridge University Press.

Melbin, Murray. (1987). *Night as Frontier: Colonizing the World After Dark*. New York: Free Press.

Menut, Albert D. (1968). Introduction: Origin of the Translation. In Albert D. Menut and Alexander J. Denomy (eds.), *Le Livre du ciel et du monde* (pp. 3–9). Madison: University of Wisconsin Press.

Merton, Robert K. (1968). *Social Theory and Social Structure* (1968 enlarged ed.). New York: Free Press.

Meyer, Christopher. (1993). *Fast Cycle Time: How to Align Purpose, Strategy, and Structure for Speed*. New York: Free Press.

Michels, Robert. (1962). *Political Parties: A Sociological Study of the Oligarchical Tendencies of Modern Democracy* (Eden Paul and Cedar Paul, trans.). New York: Free Press. (Original work published in 1911.)

Miller, C. Chet, and Laura B. Cardinal. (1994). Strategic Planning and Firm Performance: A Synthesis of More Than Two Decades of Research. *Academy of Management Journal*, 37, 1649–65.

Miller, Danny. (1990). *The Icarus Paradox: How Exceptional Companies Bring About Their Own Downfall*. New York: HarperBusiness.

Minkowski, Eugène. (1970). *Lived Time: Phenomenological and Psychopathological Studies* (Nancy Metzel, trans.). Evanston, Ill.: Northwestern University Press. (Original work published in 1933.)

Mintzberg, Henry. (1973). *The Nature of Managerial Work*. Englewood Cliffs, N.J.: Prentice-Hall.

Mintzberg, Henry. (1978). Patterns in Strategy Formation. *Management Science*, 24, 934–48.

Mintzberg, Henry. (1990). The Design School: Reconsidering the Basic Premises of Strategic Management. *Strategic Management Journal*, 11, 171–95.

Mitchell, Terence R., and Lawrence R. James. (2001). Building Better Theory: Time and the Specification of When Things Happen. *Academy of Management Review*, 26, 530–47.

Monk, Timothy H. (1980). Traffic Accident Increases as a Possible Indicant of Desynchronosis. *Chronobiologia*, 7, 527–29.

Monk, Timothy H., and Lynne C. Aplin. (1980). Spring and Autumn Daylight Saving Time Changes: Studies of Adjustment in Sleep Timings, Mood, and Efficiency. *Ergonomics*, 23, 167–78.

Monk, Timothy H., and Simon Folkard. (1976). Adjusting to the Changes to and from Daylight Saving Time. *Nature*, 261, 688–89.

Montaigne, Michel de. (1892). *The Essays of Michel de Montaigne* (vol. 1) (Charles Cotton, trans.; W. Carew Hazlitt, ed.). New York: A. L. Burt (Original work published in 1580.)

Moore, Ronald D., and Brannon Braga. (1995). All Good Things . . . (Winrich Holbe, director). In Michael Piller, Rich Berman, and Jeri Taylor (executive producers), *Star Trek: The Next Generation*. Hollywood, Calif.: Paramount Pictures.

Moore, Wilbert E. (1963). *Man, Time, and Society*. New York: Wiley.

Morgan, Gareth. (1997). *Images of Organization* (2nd ed.). Thousand Oaks, Calif.: Sage Publications.

Mumford, Lewis. (1963). *Technics and Civilization*. San Diego: Harcourt Brace. (Original work published in 1934.)

Mumford, Lewis. (1970). *The Myth of the Machine: The Pentagon of Power*. New York: Harcourt, Brace, Jovanovich.

Murray, Alan, and Urban C. Lehner. (1990, June 25). What U.S. Scientists Discover, the Japanese Convert—Into Profit. *Wall Street Journal*, A1, A16.

Nelson, Emily. (2001, February 21). How Women Warriors Replaced Gardeners in P&G Ad Campaign. *Wall Street Journal*, A1, A10.

Neustadt, Richard E., and Ernest R. May. (1986). *Thinking in Time: The Uses of History for Decision-Makers*. New York: Free Press.

Nevins, Allan. (1954). *Ford: The Times, the Man, the Company*. New York: Charles Scribner's Sons.

Newton, Isaac. (1999). *The Principia: Mathematical Principles of Natural Philosophy* (I. Bernard Cohen and Anne Whitman, trans., assisted by Julia Budenz; preceded by *A Guide to Newton's Principia* by I. Bernard Cohen). Berkeley: University of California Press. (Original work published in 1687.)

Nietzsche, Friedrich. (1968). *Twilight of the Idols or, How One Philosophizes with a Hammer* (Walter Kaufmann, trans.). In Walter Kaufmann (ed.), *The Portable Nietzsche* (pp. 463–563). New York: Penguin Books. (Original work published in 1889.)

North, J. D. (1975). Monasticism and the First Mechanical Clocks. In J. T. Fraser and N. Lawrence (eds.), *The Study of Time II* (pp. 381–98). New York: Springer-Verlag.

North, John D. (1994). Clocks: The First Mechanical Clocks. In Samuel L. Macey (ed.), *Encyclopedia of Time* (pp. 127–33). New York: Garland.

Novikov, Igor D. (1998). *The River of Time* (Vitaly Kisin, trans.). Cambridge: Cambridge University Press.

Nunnally, Jum C. (1978). *Psychometric Theory* (2nd ed.). New York: McGraw-Hill.

Oakley, Kenneth. (1964). *Frameworks for Dating Fossil Man*. Chicago: Aldine Publishing.

O'Connor, Ellen. (1998, August 7–12). The Past Recaptured: Narratives of Nostalgia as Constructing the Future. Presentation in the symposium "Stor(i)ed Knowledge: The Narrative Basis of Organizational Memory," at the Annual Meeting of the Academy of Management, San Diego, Calif.

O'Connor, Ellen. (2000, August 4–9). From Sources to Resources: Using the Past as a Strategic Asset. Presentation in the symposium "The Strategic Use of the Past for the Present and Future: Organizational History and Changes in Image, Identity, and Reputation," at the Annual Meeting of the Academy of Management, Toronto, Canada.

O'Malley, Michael. (1990). *Keeping Watch: A History of American Time*. New York: Viking.

Onken, Marina H. (1999). Temporal Elements of Organizational Culture and Impact on Firm Performance. *Journal of Managerial Psychology*, 14, 231–43.

Oppenheimer, Mark. (2000, October 25). True Belief Buried by Films, Fads. *Columbia Missourian*, 8A.

Oresme, Nicole. (1968). *Le Livre du Ciel et du Monde* (Albert D. Menut, trans.; Albert D. Menut and Alexander J. Denomy, eds.). Madison: University of Wisconsin Press. (Original work published in 1377.)

Orlikowski, Wanda J. (1992). The Duality of Technology: Rethinking the Concept of Technology in Organizations. *Organization Science*, 3, 398–427.

Orlikowski, Wanda, and JoAnne Yates. (1999, August 6–11). It's About Time: An Enacted View of Time in Organizations. Paper presented at the Annual Meeting of the Academy of Management, Chicago.

Ornstein, Robert E. (1997). *On the Experience of Time*. Boulder, Colo.: Westview Press.

Orwell, George. (1961). *1984*. New York: Signet Classics. (Original work published in 1949.)

Ostrom, Elinor, Joanna Burger, Christopher B. Field, Richard B. Norgaard, and David Policansky. (1999). Revisiting the Commons: Local Lessons, Global Challenges. *Science*, 284, 278–82.

Ouchi, William G. (1981). *Theory Z: How American Business Can Meet the Japanese Challenge*. New York: Avon.

Paese, Paul W. (1995). Effects of Framing on Actual Time Allocation Decisions. *Organizational Behavior and Human Decision Processes*, 61, 67–76.

Palmer, David K., and F. David Schoorman. (1999). Unpackaging the Multiple Aspects of Time in Polychronicity. *Journal of Managerial Psychology*, 14, 323–44.

Panshin, Alexei. (1968). *Heinlein in Dimension, A Critical Analysis*. Chicago: Advent.

Parkinson, C. Northcote. (1957). *Parkinson's Law and Other Studies in Administration*. Cambridge, Mass.: Riverside Press.

Pearce, John A., II, Elizabeth B. Freeman, and Richard B. Robinson Jr. (1987). The Tenuous Link Between Formal Strategic Planning and Financial Performance. *Academy of Management Review*, 12, 658–75.

Pedhazur, Elazar J. (1982). *Multiple Regression in Behavioral Research: Explanation and Prediction* (2nd ed.). New York: Holt, Rinehart and Winston.

Penn, Michael. (1999). Romancing the Word: The Unfinished Adventure of a Dictionary Maker's Life. *On Wisconsin*, 100(3), 24–27, 52.

Perlow, Leslie A. (1997). *Finding Time: How Corporations, Individuals, and Families Can Benefit from New Work Practices*. Ithaca, N.Y.: Cornell University Press, ILR Press.

Perlow, Leslie A. (1998). Boundary Control: The Social Ordering of Work and Family Time in a High-Tech Corporation. *Administrative Science Quarterly*, 43, 328–57.

Perlow, Leslie A. (1999). The Time Famine: Toward a Sociology of Work Time. *Administrative Science Quarterly*, 44, 57–81.

Perrow, Charles. (1984). *Normal Accidents: Living with High-Risk Technologies*. New York: Basic Books.

Perry, Clay, and Maggie Perry. (2000). *A World of Flowers*. New York: St. Martin's Press.

Persing, D. Lynne. (1992). *The Effect of Effort Allocation Information on Percep-*

tions of Intellectual Workers and Evaluations of Their Products. Dissertation Abstracts International, 52(09A), 3350–51.

Persing, D. Lynne. (1999). Managing in Polychronic Times: Exploring Individual Creativity and Performance in Intellectually Intensive Venues. *Journal of Managerial Psychology,* 14, 358–73.

Pierce, Jon L., John W. Newstrom, Randall B. Dunham, and Alison E. Barber. (1989). *Alternative Work Schedules.* Boston: Allyn & Bacon.

Pittendrigh, Colin S. (1981). Circadian Systems: General Perspective. In Jürgen Aschoff (ed.), *Handbook of Behavioral Neurobiology* (vol. 4: *Biological Rhythms*) (pp. 57–80). New York: Plenum Press.

Pittendrigh, Colin S., and Victor G. Bruce. (1959). Daily Rhythms as Coupled Oscillator Systems and Their Relation to Thermoperiodism and Photoperiodism. In Robert B. Withrow (ed.), *Photoperiodism and Related Phenomena in Plants and Animals* (pp. 475–505). Washington, D.C.: American Association for the Advancement of Science.

Plautus. (1902). From the "Baccharia." In Henry T. Riley (trans.), *The Comedies of Plautus* (vol. 2) (pp. 517–18). London: George Bell & Sons. (Original work published circa –200.)

Pliny, the Younger. (1969). To Titius Aristo. In Betty Radice (trans.), *Pliny: Letters and Panegyricus* (vol. 2: *The Letters of Pliny, Book VIII,* Letter XIV) (pp. 33–46). Cambridge: Harvard University Press. (Original work published in 105.)

Polanyi, Michael. (1966). *The Tacit Dimension.* Garden City, N.Y.: Doubleday.

Porter, Lyman W., and John Van Maanen. (1970). Task Accomplishment and the Management of Time. In Bernard M. Bass, Robert Cooper, and John A. Haas (eds.), *Managing for Accomplishment* (pp. 180–92). Lexington, Mass.: Heath Lexington Books.

Price, James L. (1972). *Handbook of Organizational Measurement.* Lexington, Mass.: D. C. Heath.

Price, James L., and Charles W. Mueller. (1986). *Handbook of Organizational Measurement.* Marshfield, Mass.: Pitman Publishing.

Price, Niko. (2000, July 4). Fox Will Stick to Reform Promise. *Columbia Missourian,* A1.

Prigogine, Ilya. (1997). *The End of Certainty: Time, Chaos, and the New Laws of Nature.* New York: Free Press.

Puffer, Sheila M. (1989). Task-Completion Schedules: Determinants and Consequences for Performance, *Human Relations,* 42, 937–55.

Quinn, Robert E., and Michael R. McGrath. (1985). The Transformation of

Organizational Cultures: A Competing Values Perspective. In Peter J. Frost, Larry F. Moore, Meryl R. Louis, Craig C. Lundberg, and Joanne Martin (eds.), *Organizational Culture* (pp. 315–34). Beverly Hills, Calif.: Sage Publications.

Quinones, Ricardo J. (1972). *The Renaissance Discovery of Time*. Cambridge: Harvard University Press.

Redelmeier, Donald A., and Robert J. Tibshirani. (1997). Association Between Cellular-Telephone Calls and Motor Vehicle Collisions. *New England Journal of Medicine*, 336, 453–58.

Reynolds, Helen, and Mary E. Tramel. (1979). *Executive Time Management: Getting 12 Hours' Work Out of an 8-Hour Day*. Englewood Cliffs, N.J.: Prentice-Hall.

Rhyne, Lawrence C. (1986). The Relationship of Strategic Planning to Financial Performance. *Strategic Management Journal*, 7, 423–36.

Richards, E. G. (1998). *Mapping Time: The Calendar and Its History*. Oxford: Oxford University Press.

Robinson, John P., and Geoffrey Godbey. (1997). *Time for Life: The Surprising Ways Americans Use Their Time*. University Park: Pennsylvania State University Press.

Robinson, W. S. (1950). Ecological Correlations and the Behavior of Individuals. *American Sociological Review*, 15, 351–57.

Rogers, Carl R., and B. F. Skinner. (1956). Some Issues Concerning the Control of Human Behavior: A Symposium. *Science*, 124, 1057–66.

Rollier, Bruce, and Jon A. Turner. (1994). Planning Forward by Looking Backward: Retrospective Thinking in Strategic Decision Making. *Decision Sciences*, 25, 169–88.

Roth, Julius A. (1963). *Timetables: Structuring the Passage of Time in Hospital Treatment and Other Careers*. Indianapolis, Ind.: Bobbs-Merrill.

Rothwell, W. (1959). The Hours of the Day in Medieval France. *French Studies*, 13, 240–51.

Rotter, Julian B. (1966). Generalized Expectancies for Internal Versus External Control of Reinforcement. *Psychological Monographs: General and Applied*, 80 (no. 1, whole no. 609), 1–28.

Roy, Donald F. (1959–60). "Banana Time": Job Satisfaction and Informal Interaction. *Human Organization*, 18, 158–68.

Sandburg, Carl. (1954). *Abraham Lincoln: The Prairie Years and the War Years* (3 vols.). New York: Dell Publishing.

Savitt, Steven F. (ed.). (1995). Introduction. In Steven F. Savitt (ed.), *Time's*

Arrows Today: Recent Physical and Philosophical Work on the Direction of Time (pp. 1–19). Cambridge: Cambridge University Press.

Schein, Edgar H. (1992). *Organizational Culture and Leadership* (2nd ed.). San Francisco: Jossey-Bass.

Schmenner, Roger W. (2001). Looking Ahead by Looking Back: Swift, Even Flow in the History of Manufacturing. *Production and Operations Management*, 10, 87–96.

Schmenner, Roger W. and Morgan L. Swink. (1998). On Theory in Operations Management. *Journal of Operations Management*, 17, 97–113.

Schneider, Benjamin. (1987). The People Make the Place. *Personnel Psychology*, 40, 437–53.

Schriber, Jacquelyn B., and Barbara A. Gutek. (1987). Some Time Dimensions of Work: Measurement of an Underlying Aspect of Organization Culture. *Journal of Applied Psychology*, 72, 642–50.

Schutz, Alfred. (1967). *The Phenomenology of the Social World* (George Walsh and Frederick Lenhert, trans.). Evanston, Ill.: Northwestern University Press.

Schwartz, Delmore. (1959). Calmly We Walk Through This April Day. In *Selected Poems (1938–1958): Summer Knowledge* (pp. 66–67). New York: New Directions.

Schwartz, Peter. (1991). *The Art of the Long View: Planning for the Future in an Uncertain World*. New York: Currency/Doubleday.

Schwartzman, Helen B. (1986). The Meeting as a Neglected Social Form in Organizational Studies. *Research in Organizational Behavior*, 8, 233–58.

Scott, W. Richard. (1975). Organizational Structure. *Annual Review of Sociology*, 1, 1–20.

Seeman, Melvin. (1959). On the Meaning of Alienation. *American Sociological Review*, 24, 783–91.

Seers, Anson, and Steve Woodruff. (1997). Temporal Pacing in Task Forces: Group Development or Deadline Pressure? *Journal of Management*, 23, 169–87.

Seiwert, Lothar J. (1989). *Time Is Money: Save It* (Edward J. Zajac and Linda I. Zajac, trans.). Homewood, Ill.: Dow Jones-Irwin.

Seneca, Lucius Annaeus. (1834). *Medea* (Charles Beck, ed.). Cambridge and Boston, Mass.: James Munroe. (Original work published circa 50.)

Sharplin, Arthur. (1998). The Lincoln Electric Company. In Gareth R. Jones, *Organizational Theory: Text and Cases* (2nd ed.) (pp. 640–51). Reading, Mass.: Addison-Wesley.

Shirer, William L. (1941). *Berlin Diary: The Journal of a Foreign Correspondent, 1934–1941.* Boston: Little, Brown.

Simonds, William A. (1943). *Henry Ford: His Life—His Work—His Genius.* Indianapolis, Ind.: Bobbs-Merrill.

Slater, Robert O. (1985). Organization Size and Differentiation. *Research in the Sociology of Organizations,* 4, 127–80.

Slocombe, Thomas E. (1999). Applying the Theory of Reasoned Action to the Analysis of an Individual's Polychronicity. *Journal of Managerial Psychology,* 14, 313–22.

Slocombe, Thomas E., and Allen C. Bluedorn. (1999). Organizational Behavior Implications of the Congruence Between Preferred Polychronicity and Experienced Work-Unit Polychronicity. *Journal of Organizational Behavior,* 20, 75–99.

Smith, Carlla S., Christopher Reilly, and Karen Midkiff. (1989). Evaluation of Three Circadian Rhythm Questionnaires with Suggestions for an Improved Measure of Morningness. *Journal of Applied Psychology,* 74, 728–38.

Smith, Ken G., and Curtis M. Grimm. (1991). A Communication-Information Model of Competitive Response Timing. *Journal of Management,* 17, 5–23.

Smith, Robert J. (1961). Cultural Differences in the Life Cycle and the Concept of Time. In Robert W. Kleemeier (ed.), *Aging and Leisure* (pp. 83–112). New York: Oxford University Press.

Sobel, Dava, and William J. H. Andrewes. (1998). *The Illustrated Longitude.* New York: Walker.

Solzhenitsyn, Aleksandr I. (1974). *The Gulag Archipelago, 1918–1956* (Thomas P. Whitney, trans.). New York: Harper & Row.

Sorokin, Pitirim A. (1943). *Sociocultural Causality, Space, Time: A Study of Referential Principles of Sociology and Social Science.* New York: Russell & Russell.

Sorokin, Pitirim A., and Robert K. Merton. (1937). Social Time: A Methodological and Functional Analysis. *American Journal of Sociology,* 42, 615–29.

Spencer, Herbert. (1899). *The Principles of Sociology* (vol. 1). New York: D. Appleton.

Spradley, James P., and Mark Phillips. (1972). Culture and Stress: A Quantitative Analysis. *American Anthropologist,* 74, 518–29.

Stalk, George, Jr., and Thomas M. Hout. (1990). *Competing Against Time: How Time-Based Competition Is Reshaping Global Markets.* New York: Free Press.

Standifer, Rhetta L., and Allen C. Bluedorn. (2000). Temporal Contexts of

Boundary Spanning: A Multi-Level Consideration of Entrainment and Polychronicity. Working Paper, Department of Management, University of Missouri-Columbia, Columbia, Mo.

Starkey, Ken. (1989). Time and Work: A Psychological Perspective. In Paul Blyton, John Hassard, Stephen Hill, and Ken Starkey (eds.), *Time, Work, and Organization* (pp. 35–56). London: Routledge.

Staw, Barry M. (1981). The Escalation of Commitment to a Course of Action. *Academy of Management Review*, 6, 577–87.

Steel, Duncan. (2000). *Marking Time: The Epic Quest to Invent the Perfect Calendar*. New York: Wiley.

Steel, Piers, Thomas Brothen, and Catherine Wambach. (2001). Procrastination and Personality, Performance, and Mood. *Personality and Individual Differences*, 30, 95–106.

Stephens, Carlene E. (1994). Standard Time: Time Zones and Daylight Saving Time. In Samuel L. Macey (ed.), *Encyclopedia of Time* (pp. 575–77). New York: Garland.

Stewart, Rosemary. (1967). *Managers and Their Jobs: A Study of the Similarities and Differences in the Ways Managers Spend Their Time*. London: Macmillan.

Stinchcombe, Arthur L. (1965). Social Structure and Organizations. In James G. March (ed.), *Handbook of Organizations* (pp. 142–93). Chicago: Rand McNally.

Straithman, Alan, Faith Gleicher, David S. Boninger, and C. Scott Edwards. (1994). The Consideration of Future Consequences: Weighing Immediate and Distant Outcomes of Behavior. *Journal of Personality and Social Psychology*, 66, 742–52.

Taylor, Frederick W. (1947a). *The Principles of Scientific Management*. In Frederick W. Taylor, *Scientific Management* (pp. 1–144). New York: Harper & Row. (Original work published in 1911.)

Taylor, Frederick W. (1947b). *Shop Management*. In Frederick W. Taylor, *Scientific Management* (pp. 17–207). New York: Harper & Row. (Original work published in 1903.)

Taylor, M. Susan, Edwin A. Locke, Cynthia Lee, and Marilyn E. Gist. (1984). Type A Behavior and Faculty Research Productivity: What Are the Mechanisms? *Organizational Behavior and Human Performance*, 34, 402–18.

Tedlock, Barbara. (1992). *Time and the Highland Maya* (rev. ed.). Albuquerque: University of New Mexico Press.

Thomas, William I., and Dorothy S. Thomas. (1928). *The Child in America: Behavior Problems and Programs*. New York: Alfred A. Knopf.

Thomas, William I., and Florian Znaniecki. (1918). *The Polish Peasant in Europe and America: Monograph of an Immigrant Group* (vol. 1). Boston: Richard G. Badger.

Thompson, E. P. (1967). Time, Work-Discipline, and Industrial Capitalism. *Past and Present*, 38, 56–97.

Thoms, Peg, and David B. Greenberger. (1995). The Relationship Between Leadership and Time Orientation. *Journal of Management Inquiry*, 4, 272–92.

Thorne, Kip S. (1994). *Black Holes and Time Warps: Einstein's Outrageous Legacy*. New York: W. W. Norton.

Tinsley, Catherine. (1998). Models of Conflict Resolution in Japanese, German, and American Cultures. *Journal of Applied Psychology*, 83, 316–23.

Totterdell, Peter, Evelien Spelten, Lawrence Smith, Jane Barton, and Simon Folkard. (1995). Recovery from Work Shifts: How Long Does It Take? *Journal of Applied Psychology*, 80, 43–57.

Trask, R. L. (1999). *Key Concepts in Language and Linguistics*. London: Routledge.

Traweek, Sharon. (1988). Discovering Machines: Nature in the Age of Its Mechanical Reproduction. In Frank A. Dubinskas (ed.), *Making Time: Ethnographies of High-Technology Organizations* (pp. 39–91). Philadelphia: Temple University Press.

Tuchman, Barbara W. (1962). *The Guns of August*. New York: Dell.

Tung, Rosalie L. (1979). Dimensions of Organizational Environments: An Exploratory Study of Their Impact on Organization Structure. *Academy of Management Journal*, 22, 672–93.

Tversky, Amos, and Daniel Kahneman. (1981). The Framing of Decisions and the Psychology of Choice. *Science*, 211, 453–58.

Tyre, Marcie J., and Wanda J. Orlikowski. (1994). Windows of Opportunity: Temporal Patterns of Technological Adaptation in Organizations. *Organization Science*, 5, 98–118.

Usher, Abbott P. (1929). *A History of Mechanical Inventions*. New York: McGraw-Hill.

Usunier, Jean-Claude G. (1991). Business Time Perceptions and National Cultures: A Comparative Survey. *Management International Review*, 31, 197–217.

Usunier, Jean-Claude G., and Pierre Valette-Florence. (1994). Perceptual Time Patterns ('Time-Styles'): A Psychometric Scale. *Time & Society*, 3, 219–41.

van der Heijden, Kees. (1996). *Scenarios: The Art of Strategic Conversation*. Chichester: Wiley.

van Eerde, Wendelien. (1998). Work Motivation and Procrastination: Self-Set Goals and Action Avoidance. Doctoral thesis, University of Amsterdam. Kurt Lewin Institute dissertation series, no. 6.

van Eerde, Wendelien. (2000). Procrastination: Self-Regulation in Initiating Aversive Goals. *Applied Psychology: An International Review*, 49, 372–89.

Varadarajan, P. Rajan, Terry Clark, and William M. Pride. (1992). Controlling the Uncontrollable: Managing Your Market Environment. *Sloan Management Review*, 33(2), 39–47.

Vinton, Donna E. (1992). A New Look at Time, Speed, and the Manager. *Academy of Management Executive*, 6(4), 7–16.

Vogt, Thomas M., John P. Mullooly, Denise Ernst, Clyde R. Pope, and Jack F. Hollis. (1992). Social Networks as Predictors of Ischemic Heart Disease, Cancer, Stroke, and Hypertension: Incidence, Survival, and Mortality. *Journal of Clinical Epidemiology*, 45, 659–66.

von Bertalanffy, Ludwig. (1952). *Problems of Life: An Evaluation of Modern Biological Thought*. New York: Wiley.

Wack, Pierre. (1985). Scenarios: Shooting the Rapids. *Harvard Business Review*, 63(6), 139–50.

Waldrop, M. Mitchell. (1992). *Complexity: The Emerging Science at the Edge of Order and Chaos*. New York: Simon & Schuster, Touchstone.

Waller, Mary J. (1999). The Timing of Adaptive Group Responses to Non-routine Events. *Academy of Management Journal*, 42, 127–37.

Waller, Mary J., Robert C. Giambatista, and Mary E. Zellmer-Bruhn. (1999). The Effects of Individual Time Urgency on Group Polychronicity. *Journal of Managerial Psychology*, 14, 244–56.

Walton, Sam. (1992). *Sam Walton, Made in America: My Story*. New York: Doubleday.

Ward, C. V., M. G. Leakey, B. Brown, F. Brown, J. Harris, and A. Walker. (1999). South Turkwel: A New Pliocene Hominid Site in Kenya. *Journal of Human Evolution*, 36, 69–95.

Ward, Carol, Meave Leakey, and Alan Walker. (1999). The New Hominid Species *Australopithecus anamensis*. *Evolutionary Anthropology*, 7, 197–205.

Warner, Rebecca. (1988). Rhythm in Social Interaction. In Joseph E. McGrath (ed.), *The Social Psychology of Time: New Perspectives* (pp. 63–88). Newbury Park, Calif.: Sage Publications.

Washington State University Foundation. (2001, February 14). *Catalyst: WSU Foundation Annual Report*. http://catalyst.wsu.edu/annualreport3.asp.

Watkins, Renée N. (1969). Introduction. In *The Family in Renaissance Florence:*

A Translation by Renée Neu Watkins of I Libri Della Famiglia by Leon Battista Alberti (pp. 1–20). Columbia: University of South Carolina Press.

Watson, Warren E., Kamalesh Kumar, and Larry K. Michaelsen. (1993). Cultural Diversity's Impact on Interaction Process and Performance: Comparing Homogeneous and Diverse Task Groups. *Academy of Management Journal*, 36, 590–602.

Waugh, Alexander. (1999). *Time: Its Origin, Its Enigma, Its History*. New York: Carroll & Graf.

Webber, Ross A. (1972). *Time and Management*. New York: Van Nostrand Reinhold.

Wedekind, Claus. (1998). Give and Ye Shall Be Recognized. *Science*, 280, 2070–71.

Wedekind, Claus, and Manfred Milinski. (2000). Cooperation Through Image Scoring in Humans. *Science*, 288, 850–52.

Weekley, Ernest. (1967). *An Etymological Dictionary of Modern English* (vol. 1). New York: Dover.

Weick, Karl E. (1979). *The Social Psychology of Organizing* (2nd ed.). Reading, Mass.: Addison-Wesley.

Weick, Karl E. (1995). *Sensemaking in Organizations*. Thousand Oaks, Calif.: Sage Publications.

Weingart, Laurie R., Rebecca J. Bennett, and Jeanne M. Brett. (1993). The Impact of Consideration of Issues and Motivational Orientation on Group Negotiation Process and Outcome. *Journal of Applied Psychology*, 78, 504–17.

Wever, R. (1965). A Mathematical Model for Circadian Rhythms. In Jürgen Aschoff (ed.), *Circadian Clocks: Proceedings of the Feldafing Summer School, 7–18 September 1964* (pp. 47–63). Amsterdam: North-Holland.

Whitehead, Alfred North. (1925a). *An Enquiry Concerning the Principles of Natural Knowledge* (2nd ed.). Cambridge: Cambridge University Press.

Whitehead, Alfred North. (1925b). *Science and the Modern World*. New York: Macmillan.

Whitehead, Alfred North. (1964). *The Concept of Nature*. Cambridge: Cambridge University Press. (Original work published in 1920.)

Whitehead, Alfred North. (1978). *Process and Reality: An Essay in Cosmology* (corrected ed.) (David R. Griffin and Donald W. Sherburne, eds.). New York: Free Press. (Original work published in 1929.)

Whitrow, G. J. (1980). *The Natural Philosophy of Time* (2nd ed.). Oxford: Oxford University Press, Clarendon Press.

Whitrow, G. J. (1988). *Time in History: The Evolution of Our General Awareness of Time and Temporal Perspective*. Oxford: Oxford University Press.

Whorf, Benjamin L. (1956). *Language, Thought, and Reality: Selected Writings of Benjamin Lee Whorf* (John B. Carroll, ed.). Cambridge: Massachusetts Institute of Technology, Technology Press.

Wildavsky, Aaron. (2001). *Budgeting and Governing* (Brendon Swedlow, ed.). New Brunswick, N.J.: Transaction.

Williams, Joseph M. (2000). *Style: Ten Lessons in Clarity and Grace*. New York: Addison Wesley Longman.

Winchester, Simon. (1998). *The Professor and the Madman: A Tale of Murder, Insanity, and the Making of the Oxford English Dictionary*. New York: HarperPerennial.

Wrege, Charles D., and Richard M. Hodgetts. (2000). Frederick W. Taylor's 1899 Pig Iron Observations: Examining Fact, Fiction, and Lessons for the New Millennium. *Academy of Management Journal*, 43, 1283–91.

Wren, Daniel A. (1994). *The Evolution of Management Thought* (4th ed.). New York: Wiley.

Wynne, Marcus. (1999). Stand and Be Counted. *Psychology Today*, 32(1), 16.

Young, Michael. (1988). *The Metronomic Society: Natural Rhythms and Human Timetables*. Cambridge: Harvard University Press.

Zaheer, Srilata, Stuart Albert, and Akbar Zaheer. (1999). Time Scales and Organizational Theory. *Academy of Management Review*, 24, 725–41.

Zaleski, Zbigniew (ed.). (1994). *Psychology of Future Orientation*. Lublin, Poland: Towarzystwo Naukowe KUL.

Zeldin, Theodore. (1994). *An Intimate History of Humanity*. New York: HarperPerennial.

Zerubavel, Eviatar. (1979). *Patterns of Time in Hospital Life*. Chicago: University of Chicago Press.

Zerubavel, Eviatar. (1981). *Hidden Rhythms: Schedules and Calendars in Social Life*. Chicago: University of Chicago Press.

Zerubavel, Eviatar. (1985). *The Seven Day Circle: The History and Meaning of the Week*. Chicago: University of Chicago Press.

Zimbardo, Philip G., and John N. Boyd. (1999). Putting Time in Perspective: A Valid, Reliable Individual-Differences Metric. *Journal of Personality and Social Psychology*, 77, 1271–88.

Index

A- and B- series, 23

Abell, George, 283n2

absenteeism and lateness: deadlines and concept of, 94; polychronicity and, 64, 68; punctuality, teaching and learning, 92

absolute/abstract, time viewed as, 5–6, 20, 22, 27, 28, 31, 39

accounting periods, 14

activity patterns: multi-tasking, *see* multi-tasking; polychronic and monochronic, *see* monochronicity; polychronicity; schedules and plans, modes of connection with, 237–39

A.D. (*Anno Domini*) system, 15–17, 18–19

Adam, Barbara, 23, 26, 37, 42

Adkins, Lesley and Roy, 298n6

"Advice to a Young Tradesman" (Franklin), 276n3

Africa, 36, 38, 40, 58, 251, 258

"age" of family and temporal depth, 123–24

age of individual: "morningness" and, 250; polychronicity, as demographic variable for, 61, 62–63, 65, 67, 68; temporal depth and, 115; transcendental future, orientation toward, 115, 140

age of national culture and temporal depth, 120

age of organization and temporal depth, 123–24, 131–32

agendas, 185

Ages, geologic and archaeologic, 33–34, 35

airlines: jet lag, 294n14; paradox of efficiency, 109–10; tempo entrainment, 175

Ajzen, Icek, 135

Alban, Billie T., 253

Albert, Stuart, 138, 173

Alberti, Leon Battista, 227, 228, 230, 232, 296–97n7, 296n6

Aldag, Ramon, 290n6, 297n10, 298n6

Alexander, Richard D., 218–20, 223, 253, 263

alienation and loss/lack of meaning, 187–89, 192

Alifas, N. P., 300n10

Alighieri, Dante, 10, 198, 212, 229, 274n4

Allen, Frederick, 104

almanacs, 165

alternative scenarios, ability to construct (scenario planning), 220–23, 226

Altman, Neil, 237, 261

Alutto, Joseph, 55, 279n4

A.M./P.M. system, 17

ambiguity, tolerance for, and polychronicity, 67, 68

Ambrose, Stephen E., 198

American Dialect Society, 214

Amiens, 90

anagram-solving experiment, 166–68

analysis of variance (ANOVA), 35

anatomy and speech, 40

ancestor worship, 192–93